IEE TELECOMMUNICATIONS SERIES 36

Series Editors: Professor J. E. Flood
Professor C. J. Hughes
Professor J. D. Parsons
Dr. G. White

TELECOMMUNICATION NETWORKS

2nd Edition

Other volumes in this series:

TELECOMMUNICATION NETWORKS

2nd Edition

Edited by
J. E. Flood

The Institution of Electrical Engineers

Published by: The Institution of Electrical Engineers, London,
United Kingdom

© 1997: The Institution of Electrical Engineers

The Institution of Electrical Engineers,
Michael Faraday House,
Six Hills Way, Stevenage,
Herts. SG1 2AY, United Kingdom

British Library Cataloguing in Publication Data

A CIP catalogue record for this book
is available from the British Library

ISBN 0 85296 884 1

Printed in England by Redwood Books, Trowbridge

Contents

List of contributors

A.R. Allwood, BSc, CEng, MIEE
British Telecom Plc

J. Atkins, BA, MSc, MIEE
British Telecom Plc

J. Bateman, BEng, MSc, CEng,
MIEE
British Telecom Plc

R.K. Bell, IEng, MIEIE, FTMA
Bell Information Technology Services Ltd

E.A.J. Boggis, MA, MIEE
Vodafone Ltd

G.J. Cook, BSc(Eng), CEng, MIEE
British Telecom Plc

EurIng M.J. Duff, MSc, CEng,
MIEE
Nortel Ltd

M. Eburne, BSc, AMIEE
British Telecom Plc

J.M. Fairley, BSc, CEng, MIEE
Stopmuch Ltd

J.E. Flood, OBE, DSc, FCGI, FInst.P,
CEng, FIEE
Aston University

J.M. Griffiths, BSc, CEng, FIEE
Queen Mary and Westfield College,
London University

E.S. Grundy, CEng, MIEE
formerly with Cable and Wireless
Ltd

M. Kelly, MSc, CEng, MIEE
Nortel Ltd

G.N. Lawrence, BSc(Eng), AKC,
CEng, FIEE
GPT Ltd

D.M. Leakey, BSc(Eng), PhD, DIC,
FCGI, FEng, FIEE
Consultant

D. Onions, MSc
British Telecom Plc

B.R.S. Panesar, BSc, MBA, DipM
Energis Communications Ltd

EurIng S.F. Smith, BSc(Eng), CEng,
FIEE
formerly with Northern Telecom
Ltd

D.J. Songhurst, MSc
Lyndwode Research

C.I. Thomas, CEng, MIEE
Vodafone Ltd

K.E. Ward, FCGI, CEng, FIEE,
MIMgt
University College, London University

N.F. Whitehead, BSc, PhD, CEng,
MIEE
formerly with British Telecom Plc

N. Winch, BSc, CEng, MIEE
GPT Ltd

Preface

The first edition of 'Telecommunication networks' was published in 1975. It is now out of date, as well as long out of print. A new edition is therefore much overdue. Inevitably, the second edition is almost a new book, rather than a revision of the previous text.

In his foreword to the first edition, James Merriman, then President of the Institution of Electrical Engineers and Board Member for Technology of the British Post Office, wrote:

'Many adjectives can today be applied to the noun "telecommunications". It is rapidly growing; it is changing under the brisk massage of internal technological change, and the adoption of quasitelecommunication disciplines in a great variety of external, human activities; it is both capital hungry and challenging to the intellect; it is inclined, at the launch of a space craft or in the wake of a cable ship, suddenly to become global, imperative and complicated; it is continuous, unresting and pervasive; and, because its externals are self-evidently simple, its infrastructure lies unrecognised in either magnitude or complexity and therefore is taken for granted. Yet, withal, in my mind the essential adjectival attributes of telecommunications are co-operative and interdependent.'

Telecommunications has, indeed, changed and grown since those words were written. Optical fibres now offer bandwidth previously unimagined. Digital technology has transformed both transmission and switching, leading to the integrated digital network and the integrated-services digital network (ISDN). Software has become indispensable and ubiquitous. The asynchronous transfer mode (ATM) offers the prospect of broadband services.

Technological changes have accompanied social, economic and political changes. In developed countries, nearly every home now has a telephone. Businesses have become more dependent than ever on effective communication and multinational companies expect to obtain it in any country where they choose to operate. Multimedia applications of telecommunications are developing. Governments are giving up their monopolies of telecommunications and allowing private companies to compete in operating rival networks.

This book discusses the structure and performance of networks in the context of the services they provide. Chapters are devoted to public and private networks, ISDN, intelligent networks, mobile radio networks and broadband networks. Other chapters explain the principles of the underlying technologies, including transmission, circuit and packet switching, signalling, stored-program control, teletraffic engineering, network management, investment appraisal and network planning. A final chapter presents some case studies which illustrate the application of these principles.

Like the first edition, this book is based mainly on material presented by its authors at vacation schools organised for the Institution of Electrical Engineers at Aston University. As editor, I am deeply indebted to the contributors for their talents and their labours. The book would not have been possible without their extensive experience in telecommunications operating and manufacturing organisations.

Acknowledgment is made to the following companies for permission to publish much information contained in the book: British Telecom, Cable and Wireless, Energis Communications, GPT, Nortel and Vodafone.

J.E. Flood
Aston University
Birmingham, UK

List of abbreviations

AAR	automatic alternative routing
ABS	alternative billing services
ACE	automatic cross-connect equipment
ACP	action control point
A/D	analogue-to-digital
ADPCM	adaptive differential pulse-code modulation
AIN	advanced intelligent network
AM	amplitude modulation
AMI	alternative mark inversion
AMPS	Advanced Mobile Phone System
ANI	automatic number indication
API	application-programming interface
ASK	amplitude-shift keying
ATD	asynchronous time division
ATM	asynchronous transfer mode
AT&T	American Telephone and Telegraph Company
Bellcore	Bell Companies Research and Engineering Organisation
BER	bit-error rate
BHCA	busy-hour call attempts
BIB	backward-indicator bit
B-ISDN	broadband ISDN
BPON	broadband passive optical network
BRL	balance return loss
BSC	base-station controller
BSN	backward sequence number
BT	British Telecom
BTS	base transceiver station
BZS	bipolar zero-substituted
CAC	customer-access connection
CCIR	Comité Consultatif International des Radiocommunications
CCITT	Comité Consultatif International Télégraphique et Téléphonique
CCS	common-channel signalling
CDMA	code-division multiple access

CIC	carrier-identification code
CLI	calling-line identity
CME	conformant management entity
Codec	coder/decoder
CPU	central processor unit
C/R	command/response bit
CRC	cyclic redundancy check
CT	centre du transit
CT	cordless telephone
DAR	dynamic alternative routing
D/A	digital-to-analogue
dB	decibel
dBm	decibels relative to 1 mW
dBm0	corresponding dBm at zero reference point
dBr	dB relative to level at zero reference point
dBW	dB relative to 1W
dBmp	dBm psophometrically weighted
DCC	data country code
DCCE	digital-cell-centre exchange
DCE	data-circuit terminating equipment
DCF	discounted cash flow
DCS	digital crossconnect system
DDD	direct distance dialling
DDF	digital distribution frame
DDI	direct dialling-in
DDSN	digital derived-services network
DDSSC	digital derived-services switching centre
DECT	Digital European Cordless Telecommunications
DID	direct inward dialling
DISU	digital international switching unit
DLCI	data-link-connection identifier
DLE	digital local exchange
DMSU	digital main switching unit
DN	directory number
DNHR	dynamic nonhierarchical routing
DNIC	data-network-identification code
DP	distribution point
DPNSS	digital private-network signalling system
DPSK	differential phase-shift keying
DQDB	distributed-queue dual bus
DSB	double sideband
DSBSC	DSB suppressed carrier
DSI	digital speech interpolation
DSS	digital subscriber signalling system

DUP	data-user part
DTE	data terminal equipment
DTMF	dual-tone multifrequency
EC	European Community
EN	equipment number
ET	exchange termination
ETSI	European Telecommunications Standards Institute
FAS	flexible access system
FAW	frame-alignment word
Fax	facsimile
FC	functional component
FCC	Federal Communications Commission
FCS	frame check sequence
FDM	frequency-division multiplexing
FDMA	frequency-division multiple access
FEXT	far-end crosstalk
FIB	forward indicator bit
FIFO	first-in/first-out
FISU	fill-in signal unit
FLMTS	Future Land Mobile Telecommunications System
FM	frequency modulation
FPS	fast packet switching
FSK	frequency-shift keying
FSN	forward sequence number
GOS	grade of service
GSM	Global System for Mobile Telecommunications
HDB	high-density bipolar
HDLC	high-level data-link control
HF	high frequency
HLR	home location register
HRX	hypothetical reference connection
IAM	initial address message
IBCN	integrated broadband communications network
IBM	International Business Machines Company
IDN	integrated digital network
IEEE	Institute of Electrical and Electronic Engineers (USA)
IFRB	International Frequency Registration Board
IN	intelligent network
I/O	input/output
IP	intelligent peripheral

ISC	international switching centre
ISD	international subscriber dialling
ISDN	integrated services digital network
ISI	intersymbol interference
ISO	International Standards Organisation
ISP	intermediate-service part
ISPBX	integrated-services PBX
ISUP	ISDN-user part
ITU	International Telecommunications Union
ITU-R	ITU—Radio communication sector
ITU-T	ITU—Telecommunications sector
IXC	interexchange carrier

LAN	local-area network
LAP	link-access protocol
LAP-B	link-access protocol—balanced-mode
LAP-D	link-access protocol—D channel
LATA	local-access and transport area
LCI	logical-channel identifier
LDDC	long-distance DC signalling
LE	local exchange
LF	low frequency
LI	length indicator
LPC	linear predictive coding
LSI	large-scale integration
LSSU	link-status signal unit
LT	line termination

MAN	metropolitan-area network
MAP	mobile application part
MDF	main distribution frame
MF	multifrequency
Modem	modulator/demodulator
Muldex	multiplexer/demultiplexer
Mux	multiplexer
MS	mobile station
MSC	mobile-service switching centre
MSU	message signal unit
MTP	message-transfer part

N	neper
NAMTS	Nippon Automatic Mobile Telephone System
NCP	network control point
NCTE	network channel-terminating equipment
NDC	network destination code

NEXT	near-end crosstalk
N-ISDN	narrowband ISDN
NMC	network-management centre
NMF	network-management forum
NMT	Nordic Mobile Telephone Service
NPV	net present value
NRM	normal response mode
NT	network termination
NTN	network terminal number
O&M	Operations and maintenance
OFTEL	Office of Telecommunications
OLR	overall loudness rating
OMC	operations & maintenance centre
ONA	Open Network Access
ONP	Open Network Provision
OSI	Open Systems Interconnection
PAD	packet assembler/disassembler
P&T	posts and telegraphs
PABX	private automatic branch exchange
PAM	pulse-amplitude modulation
PBX	private branch exchange
PC	primary centre
PCP	primary cross-connection point
PCM	pulse-code modulation
PCN	personal communication network
PDH	plesiochronous digital hierarchy
PDN	public data network
PM	phase modulation
PON	passive optical network
POP	point of presence
PRF	pulse-repetition frequency
PSK	phase-shift keying
PSS	packet-switched service
PSTN	public switched telephone network
PTDF	proportional-traffic-distribution facility
PTO	public telecommunications operator
PTT	posts, telegraphs and telephones
PVC	permanent virtual circuit
QAM	quadrature amplitude modulation
QDU	quantisation-distortion unit
QOS	quality of service
QPSK	quadrature phase-shift keying

RAM	random-access memory
RBS	radio base station
RCU	remote concentrator unit
RF	radio frequency
RLR	receive loudness rating
RMS	root mean square
ROSE	remote-operations service element
RPE-LPC	regular-pulse-excited linear-prediction coding
RSU	remote switching unit
RTNR	real-time network routing
SAP	service-access point
SAPI	service-access-point identifier
SCCP	signalling-connection control part
SCPC	single channel per carrier
SCE	service-creation environment
SCP	secondary cross-connection point
SCP	service control point
SDH	synchronous digital hierarchy
SHF	super-high frequency
SIB	service-independent building block
SIF	signalling-information field
SIO	service-information octet
SIP	SMDS interface protocol
SIR	sustained information rate
SLEE	service-logic execution environment
SLI	service-logic interpreter
SLP	service-logic program
SLR	send loudness rating
SMDS	switched multimegabit data service
SMS	service–management system
SNI	subscriber–network interface
SOH	section overhead
SONET	synchronous optical network
SPC	stored-program control
SQNR	signal/quantisation-noise ratio
SSB	single sideband
SSBSC	SSB suppressed carrier
SSP	service switching point
SS7	CCITT signalling system 7
Statmux	statistical multiplexer
STD	subscriber trunk dialling
STM	synchronous transport module
STP	signal-transfer point

S-T-S	space–time–space
SU	signal unit
SVC	switched virtual circuit
TA	terminal adapter
TACS	total-access communication system
TASI	time-assignment speech interpolation
TC	transaction capabilities
TCAP	transaction-capabilities application part
TCM	time-compression multiplexing
TDM	time-division multiplexing
TDMA	time-division multiple access
TE	terminal equipment
TEI	terminal end-point identifier
TPON	telephony on passive optical network
T-S-T	time–space–time
TU	tributary unit
TUP	telephone-user part
TV	television
UHF	ultra high-frequency
UMTS	Universal Mobile Telecommunications Service
VANS	value-added network service
VC	virtual container
VDU	visual-display unit
VF	voice frequency
VHF	very high frequency
VLR	visitor-location register
VLSI	very large-scale integration
VPN	virtual private network
VSAT	very small-aperture terminal
VSB	vestigial sideband
VT	virtual tributary
WAN	wide-area network
WDM	wavelength-division multiplexing
WDMA	wavelength-division multiple access

Chapter 1

Introduction

D.M. Leakey

1.1 The telecommunications business

To understand telecommunications, it is necessary to have at least an outline appreciation of the more basic subject of Information [1], for which telecommunications provides a transport mechanism. Information comes in various forms, for example: spoken words, written and printed documents and computer data. Information can be processed, stored and transmitted. It is unquestionably vital to modern civilization.

Telecommunications involves the following:

(*a*) Transporting information between locations, either:
   (i) one-to-one (e.g. telephony)
   (ii) one-to-many (e.g. broadcasting);
(*b*) Relaying information from one telecommunications system to another;
(*c*) Managing this transportation service. It is necessary to provide, monitor, maintain and bill for the service;
(*d*) Adding communications value, e.g.
   (i) store and forward on request
   (ii) versatile (e.g. time-dependent) routing
   (iii) customised billing; and
(*e*) Adding information value, e.g. credit-card verification and electronic Yellow Pages.

All these functions contribute to the services provided to the users of a telecommunication network. Since they must pay for them, they are known as the *subscribers* or *customers* of the network operator.

Telecommunications has been driven by powerful technological forces and by an apparently almost insatiable demand for communications in a large variety of forms [2]. Within this scenario, various players act their parts. A broad division might include:

(*a*) End users. These have an ever-increasing need for information systems of which telecommunications plays a vital constituent role;
(*b*) Enhanced service providers. With liberalisation, an ever-increasing band of competitive players is providing services beyond that of simple point-to-point communication;

(*c*) Network operators. Conventional public telecommunications operators (PTOs) are now operating in an increasingly competitive environment;

(*d*) System suppliers. The conventional telephone supply companies are now joined by a wide array of other manufacturers, particularly those associated with computing;

(*e*) Component suppliers. The traditional electronic and electromechanical component suppliers and their software equivalents;

(*f*) Regulators. Both national and international;

(*g*) Financial backers. National and private.

There is also the equally vital role of the academic institutions which provide the major source of human expertise and the underpinning basic research.

A primary interest in a business venture is the rate of change and the factors driving that change. Within telecommunications, the list includes:

(*a*) Market pull: this comes from a wide range of customers;

(*b*) Technology push: developments reduce costs and enable new services to be provided; and

(*c*) Competition: this is increasing at all levels of activity.

In principle, the drive for new services could come from a wide range of customers. However, to achieve acceptable cost levels as rapidly as possible, so that a service can become profitable, requires a demand from large businesses rather than from residential customers. As a broad generalisation, major customer pull tends to arise from large (often international) business customers to increase their internal integration, to increase their automated interaction with their (often small) suppliers and to supply their brand of services to end customers more competitively. A prime example is banking services, which include credit-card verification, cash dispensers, interbank reconciliation and, given time, home banking. In contrast, the more technically exotic requirements of the scientific community often have difficulty in achieving adequate penetration to provide acceptable price-to-service ratios in adequate timescales.

Looking at the challenge more generally, Figure 1.1 attempts to illustrate the customer pull in very broad outline. Customer classes can usefully be split as:

(*a*) residential

(*b*) small/medium business

(*c*) large international business.

The requirements split into:

(*a*) entertainment

(*b*) routine transactions

(*c*) knowledge working.

To the technically minded, knowledge working usually provides the most interest. However, it is the routine transactions which provide the real

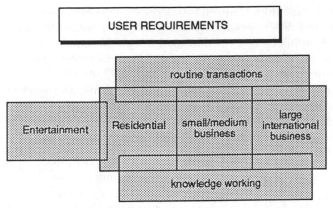

Figure 1.1 The market pull

business, although the market in this area must be closer to saturation. Entertainment, particularly in a person-to-person form of the friendly chat, should also be capable of rapid expansion. A response channel to the all-pervasive one-way television, to provide a more-interactive method of passing the time, should also provide a basis for considerable growth.

Competition, or the lack of it, is a subject well loved by politicians. At present, dogma favours national competition, without being too specific as to the international consequences. This raises many issues. Competition occurs at all levels, from customers to components. It often starts at national level, but it rapidly becomes international. This is most evident at the levels of components and large customers, but it also occurs at the network-operator level. This all leads to questions of national sovereignty and pride.

1.2 Network structures

The basic purpose of a public switched telecommunication network (PSTN) is to enable millions of independent conversations to occur simultaneously and in an ever-varying pattern, where any customer can choose and rapidly establish a connection to any other customer on the network. To fulfil this requirement, it is clearly impractical for each customer to be given a path directly to every other customer with whom he or she may wish to communicate, since the number of links required would be prohibitive. To avoid this problem, it is established practice to concentrate the routing or switching equipment at one central location, namely the exchange, to which the customers are star connected, as shown in Figure 1.2a.

As a telephone service grows, the number of customers will exceed that which can be conveniently handled by one local exchange, and the situation arises where many local exchanges are required. Since customers on one local exchange will wish to converse with customers on another local exchange, these exchanges will require interconnection. This interconnection can employ either mesh interconnection involving direct links, or star interconnection using a tandem exchange to route calls from one local exchange to another, as shown in Figure 1.2*b*. The routes between exchanges will contain many transmission links or junctions, the number being calculated on a potential-usage basis, which will depend on the community of interest between customers on the exchanges in question. In practice, it is often difficult to decide whether an exchange network should be star or mesh connected. The practical result is invariably one of compromise, and both mesh and star interconnections are employed.

The star element of the network shown in Figure 1.2*b* is representative of an elementary hierarchical structure. As the network grows in size, such groups of exchanges in different parts of the country are linked by long-distance circuits switched by trunk exchanges. Figure 1.3 gives some idea of the structures commonly in existence today. At any level, direct routes between exchanges can be included if the traffic levels justify such provision. Thus, a national network consists of the following hierarchy:

(*a*) internal lines on customers' premises that connect extension telephones to one another and to exchange lines via a private automatic branch

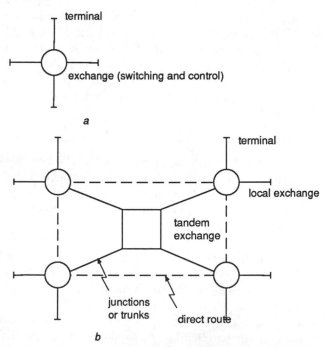

Figure 1.2   Network configurations (*a*) Customers' access network and local exchange (*b*) Multi-exchange network

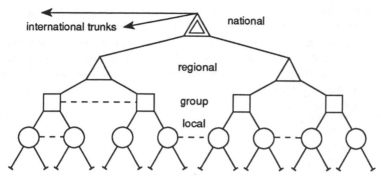

Figure 1.3   National network

exchange (PABX);*
(*b*) local networks or customers' access networks, connecting customers' telephones or PABXs to local exchanges;
(*c*) junction networks, interconnecting a group of local exchanges serving an area and connecting them to a tandem exchange or a trunk exchange;
(*d*) the trunk network which provides long-distance circuits between local areas throughout the country.

Finally, the world hierarchy is completed by the network of international circuits which provides links between the national networks of different countries. National networks are connected to the international network by international gateway exchanges in each country.

Unfortunately, the names of the different networks and their switching centres have not been standardised internationally and different names are used in different countries. For example, a switching centre is called an exchange in the UK, but a central office in the USA. An exchange on which customers' lines terminate is a local exchange in the UK but an end office or class 5 office in the USA. An exchange which switches long-distance traffic is called a trunk exchange in the UK but is a toll office in the USA. Circuits between local exchanges, or between a local exchange and a tandem or trunk exchange are junctions in the UK but are trunks in the USA. In some countries, the term 'local-area network' is used to describe the complete network of an area, including both the customer access networks and the junction network. The term 'local-area network', or LAN, is also used to denote a private data network confined to the premises of its owner.

A telecommunication network contains a large number of transmission links connecting different locations, which are known as the *nodes* of the network. Thus, each customer's terminal is a node. Switching centres form other nodes. At some nodes, certain circuits are not switched but their

* Large companies also have private networks (using circuits leased from the public telecommunications operators) linking their PABXs in different parts of a country or even across several countries.

transmission paths are joined semipermanently. Customers may require connection to services, such as recorded messages from information providers, emergency services and telephone operators for assistance in making some calls. Consequently, a telecommunication network may be considered to be the totality of the transmission links and the nodes, which are of the following types:

(*a*) customer nodes,
(*b*) service nodes,
(*c*) switching nodes, and
(*d*) transmission nodes.

To set up a connection to the required destination, and clear it down when no longer required, the customer must send information to the exchange. For a connection which passes through several exchanges, such information must be sent between all exchanges on the route. This interchange of information is called *signalling*. A telecommunication network may therefore be considered as a system consisting of the following interacting subsystems:

(*a*) transmission systems,
(*b*) switching systems, and
(*c*) signalling systems.

These systems are described more fully in later Chapters.

1.3 Network services

In practice, there are several networks, providing different services, but they use common transmission bearers, as shown in Figure 1.4. Examples include:

(*a*) the public switched telephone network (PSTN);

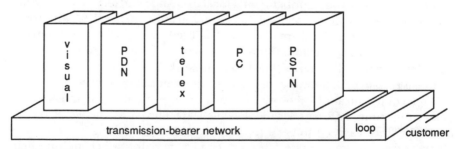

Figure 1.4   Relationship of service and bearer networks

PC = private circuits
PDN = public data network
PSTN = public switched telephone network

(*b*) private networks (using circuits leased from the public telecommunications operator);

(*c*) cellular radio networks providing mobile communications;

(*d*) data networks, usually employing packet switching; and

(*e*) special service networks, introduced to meet special demands from customers.

Although the networks might be separate to the customer, the differences should be hidden as far as is practicable. For example, service queries should be dealt with by a common service desk. Billing should be integrated where this is required. This conflict between the use of separate services networks, usually for expediency, commercial or regulatory reasons, and the need to provide at least a veneer of integration leads to an overall service and network model as illustrated in Figure 1.5.

The main PSTN plus the more specialised networks (e.g. packet, mobile etc.) share common basic trunk and junction transmission networks. Where appropriate, the separate service networks are equipped with intelligent network controls. Each element of the transmission, service and control layers is equipped with an interface-management unit which connects to the overall network-management function. This may be integrated fully, so that one system manages all the diverse service networks. Alternatively, as is more probable, some service networks have their own unique management. The service-management layer deals with those supervisory features of interest to the customer, such as service provision, service monitoring, maintenance and billing. These apparently minor activities, when multiplied by tens of millions of customers, require integrated computer systems of a size comparable with the largest in the world.

Even at the service-management level, certain specialist services might have their own unique management system. In such cases, to provide complete management integration as seen by the customer, an extra customer-contact

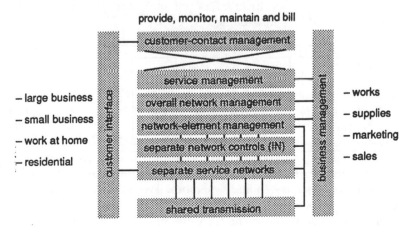

Figure 1.5   Service and network model

level is now commonly added which disguises any lack of integration at the lower levels.

To complete the picture, a customer interface or local network has to be provided, again with varying degrees of integration. There is also the overall business-management function, dealing with matters such as marketing, sales, supplies, installation and the other activities necessary for the efficient running of a modern business.

The likelihood is that complete integration is probably impracticable and possibly even undesirable. A compromise has to be reached when the number of separate service networks is tightly controlled to contain complexity, but where new network services can be introduced rapidly and at least initially on separate networks. The next technology always promises to provide full integration. Come the day, however, neglected factors seem to make the integrated ideal not quite as ideal as was initially hoped.

1.4 Network statistics

To obtain a valid comparison between one national telephone system and another, it is strictly necessary to consider relative quantities of equipment at all the major system levels. To compare complete systems on the basis of the number of telephones alone could give a false impression, because a telephone alone does not ensure success in completing a connection. However, given that the quality of service provided by countries with well-developed telephone services is reasonably similar, the number of telephones is a fair indication of the size of the telephone system as a whole.

Table 1.1 gives the basic telephone statistics of countries with more than 5 million main exchange lines [3]. An outstanding statistic is the number of telephones in the USA, which nearly equals one-half the sum total of the remaining countries listed. North America and Sweden also have the largest number of telephones per head of population, and it is evident that a saturation state could be approaching.

1.5 Regulation

The business of operating telecommunication networks has tended to be a monopoly. It is enormously expensive to dig up streets, install conduits or ducts and lay cables throughout a country. This high cost is a barrier against competitors entering the market. It is obviously desirable that the power of a monopoly should be limited to protect its customers from exploitation [4].

In most countries of the world, the telecommunications monopoly has been controlled by means of state ownership. Typically, the telecommunications operator has been a Posts and Telegraphs Department (P&T) or a public corporation. However, in the USA, the monopoly was privately owned. The Bell System of the American Telephone and Telegraph Company

(AT&T) provided service in many parts of the country and independent companies had monopolies in other areas. To protect the customers, tariffs have been regulated by the Federal Communications Commission (FCC) for long-distance traffic and by the Public Utilities Commissions of individual states for local service.

There have been great changes in recent years. In 1984, the Bell System was split into separate operating companies for different regions of the USA, with AT&T operating only a company to provide the long-distance network. Developments in technology, which had greatly reduced the cost of long-distance transmission, enabled new companies, MCI and Sprint, to enter this field. Now, the customer obtains access for long-distance calls via the local Bell or independent operating company, but can choose which of the long-distance carriers to use.

In Britain, the process of 'liberalisation' began with the telecommunications service being separated from the Post Office and made into a public corporation (British Telecom). Then customers were permitted to own their own terminal equipments (e.g. telephones and PABXs), instead of renting

Table 1.1 Countries with more than 5 million main exchange lines in 1993

	Number of main exchange lines (to nearest million)	Main exchange lines per 100 population
USA	145	56
Japan	59	47
Germany	35	48
France	30	51
UK	27	47
Italy	24	41
Russia	23	15
Canada	16	59
South Korea	16	37
Spain	14	35
Brazil	11	7
China	10	0.7
Australia	9	49
Turkey	8	14
Netherlands	7	49
India	7	0.8
Mexico	6	7
Sweden	6	68

Source: Elsevier's Yearbook of World Electronics Data, 1995.

them from the network operator. In 1984, the government partly sold British Telecom (BT) into private ownership. To provide competition, a second operator, Mercury Communications Ltd., was formed. Although it is small compared with BT, Mercury is developing local networks to provide customer access, in addition to carrying long-distance traffic. Cable television companies have also been licensed to provide telephone service to their customers. Also, competing cellular mobile radio companies have been established. Competition is still imperfect in the UK, because of the predominance of BT (at least as far as the fixed network is concerned). The Office of Telecommunications (OFTEL) was set up as the government's regulatory body.

Similar changes have occurred in other parts of the world, but operation of the telecommunication network is still a state monopoly in most countries. However, in the European Community, the EC Commission has issued an Open Network Provision (ONP) Directive which requires the telecommunication operators of member states to allow other service providers fair and equal access to leased lines in their networks. A similar requirement, known as Open Network Access (ONA), has been introduced in the USA.

The primary current trend is towards actively encouraging competition in the supply of telecommunications services. Opponents question the wisdom of this approach, suggesting that at least basic telecommunications is a natural monopoly. However, opinions are probably founded more in dogma than reason. Probably, a move from competition to monopoly and vice versa on say a thirty year cycle is about optimum. The need for regulation indicates the gross instability of competition. In more detail:

(a) Regulation distinguishes between competitive but highly regulated basic public services and competitive but more lightly regulated enhanced or added value services.
(b) In addition to the conventional fixed public-service networks, there are many basic radio-access systems and special systems (e.g. for airports) licensed as public networks.
(c) National regulation predominates in variety, making the uniform provision of international services often very difficult beyond simple interconnectivity.
(d) Commercial pressures from users, operators and suppliers will continue to ensure that regulation never approaches stability.

1.6 Standards

Successful planning and operation of the international telecommunication network depends on co-operation between all the countries involved. The standardisation which has made an effective international network possible has been carried out through the International Telecommunications Union (ITU). This was originally founded in 1865 as the International Telegraph

Union and is the oldest of the intergovernmental organisations which now form the specialist agencies of the United Nations [5].

The work of the ITU is carried out through two main bodies:

(*a*) The ITU Telecommunication Sector (ITU-T), which was formerly the Comité Consultatif Télégraphique et Téléphonique (CCITT). Its duties include the study of technical questions, operating methods and tariffs for telephony, telegraphy and data communications.

(*b*) The ITU Radiocommunication Sector (ITU-R), which was formerly the Comité Consultatif International des Radiocommunications (CCIR). It studies all technical and operating questions relating to radio communications, including point-to-point communications, mobile services and broadcasting. Associated with it is the International Frequency Registration Board (IFRB) which regulates the assignment of radio frequencies to prevent interference between different transmissions.

The ITU-R and ITU-T are composed of representatives of governments, operating administrations and industrial organisations. Each has a large number of active study groups. The recommendations of the study groups are reported to plenary sessions which meet every few years. The results of the plenary sessions are published in a series of volumes which provide authoritative records of the state of the art.

In theory, these bodies issue recommendations and these apply only to international communications. However, an international call must pass over parts of the national networks of two countries in addition to the international circuits concerned. Consequently, national standards are inevitably affected. For example, an international telephone connection could not meet the transmission requirements of the ITU-T if these were violated by the part of the national network between either the calling or called customer and the international gateway exchange. Thus, in practice, PTOs must take account of ITU-T recommendations in planning their networks and manufacturers must produce equipments which meet ITU-T specifications.

In addition, there is the International Standards Organisation (ISO). It produces standards in many fields, including information technology. Of particular importance to telecommunications is the ISO Reference Model for Open Systems Interconnection, which is described below. There are also regional standards bodies, such as the European Telecommunications Standards Institute (ETSI). In the USA, standards are produced by the American Standards Institute (ANSI) and the Institute of Electrical and Electronic Engineers (IEEE). Other national standards bodies include the Association Francaise du Normalisation (AFNOR), the British Standards Institution (BSI) and the Deutsches Institut für Normung (DIN).

The standards of individual large companies can also be influential. For example, other computer companies manufacture IBM-compatible equipment. In the USA, the Bell Companies Research and Engineering organisation (Bellcore) produces standards to facilitate communications between the different Bell regional operating companies.

Standards have many uses, which include:

(*a*) to aid interworking between networks and between features;
(*b*) to aid interchangeability between components (hardware and software);
(*c*) to aid in the overall realisation process;
(*d*) to aid module re-use (again hardware or software);
(*e*) to contain complexity; and
(*f*) to promote competition.

Most of these reasons have complex technical ramifications. However, they also have very strong commercial implications. Consequently the whole subject of standards becomes of great interest to almost all the players involved in telecommunications. Many of these interests are in opposition and commercial considerations can often override what might be considered best technical practice. Standards are also needed to timescales set by commercial needs, and severe compromises often have to be made between technical perfection and timeliness.

1.7 The ISO reference model for Open Systems Interconnection

Standards on interworking must be adhered to rigidly. These procedures are called *protocols*. Many private data networks interconnect data terminals from the same manufacturer and operate using proprietary protocols. However, the development of data communications led to the need for communication between computers and terminals from different manufacturers. This resulted in the concept of *Open Systems Interconnection*, (OSI), to enable networks to provide communication between previously incompatible machines.

The necessary specifications and protocols for OSI were developed by the International Standards Organisation (ISO). These are based on a seven-layer protocol known as the *ISO Reference Model for OSI* [6]. In this model, as shown in Figure 1.6, the process of communication is considered in seven layers. Communication appears to take place between any layer and the corresponding layer at the other end of the communication link. However, this is actually achieved by each layer at the sending end passing information down to the layer beneath it until the lowest layer is reached and then passing the information up the stack at the receiving end. The interface between each layer and the one immediately below and above it is specified. Otherwise, the actions of the layers are independent.

The layers of the OSI model are as follows:

Layer 1: The Physical Layer: This is concerned with conveying data as binary digits (bits). It specifies the form of the electrical pulses that are transmitted,

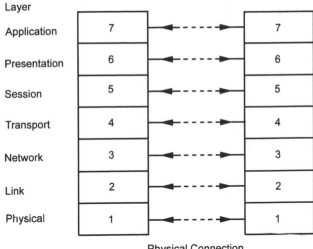

Figure 1.6   The ISO 7-layer model for Open Systems Interconnection

to ensure compatibility between the two ends of the link.

Layer 2: The Link layer: This manages the transmission of the data without error. It controls the flow of data bits, synchronises the sender and receiver and detects and corrects errors in transmission.

Layer 3: The Network Layer: This performs a routing function to set up connections across a network.

Layer 4: The Transport Layer: This establishes the end-to-end transport service for the data. It provides to layer 3 the address information for setting up connections and it determines the size of units in which data are transported.

Layer 5: The Session Layer: This arranges the appropriate type of operational session between terminals. Its protocol thus contains commands such as 'start', 'suspend', 'resume' and 'finish'.

Layer 6: The Presentation Layer: This defines the format of the data presented, to overcome differences in representation of the information as supplied to one terminal and required from the other.

Layer 7: The Application Layer: This defines the task to be performed, i.e. the applications programs needed. Examples include electronic mail, banking transactions etc.

It is useful to analyse and design telecommunications systems in terms of the OSI model [7,8]. Examples include the CCITT no. 7 signalling system, described in Chapter 8, and X.25 packet switching, described in Chapter 13. The designers of telecommunications systems are mainly concerned with layers 1–3 and the higher levels are the concern of users and the designers of information-processing systems. For telephony, level 3 only requires calling and clearing signals, address signals, ringing and tone signals. Protocols for the higher levels evolve informally between the speakers. However, for data communication, these must be specified completely and programmed in detail.

1.8 Complexity control

However a public telecommunications service is provided, in totality it is complex, indeed complex to a degree where control can be lost and chaos ensue. How can this potentially catastrophic situation be avoided? A way is to provide mechanisms by which the complexity can be understood and managed at each appropriate level in the company-management hierarchy. One such approach involves viewing the network in architectural terms, with each level of decomposition providing adequate but no more detail than is required to manage at that level. An approach which is rapidly gaining favour comprises three levels commonly referred to as:

(*a*) enterprise architecture
(*b*) functional architecture
(*c*) technical architecture

The enterprise architecture refers to the manner in which a customer's business or the operator's business is organised and run. In normal engineering terms, it is completely nontechnical but nevertheless involves considerable detail. At the other extreme, the technical architecture refers to the methods by which the supporting systems are realised. This level is very technical and riddled with appropriate jargon. The functional architecture is the middle ground. It describes (again in great detail) the customer enterprise architecture (but in terms stripped of specific reference to customer procedures) and the matching technical architecture (but again stripped of particular realisation details). It is similar to a high-level computer language, although at a level considerably higher than conventional high-level computer-programming languages.

Increasingly, systems for general service, such as virtual private networks, are conceived at this functional-architecture level. The functional-level design can then be used as a specification for the technical realisation, and as the basis for service offerings meeting specific customer needs.

Far more is likely to be heard of functional architectures and specifications in the future. Unfortunately, its understanding requires a conceptual leap which many technically-minded individuals find difficult to make.

A second relatively recent development concerns the rather obvious observation that telecommunications alone is only part of any usable information service which must also contain processing, storage and I/O functions. In addition, the word 'service' has been used to describe almost anything remotely connected with end-user service. For example the 'service' in ISDN is a most inappropriate use of the word; at this rather basic technical level, most customers would hardly apply the concept of service and in most cases would prefer to have absolutely no knowledge of what is involved.

To overcome these ambiguities another pseudohierarchy is being created, which includes:

(a) Customer applications: e.g. cash dispensers, credit verification etc;
(b) Communications features: to support the customer applications, e.g. messaging, cashless calling etc;
(c) Capabilities: the basic network attributes necessary to provide the features, e.g. digital transmission, numbering flexibility etc; and
(d) Networks: the realisation of the capabilities in a practical form.

There is no suggestion that such a classification is perfect, but it provides a considerable improvement in possible clarity of thought and the associated requirements-matching processes.

1.9 Conclusions

Some general conclusions can be drawn, which may put the whole business of telecommunications into some reasonable perspective.

(a) Telecommunications is only part of the information scene and should not be viewed separately.
(b) The real goal of a telecommunication-network operator is to meet customer needs, while making a reasonable profit.
(c) Complexity is here to stay, so we must learn to live with it and control it.
(d) Competition is also inevitable, whatever the politics and the regulatory regime.
(e) The action is rapidly becoming more international. In spite of all these influences, the right technology at the right time is a vital asset in achieving success.

1.10 References

1. PINKERTON, J.: 'Understanding information technology' (Ellis Horwood, 1990)
2. BRAY, J.: 'The communications miracle' (Plenum Press, 1995)
3. 'Yearbook of world electronics data' (Elsevier, 1995)
4. LITTLECHILD, S.C.: 'Elements of telecommunications economics' (Peter Peregrinus, 1979)
5. MICHAELIS, A.R.: 'From semaphore to satellite', International Telecommunications Union centenary volume, 1965
6. ISO 7498: 'Information processing systems: open systems interconnection basic reference model'. International Standards Organisation, 1984
7. WABAND, J.: 'Communication networks' (Aksend Associates, 1991)
8. HALSALL, F.: 'Data communications, computer networks and open systems' (Addison-Wesley, 1996), 4th edition

Chapter 2

Network structure
J. Bateman

2.1 Introduction

The purpose of a telecommunication network is to provide communication between customers. There are many examples of such networks, ranging from highly localised structures serving perhaps a single building on a factory site to large distributed structures serving the needs of an entire country. This Chapter concentrates on the latter. In providing switched and private services, it is vital to have regard for the competitive environment, which ensures that customers will only take up the services they are offered if the services represent good value for money [1]. The quality of the service that can be offered, and the costs of provision of such services, are crucially dependent upon the structure of the network [2].

The term 'network' is almost always used in the singular as a descriptor for a public telecommunications infrastructure, but it is important to recognise at the outset that such a 'network' consists of many smaller subnetworks. Furthermore, these are clearly divisible into the basic bearer networks and the services networks which are underpinned by these same bearer networks, as shown in Figure 1.4. Examples of services networks are public switched telephone networks (PSTN), public data networks (PDN), private-circuit networks, telex networks and visual-services networks.

The services and applications used by customers are treated as an entity by them. They are generally totally indifferent as to the structure of the service and bearer network, except insofar as it affects costs and quality. The network provider cannot afford the luxury of such indifference, but must understand exactly how the structure of a network impacts upon these key parameters of the product that is presented to a customer.

2.2 Network topology

To provide a network interconnecting a group of customer nodes, a circuit could be provided between each pair. This results in the fully-interconnected *mesh configuration* shown in Figure 2.1a. If there are n nodes, this requires

(a)   (b)   (c)   (d)

Figure 2.1   Basic network configurations

        a Mesh configuration
        b Bus configuration
        c Ring configuration
        d Star configuration

$n(n-1)/2$ circuits, which is clearly uneconomic if n is large. Alternatively, all the customer nodes could be connected to a single circuit, forming a *bus* or a *ring*, as shown in Figure 2.1*b* and *c*. Such configurations can be used in small local-area networks (LANs) for data communication by transmitting messages, one at a time, on the common highway at a much higher rate than they are generated by the individual nodes. However, these configurations cannot be used for normal telephony, since continuous communication is required between several nodes at the same time. Instead, a circuit is provided from each customer to a central hub node where switching equipment connects pairs of circuits together as required. This results in the *star configuration* shown in Figure 2.1*d*, which requires only n circuits instead of $n(n-1)/2$.

Consequently, call-connection switching equipment and private-circuit cross-connect equipment housed in buildings designed for easy cable access, form the heart of the public telecommunication networks. From such natural *wire centres,* cables forming *access networks* radiate out along ducts under the streets and highways to connect customers to the equipment contained in the buildings.

A national network could then be formed by interconnecting all these buildings with a long-haul transmission-bearer network. A simple example of some of the fundamental issues to be considered in determining optimum network structures can be illustrated by examining the UK network of about 6300 local exchanges. A simple calculation reveals that the number of links required fully to interconnect a 6300-node network is almost 20 million. If we were to connect the nodes together by such a network, it would give rise to a very high-cost very low-quality interconnecting network. However, an alternative would be fully to interconnect just 100 nodes, with all the rest being connected to just one of these 100 fully interconnected ones. This yields a much lower-cost network solution, and furthermore one that is of high quality. Studies can be carried out to optimise the topology by minimising the cost of transmission equipment, but there are other issues beyond reducing the costs as far as practicable. Sometimes, a higher-cost solution is accepted if the extra quality warrants it. For example, it is a general policy to divide traffic

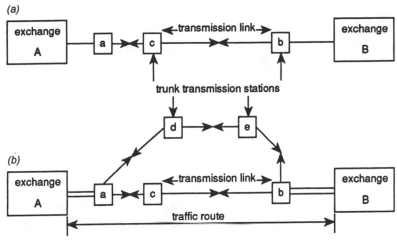

Figure 2.2 Transmission links and route diversity

    a Direct routing
    b Diverse routing

routes over one or more transmission paths in order to improve network resilience and security, as shown in Figure 2.2. This is known as *route diversity.* Although these issues of costs and quality versus structure are still of importance, the development of network structure appropriate to providing particular functionality in the network is increasingly becoming the key issue.

In practice, networks employ topologies which combine the configurations of Figure 2.1*a* and *d*. In particular, as shown in Figure 2.3*a*, stars will be concatenated into multiple-star networks, also known as *tree structures.* Trees represent a hierarchical arrangement of nodes. The lowest members of the network are connected to parents, which in turn are connected to grandparents, which in turn are connected to great grandparents and so on as shown in Figure 2.3*b*. The top level of the hierarchy consists of a single node or a fully meshed set of nodes.

In a national telecommunication network, links at the lowest level are customers' lines, links at the next level are junctions between local exchanges or between local exchanges and a tandem or trunk exchange and the higher-level links form the trunk network.

A hierarchical topology ensures that there is a fixed maximum number of links and nodes that can be involved in the interconnection of any two nodes as shown in Figure 2.3*c*. This is necessary for dimensioning the network and specifying the performance of the nodes and links. Furthermore, the discipline of this hierarchy enables different performance and capabilities to be provided at each level in the hierarchy.

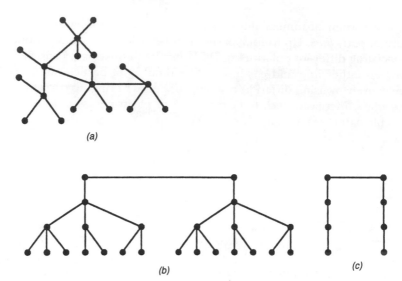

(a)

(b)   (c)

Figure 2.3   Network topologies

       a Multiple stars forming tree
       b Hierarchical structure
       c Connection having maximum number of links in tandem

2.3 Bearer networks

2.3.1 Customer-access networks

The *customer access network* or *local distribution network* [3,4,5,6] normally provides a pair of wires from each customer's premises to the local exchange. It is also sometimes called the *local loop*. As shown in Figure 2.4, each customer is served from a *distribution point* (DP), which may be overhead or underground. Multipair cables connect the DPs to the local exchange.

To enable new customers to be given service without delay, a distribution network must contain spare pairs. For economy, the number of spare pairs should be a minimum and this requires the plant to be used as flexibly as possible. Distribution points are connected by small cables to a *secondary cross-connection point* (SCP) in a street pillar. Here flexibility is obtained by rearrangeable connections between terminals. SCPs are connected by larger cables to a *primary cross-connection point* (PCP) in a street cabinet. This serves the same purpose on a larger scale. The PCPs are connected by still-larger cables to the exchange. In North America, flexibility is obtained by means of *bridge taps* [3,5]. Pairs are teed together at one or more points on a route and are thus accessible at several distribution points.

Since the cost of access networks forms a considerable proportion of the total cost of a national telecommunication network, efforts have been made to make the number of cable pairs less than the number of customers served, particularly when customers' lines are long. In the past, this has been done by the use of *party lines*. Up to four parties can be called selectively on one line by associating different polarities of DC battery with the ringing and applying it between either line and earth. A larger number of parties can be called nonselectively by using different numbers of pulses of ringing current.

Nowadays, increasing use is being made of electronic systems to enable each cable pair to serve more than one customer. Such systems are called *pair-gain systems* [3,4].

There are two types:

(*a*) multiplexers.
(*b*) concentrators.

Either frequency-division multiplexing (FDM) or time-division multiplexing (TDM) may be used to multiplex the speech signals of several customers onto a single line. Modern systems use pulse-code modulation (PCM) to enable a single four-wire circuit to serve 24 or 30 customers. The PCM systems incorporate the signalling facilities needed to enable the customers to originate and receive calls just as if they had individual lines. North American practice uses 24-channel PCM systems and European practice uses 30-channel PCM systems.

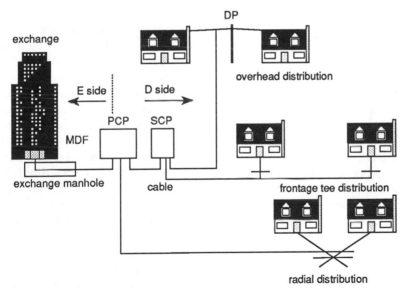

Figure 2.4   Local line network (loop)

DP = distribution point
SCP = secondary crossconnection point
PCP = primary crossconnection point
MDF = main distribution frame
D side = distribution side
E side = exchange side

A concentrator consists of switching equipment located remote from the exchange but close to its customers. Since the customers' utilisation of their telephones is normally low, a concentrator requires many fewer circuits to the exchange than the number of customers that it serves. Modern digital concentrators use PCM transmission to the exchange and thus provide pair gain both by concentration and by multiplexing.

Increasing use is now being made of optical fibres in customer access networks [5,6]. A single optical-fibre link may serve a large business customer, with a multiplexer being located on the customer's premises. Alternatively, the optical fibre may terminate at a multiplexer or concentrator in a street cabinet from which copper pairs are connected to individual customers' premises.

2.3.2 *The transmission-bearer network*

The transmission-bearer network consists of links which provide large numbers of circuits between switching nodes. Each link may convey channels over physically separate paths (space division), or over a common path which channels share by frequency-division multiplexing or time-division multiplexing. The complete transmission-bearer network contains junction networks (sometimes called the *outer core*) and the long-distance trunk network (sometimes called the *inner core*).

Traditionally, the junction network has used space division. Each junction circuit is provided by a separate pair in a multipair audio cable. These pairs are of a heavier gauge than used in the access network in order to minimise attenuation. However, increasing use has been made of PCM transmission to enable 24 or 30 circuits to be provided over two cable pairs. Optical-fibre cables have now been introduced, conveying many more channels on each fibre.

The trunk network has used large-capacity FDM carrier systems on coaxial cables and microwave radio links. Nowadays, increasing use is being made of high-capacity digital transmission systems operating over optical fibres and digital microwave radio links.

Trunk transmission nodes are known as *transmission stations* or *repeater stations*. They are usually colocated with telephone exchanges, for which they provide access to the trunk transmission network. They may also be located away from exchange buildings at places chosen as focal points of the transmission network, as determined by the topography of the country.

Repeater stations provide flexibility points at the ends of high-capacity transmission systems. In digital networks, such nodes provide interconnections at the 2 Mbit/s or 1.5 Mbit/s level between two higher-order digital line systems. The interconnections are made using coaxial jumpers across a digital distribution frame (DDF) located between two sets of back-to-back higher-order multiplexers, as shown in Figure 2.5. In a 2 Mbit/s-based network, connections may also be made via the DDF at the 8 Mbit/s and 34 Mbit/s levels. In a 1.5 Mbit/s network, they are made at the 6 Mbit/s and 45 Mbit/s levels.

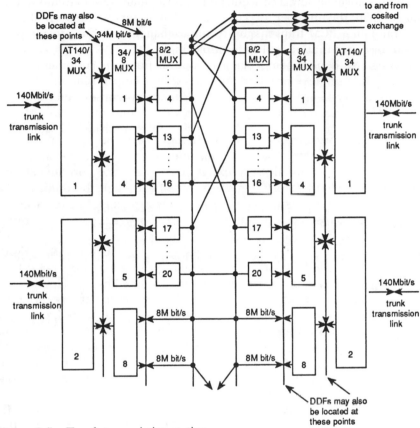

Figure 2.5   Trunk transmission station

DDF = digital distribution frame

2.4 Service networks

2.4.1 Public networks

Public telecommunication networks provide communication between all customers on that network. A public network may cover an entire country or just a specified region or regions or a class of customer within a country [7]. There may therefore be several public networks serving one country. Where this is the case, the networks are interconnected to enable the appropriate degree of access between customers on the various networks. For example, customers of a cellular mobile-radio service must be able to communicate with customers of the fixed network. Public networks may be owned and operated by private companies or local or national government departments.

The ubiquitous telephone networks throughout the world are the prime examples of public telecommunication networks. These are known as public switched telephone networks (PSTN).

2.4.2 Private networks

Private telecommunication networks provide communication between members of a single organisation. The organisation may be just one private company, a conglomerate of several companies or government departments. (The limits of such organisation groupings that may operate a private network are defined by the regulatory regime in force.) Although the domain of a private network is thus restricted, such networks can extend across a whole country or even across many countries (i.e. 'global private networks'). Private networks are considered in more detail in Chapter 11.

There is a variety of possible ownership arrangements with private networks. Usually, the links of the network are provided and operated by a public-network operator, in the form of leased private circuits. However, in certain circumstances, links may be provided by the private-network owner or a third party. The nodes of the private network are usually owned, provided and operated by the private organisation. However, in some instances, the private organisation may elect to use a public-network operator or a third party to provide and operate the private network on its behalf. Thus, the nodes and links of private network are dedicated to the customer organisation.

Private networks are making increasing use of 64 kbit/s digital circuits. In British Telecom, this is known as the 'Kilostream service' and uses the network shown in Figure 2.6. Customers' 64 kbit/s signals are multiplexed onto 2 Mbit/s links which terminate on digital crossconnects, known as *automatic crossconnect equipments* (ACE). The connections in these can be reconfigured remotely by a network controller to cater for changing customer needs. Where circuits are delivered to customers over a fibre delivery mechanism, equipment known as *flexible access system* (FAS) is also used. This uses apparatus identical to that in traditional call-connection switching equipment to provide semipermanent switched connections.

2.4.3 Virtual private networks

The term *virtual private network* (VPN) applies to the provision of private-network facilities for a customer organisation by means of the public network, usually the PSTN. Thus, a virtual private network may be viewed as a form of closed user group working within the PSTN. Although the customer organisation obtains the facilities of a private network, there are no dedicated

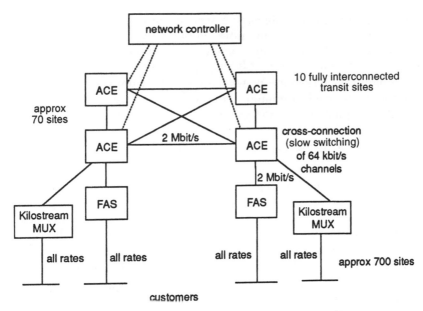

Figure 2.6   Digital private-circuit network (BT kilostream network)

ACE = automatic cross-connect equipment
FAS = flexible access system

nodes or internal links (although access from customer to the VPN may be dedicated). All connectivity within the virtual private network is provided by the public switched telephone network, with full sharing of the network elements by the 'private network' and public traffic.

The regulatory regime applying in the country defines the precise conditions for public and private networks. Public networks must adhere to a set of standards for numbering, transmission performance and interfaces between customers' attachments and the network.

Private networks have none of the above constraints applying to communication within the organisation. However, there are standards and rules applying to communication which extends to the public network (at either originating or terminating end). The provision of a virtual-private-network service by a public-network operator is complicated by the fact that VPNs do not conveniently fit within the regulatory and standards framework for public or private networks.

2.4.4 Value-added services networks

There is a further category of telecommunication network which, for convenience, can be classified as *value-added-services* (VAS) networks. These so-called networks are really public services which are provided by a value-added-services operator, using the basic infrastructure of a public network together with a dedicated value-added-services special node. In general, customers access the value-added-services nodes via the public switched telephone network or public packet network. Value-added services operating licences are only awarded for services which are more than simple conveyance of calls, i.e. value must be added. Examples of value-added services include:

(*a*) information services
(*b*) electronic mail
(*c*) voice messaging
(*d*) paging
(*e*) protocol conversion for data communication.

2.5 Node functions

2.5.1 General

In general, nodes provide a set of one or more of the following transfer functions:

(*a*) *Switching*
This may be circuit switching (concentration, expansion, interconnection) or packet switching. Circuit-switching systems are described in Chapters 5 and 6 and packet switching in Chapter 13.
(*b*) *Crossconnection*
This may be automatic (grooming, consolidation, interconnection) or manual (grooming, consolidation, interconnection).
(*c*) *Multiplexing*
This may be primary multiplexing or higher-order multiplexing.
(*d*) *Information processing*
Examples are signal termination and transfer and alarm and control information processing.

These are considered in the following Sections.

2.5.2 Switching and automatic crossconnection

There are four basic forms of automatic connection in a network, namely:

(*a*) Call connection, also known as 'fast switching', in which a path is established across a switch block for the duration of the call. The connection is usually controlled, directly or indirectly, by address signalling from the calling customer. Fast switches therefore require call-control processing and a customer-to-switch signalling system.

(*b*) Automatic crossconnection, also known as 'slow switching', in which a path is established across a switch block indefinitely. The connection is controlled, directly or indirectly, by the network operator using local or remote terminals. Slow switches do not require call processing or signalling equipment and therefore have simpler control systems. The connections through slow switches are semipermanent and under fault-free conditions may last several years.

(*c*) Locked-up time-slot connection, also known as 'nailed-up connection', in which a fast switch is used to provide a semipermanent connection, in much the same manner as a slow switch. In such a case, the call-processing and signalling capability of the switch-control system is not used. Instead, the connection is controlled, directly or indirectly, via an operational control terminal associated with the switch (on a local or remote basis).

(*d*) Packet-switch connection, in which packetised traffic is switched between links by storing and routing individual packets. Packets relating to different 'calls' are interleaved on the relevant links.

Circuit-switched services, e.g. telephony, telex, circuit-switched data etc., are provided over networks employing fast switching. Private-circuit services are time-slot connections through fast switches. Virtual private circuits, and hence virtual private networks (VPN), are provided by fast switching over the public switched telephone network (PSTN).

2.5.3 Grooming

This term describes the use of a slow switch to segregate types of circuits from a number of input multiplexed bearers into appropriate groupings for onward routing in the network. For example, Figure 2.7 shows four input 2 Mbit/s bearers containing a mixture of A and B type circuits entering a slow switch. All the A type circuits are crossconnected to one output bearer for routing to another network node; similarly for the B circuits. The A type circuits could be PSTN circuits and the B type could be private circuits.

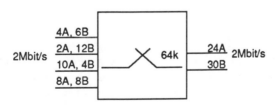

Figure 2.7   Example of grooming
A = PSTN circuit (64 kbit/s)
B = private circuit (64 kbit/s)

2.5.4 Consolidation

A slow switch performs the function of consolidation when channels of a similar type on several input multiplexed bearers (e.g. 2Mbit/s) are

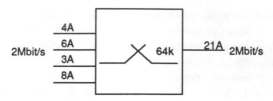

Figure 2.8   Example of consolidation

A = 64 kbit/s circuit

crossconnected to a smaller number of output multiplexed bearers. Thus, the utilisation (or loading) of the output bearers is higher than that of the input bearers, although there is no loss of traffic. Figure 2.8 illustrates the principle of consolidation.

2.5.5 Concentration

A fast switch provides concentration where the number of inlet channels is greater than the number of outlet channels. Although concentration gives an economy of equipment and circuits by increasing the traffic loading of the circuits, there is a consequential chance that traffic will be lost through the switch due to call congestion.

2.5.6 Expansion

Expansion is the opposite of concentration. It is provided by a fast switch which has more outlets than inlets. The purpose is to provide connectivity between a small number of highly-loaded trunks and a large number of lightly-
loaded lines (e.g. customers' lines).

2.5.7 Interconnection

This is the function of providing connectivity between an equal number of inlet and outlet channels. Both fast and slow switches may provide interconnection. Fast switches providing interconnection do not introduce traffic loss due to congestion. However, both fast and slow switches may exhibit blocking.

2.5.8 Multiplexing

Multiplexing enables a number of low-speed circuits to be carried over a single high-speed bearer. The total number of channels of the tributaries equals the number of channels carried over the multiplexed bearer; thus, there is no loss of traffic.

In PCM digital multiplexing, the primary multiplexer provides analogue-to-digital conversion of speech channels to give 64 kbit/s bit streams and multiplexes 24 or 30 of these into a 1.5 Mbit/s or 2 Mbit/s signal. It also provides demultiplexing and digital-to-analogue conversion for the return direction of transmission. For 64 kbit/s digital tributaries, it provides only the multiplexing and demultiplexing functions.

Higher-order multiplexers provide time-division multiplexing to combine 1.5 Mbit/s or 2 Mbit/s signals (or signals at higher rates) onto a common bearer. Normally, a channel will be multiplexed through discrete multiplex levels. For example, in the UK most telephony traffic is multiplexed initially onto a 2 Mbit/s bearer. These are then multiplexed to 8 Mbit/s, 34 Mbit/s, 140 Mbit/s and finally 565 Mbit/s which then becomes the effective line-transmission rate.

A *jump multiplexer* or *drop and insert multiplexer* inserts and extracts 1.5 Mbit/s or 2 Mbit/s tributaries from higher-order bearers without involving a chain of multiplexers and digital distribution frames.

2.5.9 Information processing

This is a category covering a wide range of functions, most of which are provided to switching nodes. The range of functions includes:

(*a*) network database (e.g. number-to-routing translation)
(*b*) alarm and control-information processing
(*c*) signal-link termination (e.g. CCITT SS No. 7 link termination)
(*d*) message switching [e.g. between CCITT SS No. 7 links at a signal-transfer point (STP)]
(*e*) synchronisation-link control (e.g. the nodal function in BT's mutual synchronisation system).

2.6 Integrated digital networks

2.6.1 General

An *integrated digital network* (IDN) is defined as a network in which all the exchanges are digital stored-program-controlled (SPC) units and all the trunk and junction traffic routes are carried on digital transmission systems [8,9]. In addition, the signalling between exchanges in an integrated digital network is assumed to be of the common-channel type [10]. Within the integrated digital network, all traffic channels are in digital (PCM) format; analogue-to-digital conversion is thus required only at the boundary of the IDN. This boundary passes through the exchange subscriber's line card for the common type of analogue subscriber line. Any boundary within the trunk or junction portion of the IDN is provided by interworking equipment at the digital exchange or, preferably, at the distant analogue exchange. This situation arises during the evolution of an IDN. It is necessary to keep the old analogue network and the new digital network operating side by side and to provide interconnections between them.

The provision of digital transmission on subscriber lines extends the integrated-digital-network boundary to the subscribers' premises. This is an essential element of an *integrated-services digital network* (ISDN) [11,12].

The optimum structure of an integrated digital network is significantly different from that of an analogue network of equivalent size. The factors contributing to this difference are:

(*a*) use of multiplexers and remote concentrator units in access networks
(*b*) size of digital transmission group
(*c*) relationship between digital transmission and switching costs
(*d*) stored-program-control flexibility
(*e*) use of common-channel signalling

Each of these factors is briefly considered in the following Sections.

2.6.2 *Use of multiplexers and remote concentrator units*

As described in Section 2.3.1, customers can now be served by multiplexers and remote concentrator units (RCUs) connected to their local exchange by optical-fibre links. As a result, the size of a local-exchange area is no longer restricted by the DC-resistance and attenuation limits of customers' lines. This enables a group of local exchanges, as shown in Figure 2.9*a* to be replaced by a single large exchange, as shown in Figure 2.9*b*. Many small exchanges can thus be replaced by multiplexers or RCUs. The remaining local exchange

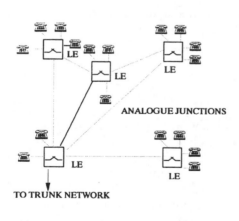

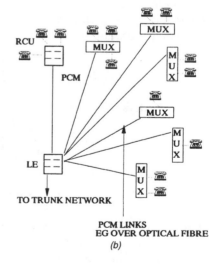

(a)

(b)

Figure 2.9   *Local-network restructuring*

a Previous analogue network
b Restructured digital network
LE = local exchange
RCU = remote concentrator unit

then serves many more customers (e.g. 50,000 instead of 10,000) over a much larger area.

2.6.3 Size of digital transmission group

An integrated digital network is based on the PCM standard group of 30 or 24 traffic circuits rather than the 12-circuit group of an analogue frequency-division-multiplex transmission network. This increase of 150% or 100% capacity in the minimun size of a transmission link raises the level of traffic needed to justify an optional route between two exchanges. Thus, IDNs have fewer optional traffic routes than analogue networks of equivalent capacity. Since the minimum route size is 30 (or 24) channels, a mandatory route, which is required irrespective of traffic level (e.g. between a local exchange and its parent trunk exchange) may have a relatively-low traffic loading per channel. Augmentation of both optional and mandatory routes requires traffic justification. When a route is augmented, either 30 (or 24) channels are added as a standard unit.

2.6.4 Relationship between digital transmission and switching costs

Both analogue- and digital-transmission costs have declined in real terms over the past 30 years, due mainly to improved multiplexing technology. However, until the advent of digital exchanges, switching costs had been increasing. Digital transmission is cheaper than analogue transmission in many circum-stances, particularly if signalling costs are considered, and digital switching, particularly in large exchange units, is significantly cheaper than analogue switching. However, it is the ratio of switching to transmission costs, rather than the absolute costs, which influences the structure of a telephone network, as described in Section 2.7.1

2.6.5 Stored-program-control flexibility

Stored-program control (SPC) is described in Chapter 7. The catchment areas of digital and analogue SPC exchanges are constrained very little by the numbering and charging arrangements for the network [9]. Thus, there is much flexibility in the way that local exchanges may be parented onto trunk exchanges, unlike the case of non-SPC exchanges in which limits are imposed by the geographical topology of the charging groups and the numbering ranges allocated to local exchanges.

These factors, together with the transmission advantages of an integrated digital network, enable the coverage of trunk-exchange catchment areas to be optimised according to network economics.

2.6.6 Common-channel signalling

As described in Chapter 8, in common-channel signalling [10] (CCS) the signals between exchanges are sent over separate data links instead of over the speech circuits. This enables a single traffic route between local and trunk exchanges to be used for all classes of traffic. By comparison, exchanges without common-channel signalling rely on the identification of the point of entry to the switch block in order to discriminate between circuits requiring different call-processing or charging activities. In many cases, particularly with analogue step-by-step exchanges, this has resulted in several separate (and often small) routes being provided. Therefore, common-channel signalling enables efficient use to be made of single traffic routes between local and trunk exchanges.

2.7 IDN trunk networks

2.7.1 Network optimisation

A trunk network has two main components: the trunk distribution network consisting of the transmission links between local exchanges and their parent trunk exchanges, and the core network consisting of the trunk exchanges and the transmission links between them. Clearly, as the number of parent trunk exchanges increases, the cost of the distribution network decreases. However, the cost of the core network increases as the number of trunk exchanges increases. Thus, there is a minimum-cost structure for the trunk network dependent on the relative costs of digital switching and transmission. The use of digital SPC exchanges in an integrated digital network imposes few constraints on the network structure, so the trunk network can be structured in the cost-optimised way [13].

It is informative to consider, as an example, the BT network restructuring resulting from the introduction of an integrated digital network. This is shown in Figure 2.10. The optimum size of the new trunk network is about 60 trunk exchanges. This should be compared with the previous analogue network of over 400 trunk exchanges established by the cost and operational constraints of predominantly step-by-step (Strowger) 2-wire analogue exchanges. The number of traffic routes in the trunk core network decreased from around 17,000 on the analogue network to about 1,800 on the IDN. The traffic generated by the digital trunk exchanges is sufficient to warrant their full interconnection. This mesh structure simplifies network planning, but it also establishes an infrastructure which can be exploited by automatic alternative routings as well as providing a high level of network resilience.

2.7.2 Network resilience

The result of planning an integrated digital network, based on cost optimisation and the exploitation of stored-program control and digital technology, is a network design comprising few, but large, switching centres and few, but large, traffic routes between them. In addition, many of the large local exchanges act as parents (or 'hosts') to several dependent remote

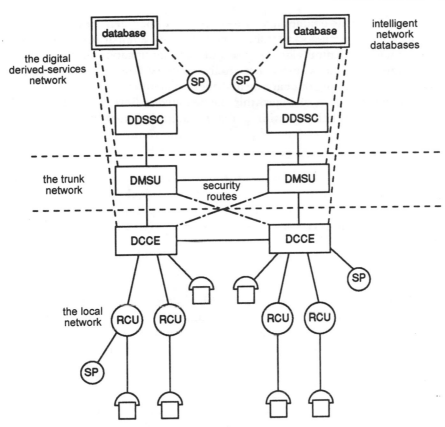

Figure 2.10 Hierarchy of BT digital PSTN

DCCE = digital cell centre exchange (local exchange)
DDSSC = digital derived-services-switching centre
DMSU = digital main switching unit (trunk exchange)
RCU = remote concentrator unit
SP = service provider

concentrator units. The security of an exchange and transmission equipment in an IDN is thus even more important than in an analogue network. The effect of route congestion and exchange-processor overload will clearly cause loss of traffic at the exchange. However, the use of automatic congestion control will throw additional traffic onto other routes, possibly causing them to become congested. Thus, there is a danger that route congestion may propagate through the network. The term *network resilience* is used to describe the ability of a network to cope with transmission and switching failure, route congestion and processor overload.

The planning of an integrated digital network must ensure an adequate level of network resilience. This is achieved using high-grade components and systems with appropriate degrees of redundancy in the hardware, together with resilient software. In addition, the capabilities of digital SPC exchanges facilitate the use of network traffic management to monitor the network performance and centrally control the initiation of remedial action [15,16]. However, despite such features, the resilience of an IDN is also very much influenced by the network structure adopted.

The various methods of improving the resilience of an integrated digital network by use of suitable network structures are briefly described below.

2.7.3 Mesh-routing strategy

A mesh structure is superimposed on the strictly-hierarchical traffic-routing patterns in both the trunk and local switching networks. This involves the provision of routes from local exchanges to nonparent trunk exchanges and to other local exchanges (sideways routes), even though not justified by traffic levels.

2.7.4 Network redundancy

Network redundancy relates to a suitable degree of overprovision of transmission and switching capacity. For example, subsequent PCM modules (30-channel or 24-channel) are added to traffic routes in advance of additional capacity being warranted by forecast traffic growth.

2.7.5 Service-protection network

A service-protection network is a special network, superimposed on the IDN to provide reserve capacity. It uses transmission capacity, e.g. 140 Mbit/s blocks, extending over the major links of the transmission-bearer network. These are provided on the basis of one service-protecting system per N parallel traffic-carrying systems ('1-in-N protection'). More elaborate methods include automatic rerouting of transmission links, thus achieving 'M-in-N protection' levels. The principal use of a service-protection network is the substitution of its capacity for transmission links which have failed. This requires the provision of broadband switches at all points of interconnection between the service-protection network and the IDN. Such switching may be automatic, initiated on detection of transmission-system failure, or manual, under local or remote control. A service-protection network may also be used to offload transmission capacity during periods of planned outages to enable work on the network to be undertaken, e.g. during transmission-link augmentation.

2.7.6 Transmission-routing diversity

The vulnerability of large traffic routes to line-plant failures can be minimised by the use of route diversity. This involves the spreading of traffic routes over two or more separate transmission paths, as described in Section 2.2.

2.7.7 Automatic alternative routing

The managed use of automatic alternative routing (AAR) allows exchanges to avoid congestion and circumvent link failure within the network. As mentioned above, the use of this technique must be bounded to ensure that it improves rather than degrades performance. One method of constraining AAR is that of partitioning traffic routes, reserving certain channels for incoming, outgoing and bothway use. Thus, large surges of traffic onto a route in one direction, resulting from automatic alternative routing, will not totally block other traffic from using that route.

The implementation of any of the above means of increasing network resilience will increase the network cost. Clearly, administrations must balance the cost of achieving a certain level of resilience against the cost of failure. The latter covers not only the operational cost but also the loss of call revenue (perhaps to a competitor) and the consequences of not meeting service obligations to customers (e.g. financial damages) or regulatory requirements.

2.7.8 Digital derived services network

As shown in Figure 2.10 the top levels of the BT integrated-digital-network hierarchy consist of a layer of just nine fully-interconnected switches which provide special functionality which is made available to all users on the system in a very economical manner [17]. In particular, services such as automatic freefone and premium-rate recorded-announcement services are provided by parenting service providers directly on those nodes and then providing access to them via the rest of the IDN. This 'overlay' network is known as the digital derived-services network.

A typical automatic freefone call would route via a local exchange and a trunk exchange to the derived-services switching exchange on which the trunk exchange is parented and then, following interrogation of the intelligent-network database, would route to the appropriate target service provider via the appropriate derived-services switch.

2.8 Multilayered model of a digital network

2.8.1 General

It is useful to describe the complex relationship between the component subnetworks of an integrated digital public switched telecommunication network (PSTN) by reference to a multilayered model, as shown in Figure 2.11. The model presents a composite of switching and transmission basic networks supported by a number of auxiliary networks.

Some of the auxiliary networks are still in the formative stages in many countries. Therefore, they may not exist as identifiable networks for many years. Although the model is described in terms of a digital stored-program-controlled telephone network, it is also applicable to an analogue

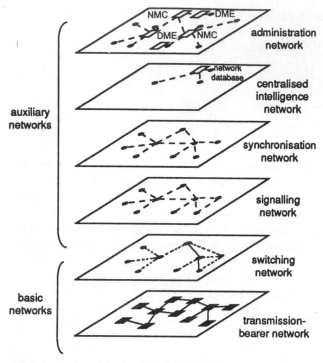

Figure 2.11   Multilayered model of a digital SPC telephone network

stored-program-controlled network, with the exception of the synchronisation auxiliary layer.

2.8.2 *Transmission-bearer networks*

The foundation of the model is the transmission-bearer network, which provides the transmission paths between all the nodes of the other layers. This network consists of the subscriber-line network between the telephone (or data terminal for an integrated-services digital network) and the main distribution frame at the local exchange, and all the transmission links between the local, junction tandem, trunk and international exchanges. The transmission-bearer network is constructed from a pattern of links and flexibility points, the latter being the nodes of the network where semi-permanent connections are made between the channels in the transmission links. In the local network, the nodes are formed by local and area distribution points where connections are made between cable pairs using jumper wire or metallic straps.

In the junction, trunk and international parts of the transmission-bearer network, the flexibility points (nodes) are located at the ends of high-capacity transmission systems in repeater stations. Such nodes provide interconnections at the 2 Mbit/s or 1.5 Mbit/s level between higher-order digital-line systems. The interconnections are achieved using coaxial jumpers across a digital distribution frame (DDF) located between two sets of digital higher-order multiplexers in back-to-back formation, as shown in Figure 2.5. Interconnection may also be provided by the DDF at 8 Mbit/s or 34 Mbit/s level, in 2 Mbit/s networks.

2.8.3 Switching network

The second layer of the model is the basic switching network, containing exchanges at all the levels of the hierarchy. The links between exchanges, in the form of bundles of circuits terminating on the corresponding switch blocks, are known as traffic routes and are shown as dotted lines in Figure 2.11. The transmission paths for the traffic routes are provided by the transmission-bearer network, shown in solid lines in Figure 2.11.

There is not necessarily a one-to-one correspondence between a traffic route and its transmission links. This concept is illustrated in Figure 2.2, where a traffic route between exchanges A and B is provided by transmission links between transmission station *a* (colocated with A) and transmission station *b* (colocated with B).

The traffic route AB is provided via one or more 2 Mbit/s paths routed over high-order transmission links *ac* and *cb*, with 2 Mbit/s or 8 Mbit/s interconnections provided via the transmission station at *c*, as shown in Figure 2.2*a*. Note that the traffic is not switched at *c*. With this arrangement, the traffic route AB will fail if either transmission link *ac* or *cb* fails. An improvement in network security is achieved by providing traffic route AB over two (or more) separate transmission routings, e.g. one 2 Mbit/s path routed *acb* and the other routed *adeb* as shown in Figure 2.2*b*. This splitting of traffic routes over parallel and separate transmission-link routings is known as *route diversity*.

2.8.4 Signalling network

The signalling network consists of the links and nodes of the common-channel signalling (CCS) systems (e.g. CCITT no. 7) within the integrated digital network [9,10]. The nodes, which are normally formed by digital SPC exchanges, may be terminal nodes (originating and terminating common-channel-signalling messages) or transit nodes, known as *signal-transfer points* (STP). A signalling link which joins two exchanges directly is called an *associated signalling link*. One which is routed via an STP is called a *quasi-associated signalling link*.

The transmission path for a common-channel-signalling link is usually provided by one or more of the PCM systems carrying traffic between two exchanges (e.g. in time slot 16 of 2 Mbit/s systems). When digital paths are not available, the signalling link may be provided at the lower rate of 4.8 kbit/s via voice-frequency modems over analogue circuits. A CCS network plan is needed to ensure that an adequate network of signalling links is available to support its dependent integrated digital network.

The plan of the common-channel-signalling network is based on the practical constraints of avoiding the need to provide additional line plant whilst meeting reliability and cost criteria. The network structure is based on the following routing rules:

(*a*) Quasi-associated routing is used if a common-channel-signalling link is required between two exchanges which are not interconnected by a traffic route.

(*b*) A single associated-signalling link is used with a traffic route of one PCM-module capacity.

(*c*) A single duplicated-signalling link is used with traffic routes of two, but less than N, PCM models. (The value of N is set by reliability considerations, but is typically 20.) Where possible, the duplicated common-channel-signalling links should be diversely routed over the transmission-bearer network.

(*d*) Traffic routes consisting of N or more PCM modules should have two duplicated, diversely routed, signalling links.

(*e*) Quasi-associated-signalling links may be used in the event of failure of associated-signalling links.

2.8.5 Synchronisation network

The synchronisation network [18] disseminates the timing from one or more reference sources to all the digital exchanges in the switching network (and to digital crossconnects). Figure 2.12 shows a four-level synchronisation hierarchy with unilateral links between levels (i.e. effective at the lower level) and bilateral links within levels (i.e. effective at both ends). The structural characteristics of a synchronisation network are:

(*a*) Line plant is not provided specifically for synchronisation purposes. Synchronisation links should preferably be derived from normal traffic-carrying PCM modules.

(*b*) The second level of the synchronisation-network hierarchy (i.e. the level below the reference node) should be well interconnected. Thus, in the event of failure of the reference node, the rest of the network is synchronised to the common frequency of the level-2 nodes.

(*c*) The number of effective links at any node should be sufficient to ensure that there is a prescribed probability that at least one link is operative despite line-plant failures on the other links.

(*d*) Nodes can have their hierarchical status changed by appropriate changes to the number, and control direction, of their synchronisation links.

(*e*) Synchronisation links should be organised so that each node can 'see' the reference node via at least two different paths via different parent nodes.

(*f*) For satisfactory operation, the number of links between a synchronisation node and the reference node should not exceed a set limit.

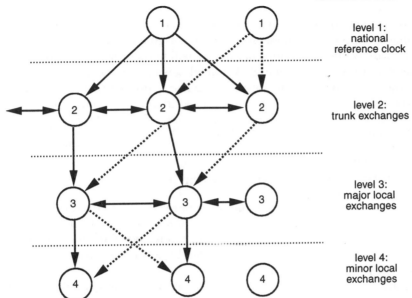

level 1:
national
reference clock

level 2:
trunk exchanges

level 3:
major local
exchanges

level 4:
minor local
exchanges

Figure 2.12   Network-synchronisation hierarchy
   —   *main*
   ...   *standby*
   →   *unilateral links*
   ‹ →   *bilateral links*

2.8.6 Administration network

The administration network [16] has a number of differing components, each fulfilling a specific aspect of operations and maintenance support to the exchanges and transmission links of the integrated digital network. These components include operations and maintenance centres, network-management centres and billing centres.

The links between these various centres and the integrated digital network may be direct or indirect. Where the level of data flow between an IDN node (i.e. exchange or transmission station) and the administration-network centre is sporadic and of low volume, modem links achieved via the PSTN may be suitable. Alternatively, access via a packet-switched service may be used. Leased-line connections are necessary where the level of data is high. Alternatively, bulk data such as the periodic exchange output for billing may be transported manually (off-line) using magnetic tapes or cartridges and so may not involve a telecommunication link.

2.8.7 Centralised-intelligence network

In a network with centralised intelligence, databases are located at a number of centres, either at key exchanges or at network-management centres

[13,19]. Local and trunk exchanges may be connected to their databases by a variety of means, according to the level of traffic involved. Where the flow of data between an exchange and a database is low, access may be provided via a public packet-switched network. Very high levels of data flow warrant the use of leased-line connections to the database. Alternatively, the common-channel-signalling network may be used for all links between exchanges and the database, using the noncircuit-related signalling capability of CCITT SS no. 7. Intelligent networks are described in Chapter 16.

2.9 Network and service integration

Historically, telecommunication networks have been built on the basis of the ubiquitous telephone network (PSTN). New services have been provided by special service-specific nodes within the network. This has led to the establishment of series of service-specific dedicated nodes and supporting infrastructure, in addition to the base public switched telephone network. Recently, the advent of digital transmission, digital stored-program-controlled switching and common-channel signalling in the network has offered the possibility of integrating many services onto the one common network [20] as shown in Figure 2.13 Briefly, the advantages of such 'multiservice' networks are:

(*a*) the common infrastructure offering the potential for economies of scale in both equipment and operational costs;

(*b*) immunity to forecasting errors in individual services due to the pooling of component traffic on common bearers and switches; and

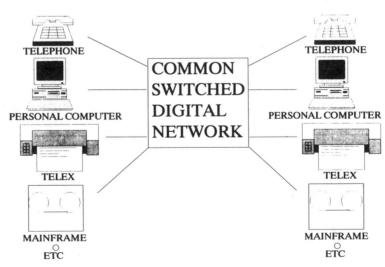

Figure 2.13   Multiservice digital network

(*c*) economies in planning efforts.

However, such service integration also has the following disadvantages:

(*a*) All services are subjected to the same network performance, which makes the provision of premium service levels difficult
(*b*) The introduction of new service capabilities on to multiservice network may be disruptive to existing services.

The terms 'integrated network' and 'separate network' when describing service provision can often be vague in that they do not fully describe the true extent of integration. Provision of an integrated-services-network service may be realised with dedicated facilities at each of the network layers:

(*a*) transmission
(*b*) switching/cross-connection
(*c*) signalling
(*d*) synchronisation
(*e*) intelligence
(*f*) admin/operation and maintenance.

In addition to the above, there are customer-facing service-management systems, dealing with such items as order handling, fault recording/progressing, account-enquiry handling and management of customer-site-visit diaries. The fuctionality of these systems influences the extent to which a customer regards the service as 'integrated'.

Integration has always existed in customer-access networks and in the main transmission-bearer network. However, the advent of digital transmission over customers' lines gives the opportunity for the capabilities of an IDN to be extended to provide customers with every service which can use a 64 kbit/s stream. These services may include, for example, high-speed data, high-speed facsimile and slow-scan television, in addition to telephony. Such networks are called *integrated-services digital networks* (ISDN) and are described in Chapter 14.

In future, there may be an increasing demand for services, such as high-definition television and colour graphics, which require greater bandwidths than can be provided by 64 kbit/s channels. Introduction of such services will require the provision of a *broadband integrated-services digital network* (B-ISDN) [20]. Such networks are the subject of Chapter 15.

2.10 References

1. CHIDLEY, J.: 'A market view of the network', *Br. Telecom Eng. J.*, 1990, **8**, pp. 200–207
2. BROWN, D.W., BELL, R.C., and MOUNTFORD, J.A.: 'Strategic planning of business operations and information systems', *Br. Telecom Eng. J.*, 1990, **9**, pp. 16–19
3. GRIFFITHS, J.M. (Ed.): 'Local telecommunications' (Peter Peregrinus, 1983)
4. GRIFFITHS, J.M. (Ed.): 'Local telecommunications II' (Peter Peregrinus, 1986)

5. ADAMS, P.F., ROSHER, P.A., and COCHRANE, P.: 'Customer access', in FLOOD, J.E., and COCHRANE, P. (Eds.): 'Transmission systems' (Peter Peregrinus, 1991), chap. 15
6. 'The access network', *Br. Telecom Eng. J.,* 1991; **10:** special issue, pp. 1–99
7. HARTSTOCK, P.E.: 'Long distance network services'. National Telecoms. and Information Administration (US Dept. of Commerce), special publication 88–21, 1988, p. 233
8. INOSE, H.: 'An introduction to digital integrated communication systems' (Peter Peregrinus, 1979)
9. REDMILL, F.J., and VALDAR, A.R.: 'SPC digital telephone exchanges' (Peter Peregrinus, 1990)
10. MANTERFIELD, R.J.: 'Common channel signalling' (Peter Peregrinus, 1991)
11. STALLINGS, W.: 'ISDN and broadband ISDN' (Macmillan, 1992), 2nd edn.
12. GRIFFITHS, J.M.: 'ISDN explained' (Wiley, 1992), 2nd edn.
13. 'Stored program controlled network', *Bell Syst. Tech. J.,* 1982, **61,** pp. 1575–1815
14. BARRON, D.A.: 'Subscriber trunk dialling', *Proc. IEE,* 1959, **106B,** pp. 341–354
15. HARA, J.A., KEYS, C.T., and WINDEKER, R.C.: 'Network management of evolving public switching network'. *Proceeding of International Switching Symposium,* 1987, p. 331
16. 'Network management', *Br. Telecom Technol. J.,* 1991, **9,** (3) special issue
17. ROBERTS, G.J.: 'The digital derived services network', *Br. Telecom Eng. J.,* 1987, **6,** pp. 105–119
18. BOULTER, R.A., and BUNN, W.: 'Network synchronisation', *Post Off. Electr. Eng. J.,* 1980, **73,** pp. 19–26
19. WEBSTER, S.: 'The digital derived services intelligent network', *Br. Telecom Eng. J.,* 1989, **8,** pp. 144–149
20. ROBERTS, J.W., and VIEVO, B.: 'Methods for the planning and evolution of multiservice telecommunications networks'. European Cooperation in the Field of Scientific and Technical Research, COST 214, report TELE/214/FR/88, European Commission, Brussels, 1988

Chapter 3

Transmission principles

J.E. Flood

3.1 Introduction

Present-day transmission systems [1,2,3] range in complexity from simple unamplified audio-frequency lines to satellite radiocommunication systems. The transmission channels they provide and the signals they convey may be classified in two broad classes: *analogue* and *digital*. An analogue signal is a continuous function of time; at any instant it may have any value between limits set by the maximum power that can be transmitted. Speech signals are an obvious example. A digital signal can only have discrete values. The commonest digital signal is a binary signal, having only two values (e.g. 'mark' and 'space' or '1' and '0'). Telegraph signals and outputs of binary-coded data from computers are thus digital signals. A television waveform is a mixture of analogue and digital signals, since it transmits both the picture contents and synchronising pulses. Some analogue and digital signals that are transmitted in telecommunication networks are listed in Table 3.1

A signal consisting of a single sinusoidal waveform is completely predictable; thus it conveys no information. A useful analogue signal must therefore contain a range of frequencies; this is known as its *bandwidth*. For a digital signal, the number of signal elements transmitted per second is called the signalling rate in *bauds*. If a nonredundant binary code is used, the rate of transmission of information (in bits per second) equals the signalling rate in bauds. If the coding contains redundancy, the bit rate is less than the number of bauds. If a multilevel signal is used (e.g. ternary or quaternary), each element conveys more than one bit of information; the bit rate is thus greater than the number of bauds.

To transmit an analogue signal without distortion, the channel must be a linear system. Cable systems and radio-relay systems equipped with linear amplifiers are examples of analogue channels. A digital channel does not require to be linear, since its output provides a number of discrete conditions corresponding to the input signal. An example of a digital channel is a telegraph circuit, whose output signal is provided by the operation of a relay. It does not follow that analogue signals must always be transmitted over analogue channels and digital signals over digital channels. Data communication and voice-frequency telegraphy over telephone lines are examples of

Table 3.1 Typical signals transmitted in telecommunication networks

(*a*) Analogue signals

Type of signal	Bandwidth
Telephone speech	300 Hz to 3.4 Hz
Facsimile Group 2*	0.3 kHz to 2.7 kHz
Broadcast programmes (e.g. music)	50 Hz to 15 kHz
Colour television (625 lines)	0 to 5.5 MHz

(*b*) **Digital signals**

Type of signal	Digit rate
Teleprinter	50 bauds
Facsimile Group 3	2400 and 4800 bit/s
Fascimile Group 4	64 kbit/s
Data	200, 600, 1200, 2400, 4800, 9600 bit/s and 48 000 and 64 000 bit/s
PCM telephony (per channel)	64 kbit/s
ADPCM telephony	32 kbit/s
Video conferencing	2 Mbit/s and lower

* Vestigial sideband transmission on 2.1 kHz carrier.

transmitting digital signals over analogue channels. Analogue signals may be coded for transmission over digital channels by means of analogue-to-digital convertors. An example is the transmission of speech by means of pulse-code modulation over lines equipped with regenerators.

An advantage of digital transmission over analogue transmission is its relative immunity to interference. For example, error-free transmission of a binary signal only requires detection of the presence or absence of each pulse, and this can be done correctly in the presence of a high level of noise. It is also possible to employ *regeneration*. Provided that a received signal is not so corrupted that it is detected erroneously, it can cause the generation of an almost perfect signal for retransmission. The use of regenerative repeaters enables the transmission performance of digital circuits to be almost independent of their length, whereas analogue signals deteriorate progressively with distance.

If a link can provide adequate transmission over a band of frequencies which is wider than that of the signals to be sent, it can be used to provide a number of channels. At the sending terminal, the signals of different channels are combined to form a composite signal of wider bandwidth. At the receiving terminal, the signals are separated and retransmitted over separate channels. This process is known as *multiplexing*. The separate channels that enter and leave the terminal stations are called *baseband channels* and the transmission link, which carries the multiplex signal, is called a *broadband channel* or a *bearer channel*.

The principal multiplexing methods are *frequency-division multiplexing* (FDM) and *time-division multiplexing* (TDM). In FDM transmission, each baseband channel uses the bearer channel for all of the time, but it is allocated only a fraction of the bandwidth. In TDM transmission, each baseband channel uses the entire bandwidth of the bearer channel, but only for a fraction of the time.

3.2 Power levels

A wide range of power levels is encountered in telecommunication transmission systems. It is therefore convenient to use a logarithmic unit for powers. This is the *decibel* (dB), which is defined as follows:

If the output power P_2 is greater than the input power P_1, then the gain G in decibels is

$$G = 10 \log_{10} \frac{P_2}{P_1} \text{ dB} \tag{3.1a}$$

If, however, $P_2 < P_1$, then the loss or attenuation in decibels is

$$L = 10 \log_{10} \frac{P_1}{P_2} \text{ dB} \tag{3.1b}$$

If the input and output circuits have the same impedance, then $P_2/P_1 = (V_2/V_1)^2 = (I_2/I_1)^2$ and

$$G = 20 \log_{10} \frac{V_2}{V_1} = 20 \log_{10} \frac{I_2}{I_1} \text{ dB} \tag{3.2}$$

In some countries, the unit employed is the *neper* (N), defined as follows:

$$\text{gain in nepers} = \log_e \frac{I_2}{I_1} \text{ N} \tag{3.3}$$

Thus, if the input and output circuits have the same impedance, a gain of 1 N corresponds to 8.69 dB.

A logarithmic unit of power is convenient when a number of circuits having gain or loss are connected in tandem. The overall gain or loss of a number of circuits in tandem is simply the algebraic sum of their individual gains and losses measured in decibels or nepers.

If a passive network, such as an attenuator pad or a filter, is inserted in a circuit between its generator and load, the increase in the total loss of the circuit is called the *insertion loss* of the network. If an active network, such as an amplifier, is inserted, the power received by the load may increase. There is thus an *insertion gain*.

The decibel, as defined above, is a unit of relative power level. To measure absolute power level in decibels, it is necessary to specify a *reference level*. This is usually taken to be 1 mW and the symbol dBm is used to indicate power levels relative to 1 mW. For example, 1 W=+30 dBm and 1 μW=−30 dBm. Sometimes (e.g. in satellite systems) the reference level is taken to be 1 W. The symbol used is then dBW.

Since a transmission system contains gains and losses, a signal will have different levels at different points in the system. It is therefore convenient to express levels at different points in the system in relation to a chosen point called the *zero reference point*. The *relative level* of a signal at any other point in the system with respect to its level at the reference point is denoted by dBr. This is, of course, equal to the algebraic sum of the gains and losses between that point and the reference point, as shown in Figure 3.1. For a 4-wire circuit (see Section 3.4), the zero-level reference point is usually taken to be the 2-wire input to the hybrid transformer.

It is often convenient to express a signal level in terms of the corresponding level at the reference point; this is denoted by dBm0. Consequently,

$$dBm0 = dBm - dBr$$

For example, if a signal has an absolute level of −6 dBm at a point where

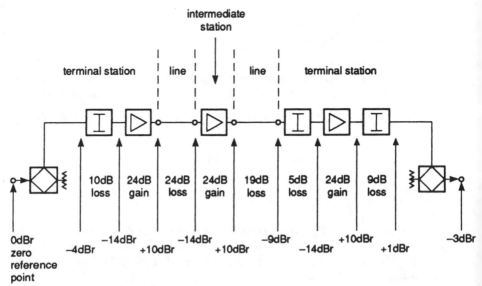

Figure 3.1   Example of relative levels in an analogue transmission system

the relative level is -10 dBr, the signal level referred back to the zero reference point is $+4$ dBm0.

3.3 Distortion

3.3.1 Nonlinear distortion

When a transmission path is nonlinear, the amplitude of the output signal is no longer directly proportional to the amplitude of the input signal. The effect is sometimes called *amplitude distortion*. The output signal also contains frequency components which are not present in the input signal. If a sinusoidal signal $v_i = V \cos \omega t$ is applied to a nonlinear transmission path, the output signal contains components at ω, 2ω, 3ω, ... Thus, the nonlinearity generates harmonics of the input frequency. The effect is therefore called *harmonic distortion*.

If the input signal contains two or more frequencies, the output signal does not only contain the fundamental frequencies and their harmonics; it contains sum and difference frequency components. The effect is called *intermodulation distortion*.

If the input signal contains more than two frequencies, the output signal contains many more intermodulation components. If the input signal occupies a wide frequency band, the spectrum of the intermodulation products is extremely complex [1]. The result is usually called *intermodulation noise*.

3.3.2 Attenuation and delay distortion

When a signal is transmitted from one point to another, there is inevitably attenuation due to energy losses in the transmission medium and delay due to the distance travelled at a finite velocity of propagation. The effect of the channel may be represented by its gain/frequency response $H(\omega)$ and its phase/frequency response $\phi(\omega)$. If the output signal is to be undistorted, then the channel must have a constant gain g and delay T for every frequency component contained in the input signal S_i. If the spectrum of S_i extends from ω_1 to ω_2, it is clearly required that

$$H(\omega) = g \quad \text{for } \omega_1 \leqq \omega \leqq \omega_2 \tag{3.4}$$

Any departure from this ideal condition is termed *attenuation distortion*.

Attenuation distortion may be corrected by inserting an *attenuation equaliser* [1]. This is a network designed to have a gain/frequency response $E(\omega)$ which is the inverse of that of the channel,

$$\text{i.e.} \quad E(\omega) \, H(\omega) = g \quad \text{for } \omega_1 \leqq \omega \leqq \omega_2$$

A flat overall gain/frequency response is thus obtained.

If a signal component $V \cos \omega t$ is delayed by a time T, the output is

$$gV \cos \omega(t - T) = gV \cos\{\omega t - \phi(\omega)\}$$

where

$$\phi(\omega) = \omega T \tag{3.5}$$

Thus, if the delay T is to be the same for all frequencies in the signal, we require $\phi \propto \omega$ for $\omega_1 \le \omega \le \omega_2$.

Any departure of the phase shift from the linear law is termed *phase distortion* or *delay distortion*. To obtain distortionless transmission, the phase/frequency characteristic should be a straight line through the origin [i.e. $\phi(0) = 0$]. This is not possible when the circuit contains transformers or coupling capacitors. Fortunately, it is not necessary for speech transmission, since the ear is insensitive to differences in phase. However, for a signal whose waveshape must be preserved (e.g. television and data signals), phase distortion must be minimised. It is then necessary to use a *phase equaliser*. This is a network designed to have a phase/frequency response $\theta(\omega)$ such that

$$\theta(\omega) + \phi(\omega) = \omega T \text{ for } \omega_1 \le \omega \le \omega_2$$

The quantity $d\phi/d\omega$, which is the slope of the phase/frequency characteristic, is called the *group delay* of the channel. A convenient measure of delay distortion is the difference between the maximum and minimum values of group delay occurring within the band ω_1 to ω_2. This is called the *differential delay* of the channel.

An equaliser may also be designed in the time domain. The equaliser is designed so that the combination of channel and equaliser in tandem has an impulse response which is within the limits specified. Now, the transfer function $H(\omega) \angle \phi(\omega)$ is the Fourier transform of the impulse response. Consequently, improving the impulse response is equivalent to improving the gain/frequency and phase/frequency responses.

3.3.3 Multipath transmission

One cause of severe attenuation and delay distortion is the *multipath effect*, shown in Figure 3.2a. This effect arises when the signal arrives at the receiving end of a channel over two or more paths having different delays. Examples of this are as follows:

(a) In a cable, there may be multiple reflections due to impedance irregularities.
(b) In a 4-wire circuit there may be echoes due to imperfect balances at the 4-wire/2-wire terminations.
(c) In long-distance HF radio transmission, the signal reflected from the ionosphere may be received both over a single-hop path and a multihop path.
(d) In VHF and UHF radio transmission, the signal may be received both by direct transmission and by reflection from the ground or from an obstacle.

If the difference between the lengths of the two paths is an exact multiple of is an odd number of half wavelengths, the signals arrive in antiphase and cancel. However, $\lambda = v/f$ (where v is the velocity of propagation and f the frequency), so the phase difference varies with frequency. Interference between the two signals thus causes the amplitude of the received signal to vary with frequency, as shown in Figure 3.2b. The phase/frequency characteristic contains ripples with the same periodicity.

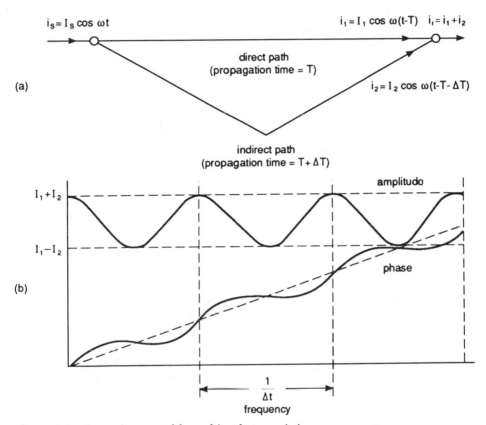

Figure 3.2 *Distortion caused by multipath transmission*

      a Interference between signals received over direct and indirect paths
      b Variation with frequency of amplitude and phase of received signal

3.4 Four-wire circuits

3.4.1 Principle of operation

It is frequently necessary to use amplifiers to compensate for the attenuation of a transmission path. Since most amplifiers are unidirectional, it is usually

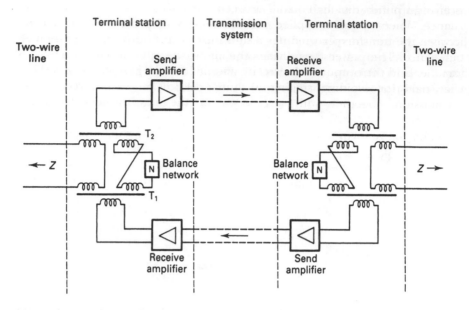

Figure 3.3   Four-wire amplified circuit

necessary to provide separate channels for the 'go' and 'return' directions of transmission. This results in a *four-wire circuit*, as shown in Figure 3.3. In practice, the go and return paths may be provided by channels in a high-capacity transmission system instead of on physical cable pairs.

At each end, the four-wire circuit must be connected to a two-wire circuit leading to an exchange. If both parts of the four-wire circuit were connected directly to the two-wire circuit at each end, a signal could circulate round the complete loop thus created. This would result in continuous oscillation or *singing*, unless the sum of the gains in the two directions was less than zero. To avoid this possibility, the two-wire line at each end is connected to the four-wire circuit by means of a *four-wire/two-wire terminating set*. This contains a *hybrid transformer** (consisting of two crossconnected transformers, as shown in Figure 3.3), and a *line-balance network* whose impedance is similar to that of the two-wire circuit over the required frequency band.

The output signal from the 'receive' amplifier causes equal voltages to be induced in the secondary windings of transformer T_1. If the impedances of the two-wire line and the line balance are equal, then equal currents flow in the primary windings of transformer T_2. These windings are connected in antiphase; thus, no EMF is induced in the secondary winding of T_2 and no signal is applied to the 'send' amplifier. Note that the output power from the

---

* The antisidetone transformer (induction coil) in a telephone also acts as a hybrid transformer, connecting the two-wire customer's line to the four-wire circuit consisting of the microphone and receiver.

receive amplifier divides equally between the two-wire line and the line balance. When a signal is applied from the two-wire line, the crossconnection between the transformer windings results in zero current in the line-balance impedance. The power thus divides equally between the input of the send amplifier and the output of the receive amplifier, where it has no effect. The price paid for avoiding singing is thus 3 dB loss in each direction of transmission, together with any losses in the transformers (typically 0.5–1.0 dB).

An alternative form of four-wire/two-wire termination, which uses a bridge circuit, is shown in Figure 3.4. If the balance impedance equals the impedance of the two-wire line, the bridge is balanced. The output of the receive amplifier then produces zero voltage at the input terminals of the transmit amplifier.

The impedance of the two-wire line varies with frequency. To achieve correct operation of four-wire/two-wire terminations, it would be necessary to measure the impedance/frequency characteristic of each line and to design a balance network to match it closely. In practice, the balance usually consists of resistor of value equal to the nominal impedance of the line (e.g. 600Ω). This is known as a *compromise balance*. Thus, a small fraction of the power received from one side of the four-wire circuit will pass through the hybrid transformer and be retransmitted in the other direction.

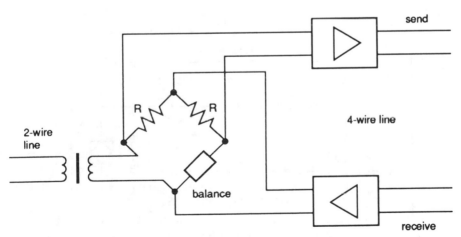

Figure 3.4   Four-wire/two-wire termination using bridge circuit

3.4.2 Echoes

In a four-wire circuit, such as that shown in Figure 3.3, an imperfect line balance causes part of the signal energy transmitted in one direction to return in the other. Both the talker and the listener may be able to hear the reflected signal and the effect is termed *echo*. The signal reflected to the speaker's end of the circuit is called *talker echo* and that at the listener's end is called *listener echo*. The paths traversed by these echoes are shown in Figure 3.5.

The attenuation between the four-wire line and two-wire line or between the two-wire line and four-wire line at each hybrid coil has been shown in Section 3.4.1 to be 3 dB. Thus, the total attenuation from one two-wire circuit to the other is

$$L_2 = 6 - G_4 \text{ dB} \tag{3.6}$$

where G_4 is the net gain of one side of the four-wire circuit (i.e. total amplifier gain minus total line loss).

The attenuation through the hybrid transformer from one side of the four-wire circuit to the other is called the *transhybrid loss*. It is equal to $6 + B$ dB [2] where

$$B = 20 \log_{10} \left| \frac{N+Z}{N-Z} \right| \text{ dB} \tag{3.7}$$

where

$$Z = \text{impedance of the two-wire line}$$

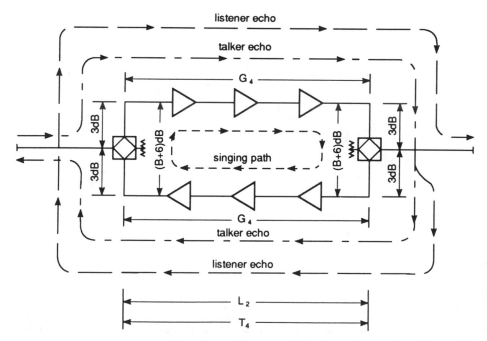

Figure 3.5   Echo and singing paths in a four-wire circuit

N=impedance of the balance network

The loss B represents that part of the trans-hybrid loss which is due to the impedance mismatch between the two-wire line and the balance network (see Section 3.11.1). It is known as the *balance-return loss* (BRL).

The attenuation, L_t, of the echo that reaches the talker's two-wire line, round the path shown in Figure 3.5, is

$$L_t=3 - G_4+(B+6) - G_4+3 \text{ dB}$$

$$=2L_2+B \text{ dB}$$

This echo is delayed by a time $2T_4$, where T_4 is the delay of the four-wire circuit (between its two-wire terminations). The attenuation L_1 of the echo which reaches the listener's two-wire line (relative to the signal received directly) is

$$L_1=(B+6) - G_4+(B+6) - G_4 \text{ dB}$$

$$=2L_2+2B \text{ dB}$$

and this is delayed by a time $2T_4$ relative to the signal received directly.

The effect of an echo is different for the speaker and the listener. For the speaker it interrupts speaking and for the listener it reduces the intelligibility of the received speech. Talker echo is the more troublesome because it is louder (by an amount equal to the BRL). The annoying effect of echo increases with its magnitude and the delay. The longer the circuit, the greater is the echo attenuation L_t required. This can be achieved by making the overall loss L_2 increase with the length of the circuit.

There is a limit to which the loss of connections can be increased to control echo. This is usually reached when the *round-trip delay* $2T_4$ is about 40 ms. This delay is exceeded on very-long transcontinental and intercontinental circuits, so it is impossible to obtain both an adequately low transmission loss and an adequately high echo attenuation. On such circuits, it is necessary to control echo by fitting devices called *echo suppressors* or *echo cancellers*.

An echo suppressor consists of a voice-operated attenuator, which is fitted in one path of the four-wire circuit and operated by speech signals on the other path. Whenever speech is being transmitted in one direction, transmission on the opposite direction is attenuated, thus interrupting the echo path. There is one such suppressor (called a 'half echo suppressor') at each end of the circuit.

A number of difficulties arise with simple echo suppressors of this type. In a very long-distance switched connection, it is possible to have a number of circuits fitted with echo suppressors connected in tandem; if these operate independently, 'lock-out' conditions can arise. It is therefore necessary to disable the echo suppressors on intermediate links in the connection. It is also necessary to disable echo suppressors during data transmission, since data-transmission systems often use a return channel to request retransmission of blocks of information when errors are detected. More-sophisticated echo suppressors have been designed [4] both to provide these facilities and to cater for the very long propagation times (250 ms each way) encountered on geosynchronous satellite links.

Echo cancellers are now also used [5]. The echo is cancelled by subtracting a replica of it from the received signal. This replica is synthesised by means of a filter, controlled by a feedback loop, which adapts to the transmission characteristic of the echo path and tracks any variations in it which may occur during a conversation.

3.4.3   Stability

If the balance-return losses of the terminations of a four-wire circuit are sufficiently small and the gains of its amplifiers are sufficiently high, the net gain round the loop may exceed zero and singing will occur. The net loss L_s of the singing path shown in Figure 3.5 is

$$L_s = 2(B+6-G_4) \text{ dB} \tag{3.8}$$

Substituting from eqn. 3.6 into eqn. 3.8 gives

$$L_s = 2(B+L_2) \text{ dB} \tag{3.9}$$

Thus, the loss of the singing path equals the sum of the two-wire-to-two-wire losses in the two directions of transmission and the BRLs at each end. The necessary condition for stability is $L_s > 0$. This requires that $L_2 + B > 0$, i.e.

$$G_2 < B \text{ (where } G_2 = -L_2) \tag{3.10}$$

The gain G_2 which can be obtained over a four-wire circuit is thus limited by the BRL. Eqn. 3.7 shows that, if $N = Z$, the balance return loss is infinite. In the limiting cases when either Z or N is zero or infinite, the balance return loss is zero. The loss between the return and go channels is then only 6 dB (plus any loss due to transformer inefficiencies).

In some cases, the two-wire line may be open-circuited for long periods because it is waiting for a call to be connected to it at the exchange. It may be short-circuited by a line or equipment fault. Although the circuit is not carrying traffic under these conditions, it is still undesirable for singing to occur. Because the magnitude of the oscillation increases until it is limited by the overloading of an amplifier, the power level of a channel which is singing greatly exceeds that for which transmission systems are designed. In a frequency-division-multiplex system, this increase in loading can cause interference with every other circuit routed over the system (see Section 3.5). Four-wire circuits are therefore usually set up to be stable even when the two-wire lines at each end are open-circuited or short-circuited ($B=0$). This requires operation with an overall net loss ($L_2 > 0$).

In practice, the attenuation of the singing path is deliberately made greater than zero. This provides a safety margin and avoids the attenuation distortion caused by echoes when the circuit is operating close to its singing point (see Section 3.3.3). The *singing point* of a circuit is defined as the maximum gain S that can be obtained (from two-wire to two-wire line) without producing singing. Thus, from eqn. 3.9, $S=B$; i.e. the singing point is given by the BRL (or the average of the two BRLs if these are different at the two ends of the circuit). The *stability margin* is defined as the maximum amount of additional gain M that can be introduced (equally and simultaneously) in each direction of transmission without causing singing, i.e. $L_s - 2M = 0$. Hence, from eqn. 3.9,

$$M = B + L_2 \text{ dB} \tag{3.11}$$

Thus, the stability margin is the sum of the two-wire-to-two-wire loss and the BRL.

In practice, a stability margin of 3 dB is found to be adequate* (i.e. $L_s = 6$ dB). If the circuit is to cater for zero BRL, the overall loss from two-wire circuit to two-wire circuit is then 3 dB.

In setting up long-distance switched connections, it is often necessary to connect a number of four-wire circuits in tandem. It is advantageous to eliminate terminating sets from the interfaces between the four-wire circuits rather than interconnect them on a two-wire basis, since the overall loss is lower. Trunk transit exchanges therefore use four-wire switching. The complete connection consists of a number of four-wire circuits in tandem with a four-wire/two-wire termination at each end of the connection. It is necessary to ensure that this complete circuit has adequate stability.

The loss of a four-wire circuit may depart from its nominal value for a number of reasons. These include:

(*a*) variation of line losses and amplifier gains with time and temperature;
(*b*) gain at other frequencies being different from that measured at the test frequency (usually 800 Hz or 1600 Hz); and
(*c*) errors in making measurements and lining up circuits.

The effect of these deviations is to cause the losses of circuits to have a statistical distribution about their mean (nominal) values. The standard deviation σ is usually between 1 dB (for modern systems) and 1.5 dB (for older systems). As a result of these variations a connection may become unstable, although its nominal loss provides an adequate stability margin. The probability of this happening must be very small (e.g. 1 in 1000).

Since the standard deviation σ increases with the number of circuits in tandem, so must the overall loss. A simple rule that has been adopted by operating administrations in a number of countries is

$$L_2 = 4.0 + 0.5n \text{ dB}$$

where L_2 is the overall loss (two-wire to two-wire) and n is the number of four-wire circuits in tandem.

When a circuit incorporates an HF radio link, variations in ionospheric

---

* If the standard deviation of G_4 is 1·0 dB and a normal probability distribution is assumed, then the probability of the gain exceeding the 3 dB stability margin is 1 in 1000.

conditions cause fading which can result in short-term variations of 20 dB in the receiver output level. Although the output level of the circuit can be controlled by a constant-volume amplifier at each radio terminal, it is impossible to ensure that the circuit always operates below its singing point. To prevent singing, a device known as a *singing suppressor* is fitted at each end of the four-wire circuit. A singing suppressor contains a voice-operated attenuator in both the 'go' and 'return' channel. In the quiescent condition, the suppressor introduces minimum loss in the return path and maximum loss in the go path. When incoming speech from the two-wire circuit is detected, the loss is removed from the go direction and introduced in the return direction. A singing suppressor thus also acts as an echo suppressor.

3.5 Crosstalk

Any sound heard in the telephone receiver associated with one channel resulting from the transmission of a signal over another channel is called *crosstalk*. Even low-level crosstalk can be intelligible and give the user the impression of a lack of secrecy.

In multipair cables the greatest cause of interference is capacitance. It can be shown that the interference between two pairs is zero if each wire of a pair has an equal capacitance to earth and has equal capacitance to each wire of the other pair. Care is therefore taken in the manufacture of multipair cables to make the capacitance unbalances as small as possible. Since any induced voltage due to capacitance unbalance or mutual inductance is directly proportional to frequency, crosstalk increases with frequency and determines the upper frequency limit for use of the cable. Crosstalk due to impedance unbalance may also arise when the two wires of a line have slightly different resistances (e.g. due to poor jointing) or the impedance of the equipment terminating the line is not balanced with respect to earth.

An additional source of crosstalk which is present in frequency-division-multiplex systems (carrier systems) is produced by amplifier nonlinearities. If an amplifier is nonlinear, then two signals of frequencies f_1 and f_2 at its input produce at the output not only f_1 and f_2 but also intermodulation products of frequencies $f_1 \pm f_2$, $f_1 \pm 2f_2$, etc., as shown in Section 3.3.1. If these components lie within the frequency bands of other channels, interchannel crosstalk results. Amplifiers for carrier systems therefore have large amounts of negative feedback to obtain very low intermodulation levels in addition to gain stability. It is also necessary for amplifiers to have a sufficiently high overload level to handle the total power of a multichannel signal and to line up circuits carefully to ensure that signal levels do not exceed their nominal values by more than a small tolerance.

When crosstalk energy is transferred from one channel to another, it can usually be detected at each end of the disturbed channel, as shown in Figure 3.6. When the crosstalk is propagated over the disturbed channel in the same direction as its own signal, the crosstalk is called *far-end crosstalk* (FEXT) or

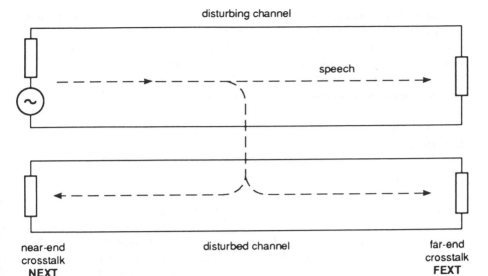

Figure 3.6 Near-end and far-end crosstalk

distant-end crosstalk. When the crosstalk is propagated over the disturbed channel in the opposite direction to its own signal, it is called *near-end crosstalk* (NEXT).

If the power of the signal entering the disturbing channel is P_1 and the power of the crosstalk received by the disturbed channel (at its near or far end) is P_2, then

$$\text{crosstalk attenuation} = 10 \log_{10} \frac{P_1}{P_2} \text{ dB}$$

If the power of the disturbed channel's own signal is P_3, then

$$\text{crosstalk ratio} = 10 \log_{10} \frac{P_3}{P_2} \text{ dB}$$

The crosstalk ratio is usually measured when test signals of the same level are applied to both the disturbing and disturbed channels.

In a multipair audio cable, power is transferred from the disturbing pair to the disturbed pair along the whole length of the cable. For FEXT, the ratio P_2/P_3 is therefore proportional to the length of the cable. For NEXT, the power transferred at a distance x from the sending end is attenuated by travelling distance x over the disturbing pair and distance x back again over the disturbed pair, a total distance of $2x$. Thus, most of the received crosstalk power is due to only a short length of the cable near the sending end and P_2 is almost independent of the total length. It can be shown [1] that for FEXT the crosstalk ratio decreases with frequency at a rate of 6 dB per octave and for NEXT it decreases by 4.5 dB per octave.

Crosstalk is most serious when it is heard during a silent interval in a conversation. The risk of crosstalk being audible is negligible if the crosstalk ratio exceeds 65 dB. A crosstalk ratio of 55 dB can often be tolerated when the crosstalk, although audible, is not intelligible.

3.6 Noise

3.6.1 General

Noise in a transmission channel may be generated internally or may be the result of interference from external sources. Interference enters a channel through unwanted electromagnetic coupling between a circuit and a source of interference such as a power line (i.e. the same mechanism as crosstalk). Internal line noise may be due to faulty joints whose contact resistance is varied by vibration caused, for example, by passing traffic. This is called *microphonic noise*. There is also an inherent source of noise due to the thermal agitation of electrons. This produces energy which is uniformly distributed over the frequency spectrum and is therefore often called *white noise*. In telephony, this is heard as a hiss. However, noise also occurs in short bursts; in telephony these are heard as clicks. This is called *impulse noise*. It is caused by electrical discharges, switching transients in electromechanical telephone exchanges and intermodulation in FDM systems.

3.6.2 Thermal noise

If the thermal noise has a power density of n watts per hertz, the noise power N in an ideal channel of bandwidth W hertz is therefore given by $N = nW$ watts. The noise voltage v_n is the result of the thermal agitation of a very large number of electrons. It therefore varies randomly with time and has a normal or Gaussian probability-density distribution, given by

$$p(v_n) = \frac{1}{\sigma\sqrt{(2\pi}} \exp(-v_n^2/2\sigma^2) \qquad (3.12a)$$

where σ is the standard deviation of v_n which is thus the RMS noise voltage.

The probability of the noise voltage being greater than $+V$ is

$$P(v_n > +V) = \int_{+V}^{\infty} p(v_n)\, dv = \tfrac{1}{2} - \int_0^v p(v_n)\, dv \qquad (3.12b)$$

and the probability of it being less than $-V$ is

$$P(v_n < -V) = \int_{-\infty}^{-V} p(v_n)\,dv = P(v_n > +V) \tag{3.12c}$$

These probabilities can therefore be obtained from standard statistical tables.

It has been shown [6,7] that the RMS noise voltage V_n produced by thermal agitation in a resistance of R ohms is given by

$$V_n^2 = 4k\ TWR\ \text{volts}^2$$

where
W = bandwidth, Hz
T = absolute temperature, K
k = Boltzmann's constant $(1.37 \times 10^{-23}\ \text{J/K})$.
If this resistance is connected to a matched load R, the noise power delivered to it is

$$P_n = \frac{(V_n/2)^2}{R} = kTW\ \text{watts} \tag{3.13a}$$

The *available noise power* is thus proportional both to bandwidth and temperature, but is independent of the resistance.

From eqn. 3.13a, the available noise power, expressed in dBW, is

$$N = 10\ \log_{10}\ k + 10\ \log_{10}\ T + 10\ \log_{10}\ W$$
$$= -228.6 + 10\ \log_{10}\ T + 10\ \log_{10} W\ \text{dBW} \tag{3.13b}$$

The available noise power is thus -228.6 dBW per hertz per kelvin. The noise power delivered into a circuit at a temperature of 290 K (i.e. 17°C) is 4×10^{-21} W per Hz of bandwidth, i.e. -174 dBm/Hz. For any specified signal-to-noise ratio, the minimum signal level that may be used is thus ultimately determined by thermal agitation noise.

When this resistive source is connected to a load, the load is also generating noise. If thermal agitation is the only source of noise, the total noise power in the matched load is the sum of the available noise powers of the source and the load (since each delivers half its total noise power to the other and dissipates half internally).

3.6.3 Noise factor and noise temperature

In an amplifier, in addition to thermal noise, there is also noise due to random fluctuations in the currents of transistors or other devices, i.e. shot noise and flicker [8,9]. The total noise present therefore exceeds that due to thermal agitation. The *noise factor* f_n of the amplifier is given by

$$f_n = \frac{\text{signal-to-noise ratio at input}}{\text{signal-to-noise ratio at output}}$$

$$\therefore f_n = \frac{\text{total noise power at output}}{\text{noise power due to thermal agitation}}$$

$$= \frac{N_0}{gkT_0 W}$$

where

N_0 = total output noise power
g = power gain of the amplifier
T_0 = reference temperature, i.e. 290 K

Since the output noise power from the amplifier is greater than that due to thermal noise alone, it is equal to the amount of thermal noise that would be present at a higher temperature than that actually present. This is known as the *noise temperature T_n*. Thus, $T_n = f_n T_0$. The difference between the noise temperature and the standard room temperature T_0 corresponds to the internally-generated noise. It is called the *excess noise temperature T_x*. Thus:

$$T_x = T_n - T_0 = T_0(f_n - 1)$$

and

$$f_n = 1 + T_x/T_0$$

The noise power at the output of an amplifier can thus be expressed in terms of an equivalent noise temperature. The noise appearing at the output terminals of a receiving antenna can also be expressed in terms of its noise temperature. In this case, the noise is due to radiation received from objects on earth, from objects in space and from interstellar gases.

If the noise factor expressed in decibels is F_n (where $F_n = 10 \log_{10} f_n$) the output noise power N_0 from an amplifier is given by

$$N_0 = F_n + G - 174 + 10 \log_{10} W \, \text{dBm}$$

where

G = power gain in decibels

If $W = 4$ kHz, then $10 \log_{10} W = 36$ dB and

$$N_0 = F_n + G - 138 \, \text{dBm}$$

3.6.4 Noise in analogue transmission systems

If an analogue transmission system contains n identical repeater sections, the total noise output power N_T is n times that of a single amplifier, i.e.

$$N_T = F_n + G - 138 + 10 \log_{10} n \, \text{dBm} \qquad (3.14)$$

If the relative output level of each repeater is L dBr, then the total output noise referred to the zero reference point is

$$N_{RO} = N_T - L \text{ dBm0}$$

Because the sensitivities of the ear and a telephone receiver vary with frequency, total noise power is not an accurate measure of its subjective effect. Therefore, noise is usually measured by a meter having a frequency-weighting network which has been standardised by the CCITT [10]. Such instruments [11] are called *psophometers*. Noise power is thus measured in units of pWp (picowatts psophometrically weighted) or dBmp (decibels relative to 1 mW psophometrically weighted). If the noise power is referred to the zero-level reference point in a system, then the units are pW0p and dBm0p. In the USA, noise power is usually referred to -90 dBm instead of 0 dBm (to make all measurements positive) and weighted noise is measured in dBrnc (where 'rn' stands for reference noise and 'c' for C message weighting). A noise level of 0 dBrnc is thus equivalent to -90 dBmp. For white noise in the band 0–4 kHz, the effect of psophometric weighting is to reduce the noise level by 3.6 dB. Thus, $-N$ dBm corresponds to $-(N+3.6)$ dBmp.

3.6.5 Companding

On some very long routes it is not possible to achieve an adeqaute signal-to-noise ratio using normal transmission techniques. For these circuits a method known as *companding* is used to reduce the effects of noise and interference [1]. At the sending end of the channel a *compressor* is used. This unit consists of an amplifier whose gain is controlled by the incoming speech signal so that low-level signals are amplified more than peak-level signals. The range of output levels (in decibels) is thus a fraction, say half, of the range of input levels. The signal-to-noise ratio for high-level input signals is unaffected; however, low-level input signals are transmitted at a relatively higher level, so they obtain an improvement in signal-to-noise ratio. At the receiving end of the channel an *expander* is used; this introduces a loss equal to the gain introduced by the expander, thus restoring the range of signal levels. The heavy attenuation of noise during quiet periods provides the major benefit of using companders. During speech, companding increases the range of levels over which a minimum signal-to-noise ratio is exceeded. This increase (in decibels) is referred to as the *companding advantage*.

The control signals applied to the compressor and expander are required to correspond to the average power of the input speech signal. They are therefore derived by a rectifier with a time constant approximating to the duration of the shortest elements of speech. This type of compandor is therefore called a *syllabic compandor* [12]. The action of an expander, in increasing the level range of output signals, magnifies the effect of any variation in the gain of the transmission channel. This can lead to instability, so syllabic compandors are only used when they are essential to make a long-distance circuit usable. For use on HF radio circuits, an improved form of compandor was developed in which a separate control signal is transmitted over the radio link to adjust the gain of the expander to suit the loss of the compressor. This system is known as Lincompex [13] (Linked Compressor and Expander). This enabled HF radio circuits to have much improved performance and removed the need for operators to make frequent adjustments.

3.7 Modulation

3.7.1 General

The processing of a signal to make it suitable for sending over a transmission medium is called *modulation* [14]. Reasons for using modulation are:

(*a*) frequency translating (e.g. when an audio-frequency baseband signal modulates a radio-frequency carrier);

(*b*) improving signal-to-noise ratio by increasing the bandwidth (e.g. using frequency modulation); and

(*c*) multiplexing (see Section 3.8).

Modulation is performed by causing the baseband modulating signal to vary a parameter of a carrier wave. A sinusoidal carrier $v_c = A \cos(\omega t + \phi)$ is defined by three parameters: amplitude A, frequency $\omega/2\pi$ and phase ϕ. Thus there are three basic modulation methods: *amplitude modulation* (AM), *frequency modulation* (FM) and *phase modulation* (PM).

When modulation is employed, a modulator is needed at the sending end of a channel and a demodulator at the receiving end recovers the baseband signal from the modulated carrier. The combination of modulator and demodulator at a terminal is often referred to as a *modem.*

3.7.2 Amplitude modulation

The simplest form of modulation is amplitude modulation. The modulator causes the envelope of the carrier wave to follow the waveform of the modulating signal and the demodulator recovers it from this envelope.

If a carrier $v_c = V_c \cos \omega_c t$ is modulated to a depth m by a sinusoidal modulating signal $v_m = V_m \cos \omega_m t$, the resulting AM signal is

$$v = (1 + m \cos \omega_m t) v_c$$

$$= (1 + m \cos \omega_m t) V_c \cos \omega_c t$$

$$= V_c\{\cos \omega_c t + \tfrac{1}{2}m \cos(\omega_c + \omega_m)t + \tfrac{1}{2}m \cos(\omega_c - \omega_m)t\}$$

If the modulating signal contains several components, $f_1, f_2, \ldots$, etc., then the modulated signal contains $f_c - f_1, f_c - f_2, \ldots$, etc. and $f_c + f_1, f_c + f_2, \ldots$, etc. in addition to f_c. If the modulating signal consists of a band of frequencies, as shown in Figure 3.7a, the modulated signal consists of two *sidebands*, each occupying the same bandwidth as the baseband signal, as shown in Figure 3.7b. In the *upper sideband*, the highest frequency corresponds to the highest

(a)

0   F_m

(b)

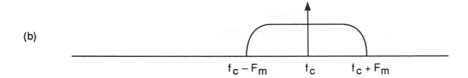

$f_c - F_m$ f_c $f_c + F_m$

(c)

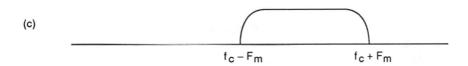

$f_c - F_m$ $f_c + F_m$

(d)

f_c $f_c + F_m$

(e)

f_c $f_c + F_m$

Figure 3.7   Frequency spectra for amplitude modulation

  a Baseband signal
  b Simple amplitude modulation (AM)
  c Double-sideband suppressed-carrier (DSBSC) modulation
  d Single-sideband suppressed-carrier (SSBSC) modulation
  e Vestigial-sideband (VSB) modulation

frequency in the baseband; this is therefore known as an *erect sideband*. In the *lower sideband*, the highest frequency corresponds to the lowest frequency in the baseband; this is known as an *inverted sideband*. Simple amplitude modulation makes inefficient use of the transmitted power, as information is transmitted only in the sidebands but the majority of the power is contained in the carrier. However, AM permits the use of a relatively simple receiver using an envelope demodulator. It is therefore used for services, such as broadcasting, where this is an important advantage.

It is possible, by using a balanced modulator [15], to eliminate the carrier and generate only the sidebands, as shown in Figure 3.7c. This is known as *double-sideband suppressed-carrier modulation* (DSBSC). To demodulate a DSBSC signal, it is necessary to use a *coherent demodulator* (consisting of a balanced modulator supplied with a locally-generated carrier) instead of the envelope demodulator used with simple AM.

If the incoming DSBSC signal is

$$v_i = \tfrac{1}{2}mV_c\{\cos(\omega_c + \omega_m)\,t + \cos(\omega_c - \omega_m)\,t\}$$

and the coherent demodulator multiplies this with a local carrier $[v_c = \cos(\omega_c t + \theta)$, its output voltage is

$$v = \tfrac{1}{4}m\,V_c[\cos\{(2\omega_c + \omega_m)\,t + \theta\} + \cos\{(2\omega_c - \omega_m)\,t + \theta\}$$
$$+ \cos(\theta + \omega_m)\,t + \cos(\theta - \omega_m)\,t]$$

The components at frequencies $(2\omega_c \pm \omega_m)$ are removed by a lowpass filter and the baseband output signal is

$$v_0 = \tfrac{1}{4}mV_c\{\cos(\theta + \omega_m t) + \cos(\theta - \omega_m t)\} \qquad (3.15a)$$
$$= \tfrac{1}{2}\,mV_c\,\cos\theta\,\cos\omega_m t \qquad (3.15b)$$

Thus v_0 represents the original baseband signal, provided that the phase θ of the local carrier is stable.

A further economy in power, and a halving in bandwidth, can be obtained by producing a *single-sideband suppressed-carrier* (SSBSC) signal, as shown in Figure 3.7d. If the upper sideband is used, the effect of the modulator is simply to produce a frequency translation of the baseband signal to a position in the frequency spectrum determined by the carrier frequency. If the lower sideband is used, the band is inverted as well as translated.

A coherent demodulator is required for demodulating a SSBSC signal. Since only one sideband is present, the demodulated output signal, from eqn. 3.15a, is

$$v_0 = \tfrac{1}{4}mV_c\cos(\omega_m t \pm \theta)$$

The SSBSC signal requires the minimum possible bandwidth for transmission. Consequently, the method is used whenever its complexity is justified by the saving in bandwidth. An error in the frequency of the local carrier at the receiver results in a corresponding shift in the frequencies of the components in the baseband output signal. For speech transmission, frequency shifts of the order of ± 10 Hz are not noticeable, but the errors that can be tolerated for telegraph and data transmission are less. The CCITT specified [16] that the frequency shift should be less than ± 2 Hz.

By using SSBSC, it is possible to transmit two channels through the bandwidth needed by simple AM for a single channel; one uses the upper sideband of the carrier and the other uses the lower sideband. This is known as *independent-sideband modulation* and is used in HF radio communication [2]. However, it is also possible to do this using DSBSC. The transmitter uses two modulators whose carriers are in quadrature. The receiver uses two coherent demodulators whose local carriers are in quadrature. Eqn. 3.15*b* shows that each demodulator produces a full output from the signal whose carrier is in phase (since cos 0=1) and zero output from the signal whose carrier is in quadrature (since cos $\pi/2=0$). This is called *quadrature amplitude modulation* (QAM). In practice, the method is not used for analogue baseband signals since small errors in the phases of the local carriers cause a fraction of the signal of each channel to appear as crosstalk in the output from the other. However, the method is used for transmitting digital signals, as described in Section 3.7.4.

If the baseband signal extends down to very low frequencies, as in television, it is almost impossible to suppress the whole of the unwanted sideband without affecting low-frequency components in the wanted sideband. Use is then made of *vestigial-sideband* (VSB) transmission instead of SSBSC. A conventional AM signal (as shown in Figure 3.7*b*) is first generated and this is then applied to a filter having a transition between its pass and stop band that is skew symmetric about the carrier frequency. This results in an output signal having the spectrum shown in Figure 3.7*e*. If a coherent demodulator is used, the original baseband signal can be recovered without distortion. It is also possible to use a simple envelope demodulator for VSB, but some nonlinear distortion then results [14]. VSB transmission does, of course, require a greater channel bandwidth than SSB. However, for a wideband signal such as television, the bandwidth saving compared with DSB is considerable.

3.7.3 Frequency and phase modulation

The instantaneous angular frequency of an alternating voltage is $\omega = d\phi/dt$ radian/s. This relationship between frequency and phase means that frequency modulation (FM) and phase modulation (PM) are both forms of *angle modulation*.

A sinusoidal carrier modulated by a sinusoidal baseband signal may be represented by

$$v = V_c \cos(\omega_c t + \beta \sin \omega_m t)$$

where

$$\beta = modulation\ index$$

The maximum phase deviation is $\Delta\phi = \pm\beta$ and, since $d\phi/dt = \omega_m\beta \cos \omega_m t$, the maximum frequency deviation is $\Delta F = \pm\beta f_m$. In PM, the phase deviation is proportional to the modulating voltage; therefore β is independent of its frequency and the frequency deviation is proportional to it. In FM, the frequency deviation is proportional to the modulating voltage; therefore the deviation frequency is independent of its frequency and β is inversely proportional to it. Thus, for FM, the modulation index may be defined as

$$\beta=\frac{\text{maximum frequency deviation of carrier}}{\text{maximum baseband frequency}}=\frac{\Delta F}{F_m}$$

In angle modulation, the information is conveyed by the instantaneous phase of the signal. Consequently, phase distortion in the transmission path causes attenuation distortion of the received signal. The differential delay of the transmission path must therefore be closely controlled over the bandwidth required to transmit the signal.

The frequency spectrum of the transmitted signal contains higher-order sideband components [14] at frequencies $f_c \pm nf_m$ (where $n=1,2,3,...$). The bandwidth required is thus greater than for AM. However, there is a little energy outside the band $(1\pm\beta)f_c$. To a first approximation the bandwidth W required [1] is

$$W=2F_m(1+\beta) \quad \text{(Carson's rule)}$$

Thus, for FM,

$$W=2(F_m+\Delta F)$$

where ΔF=maximum frequency deviation.

For low-index modulation ($\beta<1$) the bandwidth needed is little more than that for AM, but for large values of β it is much more.

In FM, the amplitude of the signal conveys no information. Thus, amplitude variations due to fading have no effect on the output-signal level. Amplitude variations caused by noise also have no effect.* However, noise also perturbs the phase of the signal and thus its instantaneous frequency. For white noise, it can be shown [14] that the improvement in signal-to-noise ratio compared with fully-modulated AM, for the same transmitter carrier power and same noise-power density at the receiver, is given by

$$\frac{\text{output signal/noise power ratio for FM}}{\text{output signal/noise power ratio for AM}}=3\beta^2$$

The improvement obtained is thus proportional to the square of the frequency deviation used.

The amplitude of the disturbance produced in the output of an FM receiver by an interfering signal is proportional to the frequency difference between the interference and the unmodulated carrier frequency. White

---

* For this assumption to be valid, the input signal-to-noise ratio must exceed a *threshold* [1, 14] of approximately 12 dB. When the input signal-to-noise ratio decreases below this, peaks of noise voltage begin to obliterate the carrier and the output signal-to-noise ratio deteriorates rapidly. Radio links used for commercial telecommunication circuits must therefore have signal-to-noise ratios well above the threshold.

noise thus produces a 'triangular' output-noise spectrum in which the RMS noise voltage (per hertz) is proportional to frequency [1]. It is therefore common practice to insert in front of the modulator at the transmitter a *pre-emphasis* network which has a rising gain/frequency characteristic across the baseband. The receiver contains a *de-emphasis* network with the inverse gain/frequency characteristic, to obtain a channel having a flat overall gain/frequency characteristic and a uniform noise spectrum. For single-channel telephony or broadcasting, this provides a better output signal-to-noise ratio. When FM is used to transmit a wideband signal consisting of a block of telephone channels assembled by frequency-division multiplexing, the use of pre-emphasis is essential [1]. Otherwise, there would be a large difference between the signal-to-noise ratios of channels at the top and bottom of the band.

Because of its good signal-to-noise ratio properties, FM is used for radio circuits whenever sufficient bandwidth can be provided. This is not possible for telephony in the more congested parts of the radio-frequency spectrum (VHF and lower frequencies), but it is standard practice at UHF and SHF. For telegraph signals, because of their narrow baseband, it is possible to use FM in the HF band [2]. This is known as *frequency-shift keying* (FSK). This form of transmission is also used for data transmission [17,18] over telephone circuits in the switched telephone network [19,20]. A typical data modem for use on telephone circuits [21] uses frequencies of 1.3 kHz and 1.7 kHz to transmit at 600 baud or 1.3 kHz and 2.1 kHz to transmit at 1200 baud.

In phase modulation (PM), the frequency deviation βf_m is proportional to the frequency of the modulating signal as well as its amplitude. Consequently, for signals (such as speech) which have the major proportion of their energy at the lower end of the baseband, PM makes inefficient use of transmission-path bandwidth in comparison with FM. Moreover, to demodulate a PM signal, the receiver must compare the phase of the incoming carrier with that of a locally generated carrier which must be extremely stable. FM is therefore preferred to PM for the transmission of analogue signals.

Phase modulation is, however, used for the transmission of digital signals [17,18]. This is called *phase-shift keying* (PSK). The carrier phase is switched to one of a number of possible values, e.g. 0° and 180° for a binary system,* or 0°, 90°, 180° and 270° for a four-level system. The receiver detects which level has been sent by comparing the phase of the incoming signal with that of a locally-generated reference carrier. The stability of this reference must be very high, so an alternative method, known as *differential phase-shift keying* (DPSK), is often used to avoid the need for a reference carrier. In DPSK the phase difference between successive intervals is used to convey the information, this difference being measured at the receiver by comparing the received waveform with the same waveform delayed by one interval [23].

---

* This is sometimes called *phase-reversal keying* (PRK). It is equivalent to DSBSC with a modulating signal that is either +1 or −1.

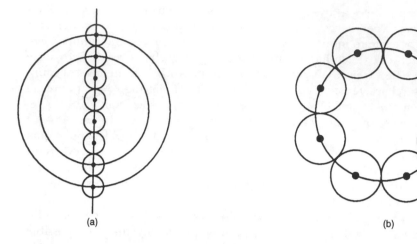

(a)   (b)

Figure 3.8   Signal-space diagrams for eight-state amplitude and phase modulation

        a Amplitude modulation
        b Phase modulation

3.7.4 Hybrid modulation

If a carrier is modulated by a digital signal, its amplitude and phase can only have a finite number of values. These can be represented in a signal–space diagram, as shown in Figure 3.8. This shows an eight-level amplitude-shift-keyed (ASK) signal and an eight-level PSK signal. The circles show the maximum permissible perturbation from the ideal signal before an error may occur in the transmission. It is obvious that, in this respect, PSK is superior to ASK.

It is possible to combine both ASK and PSK to obtain *hybrid modulation* [22], as shown in Figure 3.9. The eight-level system shown in Figure 3.9*b* has a similar error threshold voltage to the PSK system of Figure 3.8*b*. However, it corresponds to a lower mean transmitted power, because four of the eight states have a smaller carrier amplitude than those in Figure 3.8*b*. These hybrid modulation schemes can be implemented using *quadrature amplitude modulation* (QAM), i.e. DSBSC modulation of two carriers in quadrature (as discussed in Section 3.7.2). For example, if a four-level digital signal is applied to each of the two modulators, the sum of their outputs has four values of its in-phase component and four values of its quadrature component; this produces the 16-state signal constellation shown in Figure 3.9*c*. Two coherent demodulators reproduce the two four-level signals at their outputs.

16-state QAM is used to provide 2400 bit/s data transmission over switched telephone circuits with a signalling rate of only 600 baud (CCITT Recommendation V.22 *bis*) and 9.6 kbit/s over private circuits with a signalling rate of only 2400 baud (CCITT Recommendation V.32). The output of the four-state QAM system shown in Figure 3.9*a* has a constant amplitude; it therefore

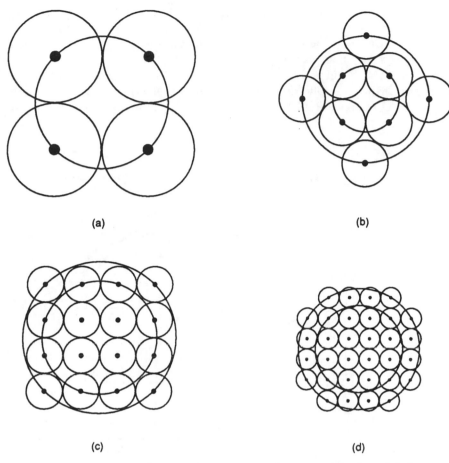

Figure 3.9   Signal-space diagrams for m-state quadrature-amplitude modulation

    a 4-state
    b 8-state
    c 16-state
    d 32-state

provides phase-shift keying and is called *quadrature phase-shift keying* (QPSK). The input is coded so that phase changes of the transmitted signal are always 90° and never 180°. If nine-state QAM is used in conjunction with partial-response coding (which is sometimes called duobinary coding), the system is called a *quadrature-partial-response system* (QPRS) [23].

3.7.5 Pulse modulation

In the examples above, the carrier waveforms are sinusoidal. It is also possible to modulate carriers having other waveforms. For example, it is possible to modulate trains of pulses [14] to produce pulse-amplitude modulation (PAM), pulse-frequency modulation (PFM) or pulse-phase modulation (PPM). It is also possible to produce pulse-length modulation (PLM), which is sometimes known as pulse-width modulation (PWM).

A basic PAM system is shown in Figure 3.10*a*. If a train of pulses as shown in Figure 3.10*c* is amplitude modulated by the baseband signal shown in Figure 3.10*b*, the resulting PAM signal is shown in Figure 3.10*d*. The baseband

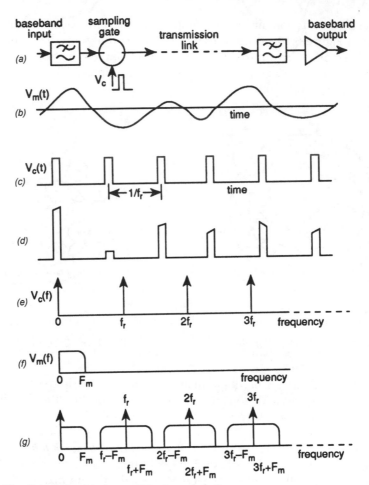

Figure 3.10   Principle of pulse-amplitude modulation

    a Basic system (one direction of transmission only)
    b Baseband signal
    c Unmodulated pulse train
    d Modulated pulse train
    e Spectrum of unmodulated pulse train
    f Spectrum of baseband signal
    g Spectrum of modulated pulse train

signal is represented by a sequence of samples of it, so the process is also called *sampling*.

A train of pulses having a pulse-repetition frequency (PRF) f_r may be represented by a Fourier series

$$v_c = \tfrac{1}{2}a_0 + \sum_{n=1}^{\infty} a_n \cos n\omega_r t$$

Its spectrum thus contains a DC component, the PRF and its harmonics, as shown in Figure 3.10e. If the pulses are of short duration, the amplitudes a_n of the harmonics are approximately equal up to large values of n.

If the pulse train is amplitude modulated by a sinusoidal baseband signal, the resulting PAM signal is

$$v = (1 + m \cos \omega_m t) v_c$$

$$= \tfrac{1}{2}a_0 + \tfrac{1}{2}a_0 m \cos \omega_m t + \sum_{n=1}^{\infty} a_n \cos n\omega_r t$$

$$+ \tfrac{1}{2}m \sum_{n=1}^{\infty} a_n \{\cos (n\omega_r + \omega_m) t + \cos (n\omega_r - \omega_m) t\}$$

Thus, the spectrum of the PAM signal contains all the components of the original pulse train, together with the baseband frequency f_m and upper and lower side frequencies $(nf_r \pm f_m)$ about the PRF and its harmonics. If the modulating signal consists of a band of frequencies, as shown in Figure 3.10f, the spectrum of the PAM signal contains the original baseband, together with upper and lower sidebands about the PRF and its harmonics, as shown in Figure 3.10g.

The PAM signal can be demodulated by means of a lowpass filter which passes the baseband and stops the lower sideband of the PRF and all higher frequencies. For this to be possible, we require

$$F_m \leqq f_0 \leqq f_r - F_m$$

where F_m is the maximum frequency of the baseband signal and f_0 is the cutoff frequency of the filter. Thus, we require

$$f_r \geqq 2F_m \qquad\qquad (3.16)$$

Eqn. 3.16 is a statement of the *sampling theorem*. This may be expressed as follows: if a signal is to be sampled and the original signal is to be recovered from the samples without error, the sampling frequency must be at least twice the highest frequency in the original signal. The sampling theorem is due to Nyquist and the lowest possible rate at which a signal may be sampled, $2F_m$, is often known as the *Nyquist rate*. If the sampling frequency is less than the Nyquist rate, the lower sideband of the PRF overlaps the baseband and it is impossible to separate them. The output from the lowpass filter then contains unwanted frequency components; this situation is known as *aliasing*.

To prevent aliasing, it is essential to limit the bandwidth of the signal before sampling. Thus, practical systems pass the input signal through an anti-aliasing lowpass filter of bandwidth $\frac{1}{2}f_r$ before sampling, as shown in Figure 3.10a. Practical filters are nonideal; it is therefore necessary to have $f_r > 2F_m$ in order that the anti-aliasing filter and demodulating filter can have both very low attenuation at frequencies up to F_m and very high attenuation at frequencies down to $f_r - F_m$. For telephony, a baseband from 300 Hz to 3.4 kHz is provided and a sampling frequency of 8 kHz is used. Thus, $f_r - F_m = 4.6$ kHz and there is a *guardband* of 1.2 kHz to accommodate the transition of the filters between their passband and their stop band.

When PFM, PPM and PLM are used, the sidebands include higher-order sideband components [14] at frequencies $nf_r \pm jf_m$ (where $j = 1,2,3, ...$), as in FM and PM. It is therefore necessary to use a sampling frequency greater than $2F_m$ to minimise aliasing.

These methods of pulse modulation all generate analogue signals. It is possible to produce a digital signal by applying a train of PAM samples to an analogue-to-digital (A/D) convertor, as shown in Figure 3.11. Each analogue sample is thus converted to a group of on/off pulses which represents its voltage in a binary code. This process is called *pulse-code modulation* (PCM) [24]. At the receiving terminal, a digital-to-analogue (D/A) convertor performs the decoding process. The combination of coder and decoder at a PCM terminal is often referred to as a *codec*. The group of bits (i.e. binary pulses) representing one sample is called a *word* or a *byte*. An 8-bit byte is sometimes called an *octet*.

For telephony, speech samples are usually encoded in an 8-bit code. Since sampling is at 8 kHz, a telephone channel requires binary digits to be sent at the rate of $8 \times 8 = 64$ baud. Nyquist has shown [25] that the minimum bandwidth required to transmit pulses is half the pulse rate. Thus, a bandwidth of at least 32 kHz is required to transmit a single telephone channel. The advantages of digital transmission are won at the expense of a much greater bandwidth requirement.

PCM introduces a transmission impairment known as *quantisation distortion*.

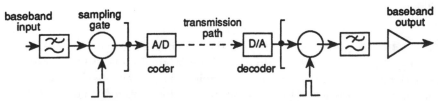

Figure 3.11   PCM transmission system (one direction of transmission only)

This arises because the system can only transmit a finite number of sample values separated by steps called *quanta*. For example, if the PCM system uses an eight-digit code, it can transmit $2^8 = 256$ different sample values. There is thus a small difference between the output signal and the original input signal. This form of nonlinear distortion is known as quantisation distortion. If the amplitude of the input signal is large compared with a quantising step, the errors in successive samples are nearly random. The spectrum of the distortion products thus approximates to that of white noise and the effect is usually called *quantisation noise.*

If the coder uses quantising steps of uniform size, then a large-amplitude signal is represented by a large number of steps and is reproduced with little distortion. However, a small-amplitude signal will range over few steps and a large percentage distortion will be present in the output signal. Thus, the signal-to-quantisation-noise ratio (SQNR) varies with the level of the input signal. This effect can be reduced by using a compandor to increase the amplitudes of low-level signals to enable them to use a larger number of quantising steps.

The required effect can be obtained by using nonuniform quantising [24], with small steps at low input voltages and large steps at large input voltages. This is known as *instantaneous companding* to differentiate it from syllabic companding. The reduction in quantising noise obtained at low signal levels (relative to a system with uniform quantisation) is called the companding advantage.

Two nonlinear encoding laws have been standardised by the CCITT [26]; these are known as the A law and the μ law. The former is used in Europe and the latter in North America and Japan. The A law gives a companding advantage of 24 dB and the μ law gives 30 dB advantage.

3.8 Multiplexing

3.8.1 Frequency-division multiplexing

In frequency-division-multiplex (FDM) transmission, a number of baseband channels are sent over a common wideband transmission path by using each channel to modulate a different carrier frequency. Systems using this process are called *multichannel carrier systems.* They are comonly employed to transmit 24 telegraph channels over a single telephone channel, to transmit 24 telephone channels over a balanced pair or many hundreds of telephone channels over a coaxial pair or a microwave radio link.

A *channel translating equipment*, or *channelling equipment*, for multiplexing 12 telephone channels is shown in Figure 3.12a. At the sending end, each incoming baseband signal $(0 \leq f_m \leq F_m)$ from an audio-frequency circuit is applied to a balanced modulator supplied with the appropriate carrier f_c. The output of this modulator is a double-sideband-suppressed-carrier signal, as shown in Figure 3.7c. This signal is applied to a bandpass filter which suppresses the upper sideband (f_c+f_m) and transmits the lower sideband (f_c-f_m). The outputs of these filters are commoned to give a composite output signal containing the signal of each telephone channel translated to a different portion of the frequency spectrum, as shown in Figure 3.12c.

At the receiving end, the incoming signal is applied to a bank of bandpass filters, each of which selects the frequency band containing the signal of one channel. This signal is applied to a modulator supplied with the appropriate carrier f_c and the output of this modulator consists of the baseband signal and unwanted high-frequency components. The unwanted components are suppressed by a lowpass filter and the baseband signal is transmitted to the audio-frequency circuit at the correct level by means of an amplifier.

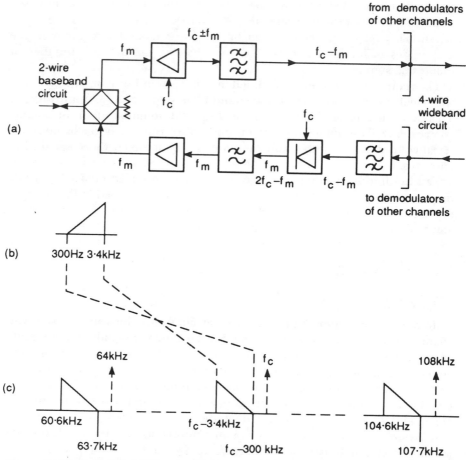

Figure 3.12   *Principle of frequency-division multiplexing*

>   *a* Channel-translating equipment
>   *b* Frequency band of baseband signal
>   *c* Frequency band of wideband signal (CCITT basic group B)

Suppressed-carrier modulation is used to minimise the total power to be handled by amplifiers in the wideband transmission system. The use of single-sideband modulation (SSB) maximises the number of channels that can be transmitted in the bandwidth available. To avoid interchannel crosstalk the sidebands of adjacent channels obviously must not overlap, so the spacing between carrier frequencies is determined by the highest frequency F_m of the baseband signal. Practical bandpass filters cannot have a perfectly sharp cutoff; so it is necessary to leave a small guardband between the frequency bands of adjacent channels. Figure 3.12c shows the standard *basic group* of 12 channels (CCITT basic group B). The carrier spacing is 4 kHz; thus 12 channels occupy the band from 60 to 108 kHz. Each channel has a baseband from 300 Hz to 3·4 kHz. The frequency guardband between adjacent channels is only 900 Hz, so crystal filters are used to obtain the necessary sharp transitions between pass and stop bands.

3.8.2 Time-division multiplexing

Another method of multiplexing the signals of a number of baseband channels to transmit them over a wideband transmission path is called *time-division multiplexing* (TDM), which is illustrated in Figure 3.13. At the sending terminal, a baseband channel is connected to the common transmission path by means of a sampling gate which is opened for short intervals by means of a train of pulses. In this way, samples of the baseband signal are sent at regular intervals by means of amplitude-modulated pulses. Pulses with the same repetition frequency f, but staggered in time, as shown in Figure 3.13b, are applied to the sending gates of the other channels. Thus the common transmission path receives interleaved trains of pulses, modulated by the signals of different channels. At the receiving terminal, gates are opened by pulses coincident with those received from the transmission path so that the demodulator of each channel is connected to the transmission path for its allotted interval and disconnected throughout the remainder of the pulse-repetition period. The combination of a multiplexer and a demultiplexer at a TDM terminal is sometimes referred to as a *muldex* or *mux.*

The pulse generator at the receiving terminal must be synchronised with that at the sending terminal to ensure that each incoming pulse train is gated to the correct outgoing baseband channel. A distinctive synchronising-pulse signal is therefore transmitted in every repetition period in addition to the channel pulses. The complete waveform transmitted during each repetition period contains a number of time slots: one is allocated to the synchronising signal and the others to the channel samples. The comlete waveform is called a *frame*, by analogy with a television-signal waveform, and the synchronising signal is called the *frame-alignment signal.*

The elementary TDM system shown in Figure 3.13 uses pulse-amplitude modulation (PAM). Other analogue methods of pulse modulation, including pulse-length modulation and pulse-position modulation can also be used [14]. However, these methods are not employed for line transmission. This is

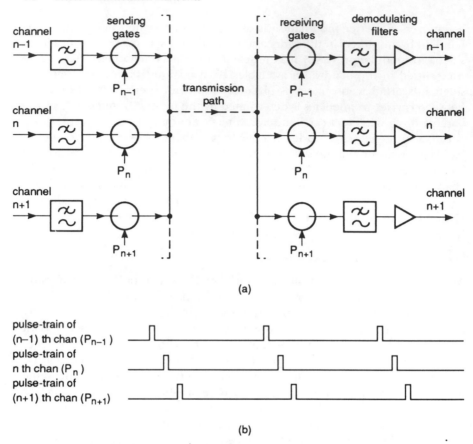

(a)

(b)

Figure 3.13   *Principle of time-division multiplexing*

        a Elementary TDM system (one direction of transmission only)
        b Channel pulse trains

because attenuation and delay distortion cause the transmitted pulses to spread in time so that they interfere with the pulses of adjacent channels and cause crosstalk. To overcome this problem, pulse-code modulation (PCM) is used for TDM transmission of telephone signals.

In PCM, the coder used for A/D conversion and the decoder used for D/A conversion are required to perform their operations within the duration of the time slot of one channel. They can therefore be common to all the channels of a TDM system, as shown in Figure 3.11.

3.8.3 Time-compression multiplexing

Time-compression multiplexing (TCM) uses time division. However, instead of sending a sample from each baseband channel sequentially, as in conventional TDM, the signal of each channel is stored for a short time, this process being carried out concurrently for all channels. The signals are then read out from each store in turn and transmitted in a much shorter time over the wideband channel. At the receiving terminal, the received signals are written into another set of stores. They are then read out in parallel and fed concurrently to the corresponding channels.

If analogue pulse modulation is used, transmission of TCM requires less bandwidth than TDM. Because most adjacent samples are from the same channel, there is no need for guard intervals between them to prevent interchannel crosstalk. The bandwidth required is thus only a few percent more than for FDM transmission. If digital transmission is used, there is no need for guard intervals between channels: thus, the bandwidths needed for TDM and TCM are the same. However, the frame structures are different. If a TCM system stores k samples of each channel before sending them, the transmitted frame contains k adjacent samples from each channel in turn. Consequently, the duration of the TCM frame is k times that of a TDM frame.

The TCM principle is used for *time-division multiple access* (TDMA) in satellite-communication systems. Because the earth stations have different locations, the propagation times between them and the satellite differ. If TDM were used, it would be impracticable to ensure correct interleaving of samples arriving at the satellite. Using TCM, if each earth station assembles k samples before sending them, the tolerances on timing are increased by a factor k. In a typical TDMA satellite system, the frame period is 2 ms, instead of the 125 μs used in TDM telephony.

A form of time-compression multiplexing is used in *packet switching*. Data from a slow-speed source are assembled to form packets; these are then transmitted over a data link at a higher speed, interleaved with packets from other sources. Since packets from different sources must be routed to different destinations, 'overheads' are incurred. Each packet contains a *header* which includes its address, together with other information required by the protocols which control the system. Packet switching is described in Chapter 13.

A special form of packet switching is used in the *asynchronous-transfer-mode* (ATM) system. This is a system designed to handle traffic inputs with different bandwidths. It uses short fixed-length packets called *cells*. The cells, which consist of five header bytes and 48 data bytes, are short in order to minimise delay (which is essential for telephony). The system operates at a fixed digit rate (approximately 150 Mbit/s) and different services are handled by changing the intervals between cells. For example, a 33 Mbit/s video codec will require many more cells in a given time than a 64 kbit/s speech codec. ATM working is described in Chapter 15.

3.9 Bandwidth compression techniques

Natural speech is a highly redundant method of encoding information. Some redundancy is removed by restricting the band of a telephone channel to

300–3400 Hz. Nevertheless, the information capacity of a normal telephone channel is much greater than the information rates of the messages conveyed. Methods have therefore been invented to convey telephone speech through analogue channels having less than 3 kHz bandwidth and digital channels conveying less than 64 kbit/s. These methods make use of the *a priori* information that the signals to be transmitted have been generated by the human-voice mechanism.

An early example was the vocoder [27]. At the sending end of a channel, the speech signal is analysed by a bank of filters into several frequency bands and the relatively slowly changing amplitudes of these components are transmitted over a narrowband channel. At the receiving end, an output signal is synthesised by generating components in these frequency bands with the same relative amplitudes. This output is intelligible but unnatural. More recent methods, which reduce the bit rate for the digital transmission of speech, are described in Reference 28.

Television signals are also highly redundant and methods have been developed for processing them in order to transmit much lower bit rates than those required for normal PCM. These methods are also discussed in Reference 28.

An obvious redundancy in telephone transmission can be removed more easily than the redundancy in speech signals themselves. Telephone connections normally provide a full-duplex circuit. Speech can be transmitted in both directions at all times, although the user at each end speaks for only about 40% of the time. The rest of the time is occupied by pauses in the speech and listening to the speaker at the other end of the connection. A system called *time-assignment speech interpolation* (TASI) [29] was developed to remove this redundancy and so reduce the number of channels required in each direction to less than half the number of telephone connections provided.

At the sending end of a TASI system, a number of input channels N is connected to a smaller number of channels M, provided over the transmission medium, by means of a high-speed switch. At the receiving end, these M channels are connected to N output channels by a similar switch. When speech activity is detected on an input channel, it is assigned to a channel of the transmission system and remains connected to it for the duration of the speaker's 'talkspurt'. A signal is transmitted to the switch at the receiving end, instructing it to connect the channel assigned in the transmission system to the output channel which corresponds to the active channel at the sending end. Since it takes a finite time to detect the presence of speech on an input channel, the first few milliseconds of the initial syllable of each talkspurt are lost. Also, when the number of active input channels exceeds the number of channels in the transmission system, then 'freeze out' occurs and complete talkspurts are lost. The TASI advantage, i.e. the ratio N/M, depends on the TASI activity factor, i.e. the percentage of time for which speech is present on an input circuit, and on the percentage of freeze out that is acceptable. A typical TASI system transmits 235 speech channels over a 96-channel transmission system [1].

It is economic to employ TASI if the cost of the necessary speech detection, switching and signalling is less than the cost of the additional telephone circuits that would otherwise be required. Considerable savings can be made if these circuits are very expensive, for example in a trans-oceanic cable or a satellite communications system.

When PCM transmission is employed, the switching and multiplexing can be combined, both using TDM. This is called *digital speech interpolation* (DSI) [30]. It is used with satellite systems that employ time-division multiple access (TDMA), as described in Section 3.8. Using DSI, 240 speech channels at 64 kbit/s are transmitted over 127 satellite 64 kbit/s channels. When the latter are all busy, freeze out is avoided by changing from 8-bit to 7-bit coding in order to use fewer digits per time slot and thus provide up to 16 more channels.

Signal interpolation can also be applied in data transmission, when a number of low-speed data terminals are connected to a high-speed data link by time-division multiplexing. If the input data traffic is 'bursty', i.e. messages are short with intervals between them, the number of terminals served can be greater than the number of time slots provided by the data link. The multiplexer assigns time slots only to those terminals actively sending data. Such a multiplexer [22] is called a *statistical multiplexer*, or 'statmux' for short. No information is lost when 'freeze-out' occurs, because data can be stored at the input ports of the multiplexer and sent when time slots become free.

3.10 Digital transmission

3.10.1 Bandwidth requirements

The minimumn bandwidth needed to transmit a digital signal at B baud has been shown by Nyquist [25] to be $W_{min} = \frac{1}{2}B$. To achieve this, the channel must have an ideal lowpass-filter characteristic, i.e. its attenuation and delay must be constant from zero frequency to frequency W_{min} and its attenuation infinite at all higher frequencies. The impulse response then has zeros at intervals of $1/2W_{min}$. Thus, if impulses are transmitted at rate $B = 2W_{min}$, the peak of each received pulse can occur when the responses to all preceding and following impulses are zero. There is then said to be zero *intersymbol interference* (ISI).

In practice, it is not possible to obtain a channel with an ideal lowpass characteristic. However, Nyquist showed [25] that zero ISI can also be obtained if the gain of the channel changes from unity to zero over a band of frequencies W with a gain/frequency response that is skew-symmetrical about $f = \frac{1}{2}W$. It is also impossible to generate a perfect impulse (since it has zero duration and infinite amplitude). The transfer characteristic of the channel should therefore be equalised so that the output signal has the required spectrum. A commonly used signal is that having a *raised cosine spectrum*.

Bandwidth is used most efficiently by making the 'roll off' of the spectrum occur over a small frequency range, but this increases problems of timing and equalisation. Commonly, a roll off between zero frequency and *B* is used. The bandwidth is then twice the theoretical minimum.

3.10.2 Equalisation

Digital transmission systems can use gain and phase equalisation to obtain an output-signal spectrum corresponding to a pulse waveform with negligible intersymbol interference, e.g. the raised-cosine spectrum. However, time-domain equalisers are often employed.

A common form of time-domain equaliser is the *transversal equaliser* (TVE) shown in Figure 3.14. This consists of a delay line tapped at intervals equal to the intersymbol interval. Each tap is connected to an amplifier (which may be an inverter to obtain negative gain). The output of the equaliser is the sum of the outputs of these amplifiers. It is possible to adjust the gains of the amplifiers (in magnitude and sign) to cancel ISI by adding appropriately weighted versions of preceding and following pulses at the time of each symbol, and thus cancel interference between them.

A TVE can be adjusted mannually. However, if the characteristics of the transmission path change with time, it is preferable for the equaliser to be adjusted automatically. A TVE which does this for itself during the course of normal operation is called an *adaptive equaliser.*

In principle, to calculate the coefficients of a TVE, one needs to know the impulse response of the transmission path. A well known method of identifying the impulse response of a linear system is to crosscorrelate its output voltage with a known input signal, such as a pseudorandom sequence of pulses. This suggests a method of implementing an adaptive equaliser. If the input signal to the transmission path is a random sequence of pulses, the input to the TVE can be crosscorrelated with the output signal from it as if the latter were correct. The result can be used to adjust the coefficients of the TVE. As a result, the output signal becomes more nearly correct. If the process is repeated, the coefficients will converge to the correct values and the output signal to the correct waveform. This process has been called 'decision directed', since the equaliser goes through a learning process based on its own decisions.

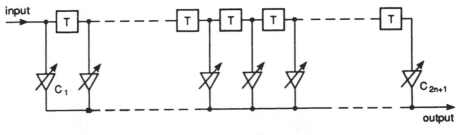

Figure 3.14   Block diagram of transversal equaliser

A number of different algorithms have been implemented to perform this process [18]. Since the process depends on the input data being random, the equaliser settings will diverge from the correct values when the input signal becomes nonrandom, e.g. a long sequence of '0's or '1's. The equaliser settings will oscillate about their correct values until the repetitive sequence ends and will then converge again to the correct values.

3.10.3 Noise and jitter

The principal advantage of digital transmission is that it is possible to obtain satisfactory transmission in the presence of very severe crosstalk and noise. For binary transmission, it is only necessary to detect the presence or absence of each pulse. Provided that the interference level is not so high as to cause frequent errors in making this decision, the output signal will be almost noise-free.

Consider an idealised train of unipolar binary pulses, as shown in Figure 3.15a. If the symbols '0' and '1' are equiprobable, i.e. $P(0) = P(1) = \frac{1}{2}$, the mean signal power corresponds to $S = \frac{1}{2}V^2$. Thus, the signal-to-noise ratio is

$$\frac{S}{N} = \frac{V^2}{2\sigma^2}$$

where σ is the RMS noise voltage.

The receiver compares the signal voltage v_s with a threshold voltage of $\frac{1}{2}V$, giving an output '0' when $v < \frac{1}{2}V$ and '1' when $v > \frac{1}{2}V$. If a noise voltage v_n is added, an error occurs if $v_n < -\frac{1}{2}V$ when $v_s = +V$, or if $v_n > +\frac{1}{2}V$ when $v_s = 0$. Thus, the probability of error P_e is given by

$$P_e = P(0)\,P(v_n > +\tfrac{1}{2}V) + P(1)\,P(v_n < -\tfrac{1}{2}V)$$

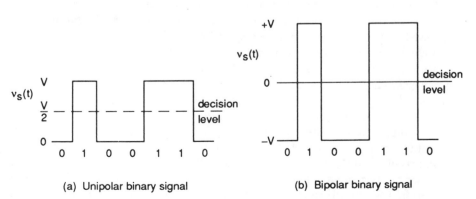

(a) Unipolar binary signal   (b) Bipolar binary signal

Figure 3.15   Detection of digital signals

But $P(0)+P(1)=1$ and, for white noise, $P(v_n<-\tfrac{1}{2}V)=P(v_n>+\tfrac{1}{2}V)$. Therefore

$$P_e=P(v_n>\tfrac{1}{2}V)$$

$$=P\{v_n/\sigma>(S/N)^{1/2}/\sqrt{2}\} \tag{3.17}$$

Hence, the error probability can be calculated from eqn. 3.12 by using a normal-probability table.

If a bipolar binary signal is used, as shown in Figure 3.15b, the mean signal power is $S=V^2$ and the signal-to-noise ratio is $S/N=V^2/\sigma^2$. The receiver gives an output '0' when $v<0$ and '1' when $v>0$. An error occurs if $v_n<-V$ when $v_s=+V$, or if $v_n>+V$ when $v_s=-V$. Thus, the probability of error is given by

$$P_e=P(0)P(v_n>+V)+P(1)P(v_n<-V)$$

$$=P(v_n>+V)$$

$$=P\{v_n/\sigma>(S/N)^{1/2}\} \tag{3.18}$$

Consequently, the same error rate is obtained with a 3 dB lower signal-to-noise ratio. Alternatively, a much lower error rate can be obtained for the same signal-to-noise ratio.

The above analysis can be extended to multilevel digital signals, by replacing the pulse amplitude with the spacing between adjacent signal levels. If the number of levels is large, the error rate is nearly doubled. This is because all the intermediate levels can be misinterpreted in either direction, owing to noise voltages of either polarity.

For a unipolar binary signal disturbed by white noise, the bit-error rate, calculated from eqn. 3.17, varies with the signal-to-noise ratio as shown in Figure 3.16. For example, if the signal-to-noise ratio is 20 dB, less than one digit per million is received in error. For telephone transmission, an error rate of 1 in 10^3 is intolerable, but an error rate of 1 in 10^5 is acceptable. Lower error rates are required for data transmission; if the error rate of the transmission link is inadequate, it is necessary to use an error-detecting or error-correcting code for the data [18,22].

On a long transmission link, it is possible to use regenerative repeaters instead of analogue amplifiers. A regenerative repeater [1] samples the received waveform at intervals corresponding to the digit rate. If the received voltage at the sampling instant exceeds a threshold voltage, this triggers a pulse generator which transmits a pulse to the next section of the line. If the received voltage is below the threshold, no pulse is generated. If both positive and negative pulses are transmitted, the regenerator is required to compare the received voltage against both a positive and negative threshold and to retransmit pulses of either polarity.

If the regenerators in a PCM link operate with a negligible error rate, the signal received at the far end of the link will be almost identical to that which was sent, regardless of the number of repeater sections. Thus, the requirements of each repeater section are no more severe than those of the overall system. This gives a considerable advantage over analogue transmis-

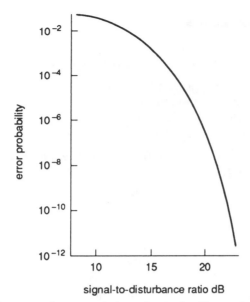

Figure 3.16 *Error rate for transmission of unipolar binary signal disturbed by white noise*

sion. For example, in an analogue system with 100 repeaters, the signal-to-noise ratio for each repeater section needs to be 20 dB better than that which would be needed if there was only one section. For a PCM system with 100 regenerators and an error rate of 1 in 10^6, the error rate per repeater section needs to be 1 in 10^8. Figure 3.16 shows that the signal-to-noise ratio for each repeater section need only be approximately 1 dB better than that required if there was only a single section. In practice, since the error rate changes so rapidly with signal-to-noise ratio, it is likely that the error performance of one regenerator in the link will be much worse than the others. Its errors will then predominate and determine the overall performance of the transmission link.

Other impairments can be caused by transmission over a link containing a number of regenerative repeaters. The instants at which pulses are retransmitted by a regenerative repeater are determined by a local oscillator synchronised to the digit rate, which must be extracted from the received waveform. Variations in the extracted frequency can cause a periodic variation of the times of the regenerated pulses, which is known as *jitter* [31–33]. This is not a great problem as long as the tolerance to jitter of any subsequent equipment in a link exceeds the amount of jitter produced by preceding equipment. However, at the end of the link it causes phase modulation of the reconstructed samples, which produces distortion products in the analogue output signal. There can also be a long-term variation in the times of the regenerated pulses due to changes in propagation time. This is known as *wander.*

3.11 Transmission media

3.11.1 Lines

The *primary coefficients* of a uniform transmission line are:

R=resistance in ohms per unit length
G=leakance in siemens per unit length
L=inductance in henries per unit length
C=capacitance in farads per unit length

A steady-state solution to the partial differential equations for an infinitely long line [34] is given by

$$V(x) = Ve^{-Px} \qquad (3.19a)$$

$$I(x) = Ie^{-Px} \qquad (3.19b)$$

$$I = \frac{V}{Z_0} \qquad (3.19c)$$

where $V(x)$ and $I(x)$ are the voltage and current at a distance x from the source feeding power into the line and

$$P = \{(R+j\omega L)(G+j\omega C)\}^{1/2} \qquad (3.20a)$$

$$Z_0 = \left(\frac{R+j\omega L}{G+j\omega C}\right)^{1/2} \qquad (3.20b)$$

The *secondary coefficients* P and Z_0 are called the *propagation coefficient* and the *characteristic impedance* of the line, respectively.

The propagation coefficient P is complex, so we can write

$$P = \alpha + j\beta$$

The real part α of the propagation coefficient thus gives the attenuation of the line in nepers per unit length; it is therefore called the *attenuation coefficient*. The imaginary part β gives the phase change in radians per unit length; it is therefore called the *phase coefficient*. The phase of the transmitted wave changes by 2π in a distance $\lambda = 2\pi/\beta$; this is thus the wavelength. The velocity of propagation is given by

$$v = f\lambda = \omega/\beta \qquad (3.21)$$

In general, the attenuation and velocity vary with frequency. The attenuation/frequency characteristic of a typical audio cable is shown in Figure 3.17a.

Since $I(x) = V(x)/Z_0$, if a line of finite length is terminated in impedance Z_0, the voltages and currents at any point on the line are identical to those in an infinite line, given by eqns. 3.19a–c. Such a line is said to be *correctly terminated*.

In practice, it is not always possible to obtain a terminating impedance matching Z_0 over the complete band of frequencies to be transmitted. A reflected wave is then produced at the termination. The ratio of reflected voltage to incident voltage and reflected current to incident current is called the *reflection coefficient* ρ. For a terminating impedance Z_L, it can be shown [34] that

$$\rho = \frac{Z_L - Z_0}{Z_L + Z_0} \tag{3.22}$$

The reciprocal of this ratio expressed in decibels (i.e. $-20 \log_{10}|\rho|$) is called the *return loss*. Obviously, for $Z_L = Z_0$, $\rho = 0$ and the return loss is infinite. For a short circuit, $\rho = -1$; thus the reflected wave is in antiphase with the incident wave and no voltage appears at the end of the line. For an open circuit, $\rho = +1$; thus the reflected wave is in phase with the incident wave and doubles the voltage at the end of the line.

Reflections may be produced not only by incorrect terminations but also by impedance irregularities at intermediate points on the line. This causes attenuation and delay distortion, as shown in Section 3.3.3. Reflections also provide a convenient method of locating line faults. In pulse-echo testing, a short pulse is sent into the line and the impedance irregularity due to the fault causes a reflection. The time interval between the pulse being sent and the reflected pulse returning to the sending end corresponds to twice the distance to the fault.

From eqn. 3.20a, the attenuation and phase coefficients are

$$\alpha = \{\tfrac{1}{2}(R^2 + \omega^2 L^2)^{1/2}(G^2 + \omega^2 C^2)^{1/2} + \tfrac{1}{2}(RG - \omega^2 LC)\}^{1/2}$$
$$\beta = \{\tfrac{1}{2}(R^2 + \omega^2 L^2)^{1/2}(G^2 + \omega^2 C^2)^{1/2} - \tfrac{1}{2}(RG - \omega^2 LC)\}^{1/2}$$

These expressions are complicated, but the following approximations are valid in different frequency regions:

(a) For low frequencies (e.g. for telegraphy):

$$\omega L \ll R \text{ and } \omega C \ll G,$$

then

$$\alpha = \surd(RG), \quad \beta = 0 \quad Z_0 = \surd(R/G)$$

(b) For audio frequencies:

$$\omega L \ll R \text{ but } \omega C \gg G$$

then

$$P = \surd(j\omega CR)$$

therefore

$$\alpha = \beta = \surd(\tfrac{1}{2}\omega CR)$$
$$Z_0 = \surd(R/j\omega C) = \surd(R/\omega C)\underline{/-\pi/4}$$

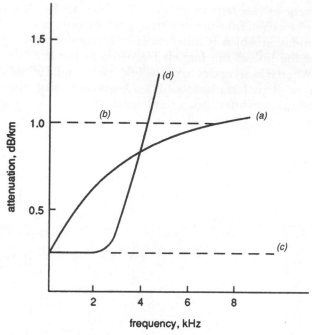

Figure 3.17   Attenuation/frequency characteristics of an audio-frequency cable

    a Unloaded line
    b Equalised line
    c Continuous loading
    d Lumped loading

Thus, the attenuation increases as the square root of frequency. The characteristic impedance is proportional to $1/\sqrt{f}$ and is capacitive.

(*c*) For radio frequencies:

$$\omega C \gg G \text{ and } \omega L \gg R$$

then

$$\alpha = \tfrac{1}{2}R\sqrt{(C/L)}$$

$$\beta = \omega\sqrt{(LC)}$$

$$Z_0 = \sqrt{(L/C)}$$

Thus, the characteristic impedance is purely resistive and independent of frequency. The velocity of propagation (ω/β) is also independent of frequency. The attenuation is not independent of frequency because skin effect reduces the depth of penetration of current and so causes R to increase with frequency. It can be shown [35,36] that $R \propto \sqrt{\omega}$, so α is also proportional to $\sqrt{\omega}$.

Attenuation increases as the square root of frequency over a very wide range of frequencies (i.e. regions b and c). The increase of attenuation with frequency can be offset, over the required band, by using an equaliser. This results in attenuation which is independent of frequency but equal to the attenuation of the line at the highest frequency in the band, as shown in Figure 3.17b. When it is necessary to preserve the waveshape of the signal (e.g. for television or digital transmission) the equaliser must correct phase distortion in addition to attenuation distortion.

A special case, which was first noted by Heaviside, arises when

$$LG = CR \qquad (3.23)$$

If this condition is substituted in eqn. 3.20a, then

$$\alpha = \sqrt{(RG)}, \quad v = 1/\sqrt{(LC)}$$

Thus, both attenuation and velocity of propagation are independent of frequency. Eqn. 3.23 is therefore known as the *distortionless condition*.

For practical cables, the left-hand side of eqn. 3.23 is much smaller than the right-hand side. It is not possible to reduce R or C, since these will already have been made as small as is practicable. It is undesirable to increase G, since this increases the attenuation. It is possible to increase L to satisfy the condition of eqn. 3.23; this is called *loading*. As shown in Figure 3.17c, it reduces the attenuation at all frequencies to that of the unloaded cable at low frequencies. (In contrast, equalisation increases the attenuation at all frequencies to that at the highest frequency in the required band, as shown in Figure 3.17b).

Loading can be done uniformly (e.g. by wrapping the conductors with tapes of magnetic material); this reduces the attenuation to a low value which is independent of frequency, as shown in Figure 3.17c. This technique, called *continuous loading*, is expensive and was rarely used. In practice, the inductance of audio-cable pairs can be increased artificially by means of coils added at regular intervals. This is known as *lumped loading*. Since the added inductance is lumped instead of being distributed, this has the effect of inserting a lowpass filter in the line, as shown in Figure 3.17d. It can be shown [1] that the cutoff frequency is $f_0 = 1/\pi\sqrt{(L_s C_s)}$, where L_s and C_s are, respectively, the total inductance and capacitance of a loading-coil section. To obtain an adequate cutoff frequency for telephony, it is only possible to increase the inductance by a factor of about 100, whereas an increase of about 1000 times would be required to satisfy eqn. 3.23.

Audio cables have extensively used 88 mH loading coils at a spacing of 1.83 km (2000 yards). This gives a cutoff frequency of 3.4 kHz for 9.9 mm conductors (20 lb/mile) and 3.9 kHz for 0.63 mm conductors (10 lb/mile). The loading reduces the attenuation at 800 Hz from 0.71 dB/km to 0.23 dB/km and from 1.05 dB/km to 0.45 dB/km, respectively. Many such circuits have now been converted to PCM transmission by replacing the loading coils with regenerative repeaters at the same locations. Thus, the traditional standard for loading has influenced the design of modern PCM systems.

Although the attenuation of a loaded cable is almost independent of frequency up to the cutoff frequency, the sharp frequency cutoff does introduce phase distortion. This has negligible effect on speech transmission, but can be harmful for data transmission [37]. The addition of inductance also reduces the velocity of propagation. Typical values are 220 000 km/s for unloaded cables and only 22 000 km/s for loaded cables. The resulting increase in propagation delay makes it undesirable to use loaded cables for very long circuits.

3.11.2 Radio

The properties of radio links are determined by the mechanisms affecting the propagation of the radio waves and these depend on the frequencies used for transmission. The different bands of radio frequencies and the standard nomenclature used for them are given in Table 3.2.

The simplest form of radio propagation is direct propagation from a transmitting antenna to a receiving antenna through free space. In practice, however, perfect direct propagation is not obtained because of the presence of the earth and imperfections of its atmosphere. Propagation between a transmitting and receiving antenna can take place over a number of different paths, which are illustrated in Figure 3.18. The received signal can therefore be the resultant of a number of the following:

(*a*) a *direct wave* (i.e. free-space propagation);

Table 3.2 Nomenclature of radio frequencies

Classification	Abbreviation	Frequency	Wavelength
		kHz	km
Very-low frequencies	VLF	3–30	100–10
Low frequencies	LF	30–300	10–1
			m
Medium frequencies	MF	300–3000	1000–100
		MHz	
High frequencies	HF	3–30	100–10
Very-high frequencies	VHF	30–300	10–1
			cm
Ultra-high frequencies	UHF	300–3000	100–10
		GHz	
Super-high frequencies	SHF	3–30	10–1
Extra-high frequencies	EHF	30–300	1–0.1

The general term 'microwave' is used to describe radio waves of frequencies higher than 1 GHz.

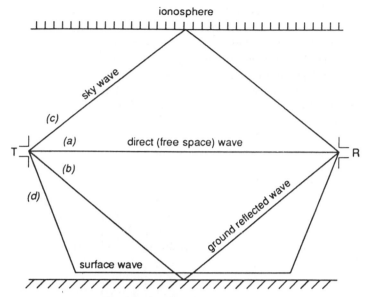

ionosphere

Figure 3.18 Transmission paths between two antennas

(*b*) a *reflected wave* (reflected from the ground);
(*c*) a *sky wave* (reflected by ionised layers above the earth, known as the *ionosphere*);
(*d*) a *surface wave* caused by diffraction round the earth associated with currents induced in the ground); and
(*e*) a wave reflected by variations in refractive index caused by local perturbations in the atmosphere (tropospheric scattering).

All these mechanisms can be present for propagation over a radio link. However, some of them are negligible in certain frequency ranges, so that very different propagation characteristics are obtained [2,38,39].

In the VLF band (i.e. below 30 kHz), the wavelengths are so long that they are comparable to the height of the lowest ionospheric layer (approximately 50 km). The ionosphere and the surface of the earth thus act as conducting planes forming a waveguide [2,40]. Consequently, VLF signals can have worldwide coverage. This band is used for telegraph transmission, for navigational aids and for distributing standard frequencies.

In the LF band (i.e. 30–300 kHz), propagation by the surface wave provides the propagation mechanism. Stable transmission is obtained over distances up to about 1500 km. This band is used for sound broadcasting.

In the MF band, there is also surface-wave propagation. However, the amount of diffraction is proportional to the wavelength λ, so the strength of the surface wave decreases with frequency. Consequently, MF transmission provides shorter ranges than LF transmission. The sky wave also has some effect and this increases the coverage, particularly at night when the absorption of the ionosphere is a minimum. This band is also used for sound broadcasting.

Since the strength of the surface wave decreases with frequency, it has negligible effect in the HF band and the propagation is mainly by means of the sky wave reflected by the ionosphere. The ionosphere is formed by ultraviolet and X-ray emission from the sun ionising the molecules of the upper atmosphere. There are several layers in which the ionisation density is a maximum; these are designated the D, E, F_1 and F_2 layers in order of height. The ionisation causes both absorption and refractive bending [2,41]. The former causes attenuation which is proportional to $1/f^2$. The latter can cause the wave to return to earth, as shown in Figure 3.18, at a distance from the transmitter known as the *skip distance*.

The properties of the ionosphere vary from hour to hour and from day to day throughout the year. Transmitting stations therefore need to be able to use several frequencies in order to provide reliable communication. Observations carried out at many stations throughout the world result in charts being prepared to enable the optimum frequency to be predicted for communication between any two points at a given time.

Fluctuations in the ionosphere cause the received signal to be made up of a number of components whose path lengths differ. When interference between these components results in a reduction of received signal strength, *fading* occurs. There are both slow fades caused by diurnal and seasonal variations in the ionosphere and rapid fades due to random fluctuations. These can cause reductions in received signal strength of up to 30 dB. Since different frequency components travel over different path lengths, modulated signals can have severe attenuation distortion, as explained in Section 3.3.3. This is known as *selective fading*. Fading is especially severe on paths exceeding 400 km, on which transmission takes place by multiple hops, as shown in Figure 3.19.

Fading can be combated by *diversity reception* [41,42]. In space-diversity systems, two or more antennas are used and the output is selected from the antenna receiving the strongest signal. The method relies on the theory that, since fading is due to small changes in the path lengths of multipath signals, it is unlikely to occur simultaneously at two separated receiving antennas.

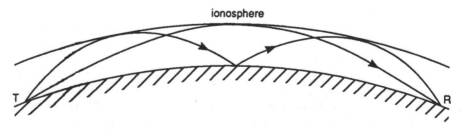

ionosphere

Figure 3.19   Multiple paths involving reflections from the ionosphere

Since the path difference producing cancellation depends on frequency, protection against fading can also be obtained by using two or more carrier frequencies received on a single antenna. This is known as *frequency-diversity* reception.

HF radio was the principal means of long-distance international communication, apart from submarine telegraph cables, until submarine telephone cables were introduced. It is still used for routes where there is insufficient traffic to justify a submarine cable or a satellite link. HF radio is still needed for long-distance communication to ships and aircraft, although satellites are now also used.

Since the sky-wave mechanism does not operate at frequencies much above 30 MHz, radio communication at VHF and UHF depends on direct free-space wave propagation. If the wave travelled in a true straight line, the range of communication would be strictly limited to 'line-of-sight' distances. However, the path of the wave is affected by variation with height of the refractive index of the troposphere [43]. The waves are refracted as they travel through the troposphere, as shown in Figure 3.20*a*, and this increases the range beyond the optical horizon. In planning routes, this effect can be allowed for by assuming that the earth has an *effective radius* which is approximately 4/3 times its actual radius [2] as shown in Figure 3.20*b*.

In addition to the direct wave, there is also a wave reflected from the surface of the earth, as shown in Figure 3.21a. The received field strength is due to the resultant of these two waves. Reflection from the earth produces a phase reversal. If the difference between the two path lengths is an odd number of half wavelengths the waves add, but for an even number they cancel. Thus, as shown in Figure 3.21*b*, the signal passes through a series of maxima and minima as the distance from the transmitter increases, until the

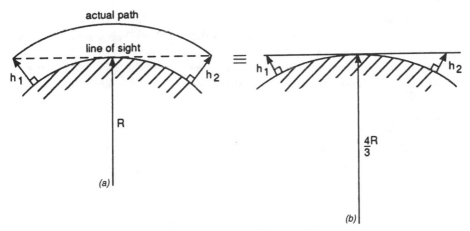

Figure 3.20   Refraction of VHF wave in earth's atmosphere

   a Actual conditions
   b Equivalent line-of-sight path with modified earth's radius

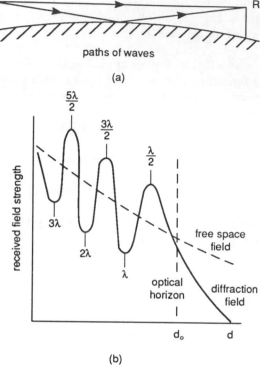

(a)

(b)

Figure 3.21   *VHF transmission by direct wave and wave reflected by ground*

    a Paths of waves
    b Variation of field strength with distance *d* from transmitter

difference between the path lengths becomes less than half a wavelength. The field strength then decays steadily as cancellation of the two waves becomes more nearly complete.

Transmission in the VHF band and the lower part of the UHF band is used mainly for television broadcasting and for short-range mobile communication.

Line-of-sight fixed radio links in telecommunication networks now use frequencies above 1 GHz, known as microwave frequencies. For these shorter wavelengths, highly directional antennas can be used. If the antennas are mounted at a sufficient height, reflection from the ground is negligible and transmission approximating to free-space propagation can be obtained. The total path attenuation is then given by [1,44]:

$$L = 20 \log_{10}(4 \pi r/\lambda) - G_T - G_R \text{ dB}$$

where G_T and G_R are the gains of the transmitting and receiving antennas in decibels and *d* is the distance between them.

To obtain effective free-space propagation, the path profile must be

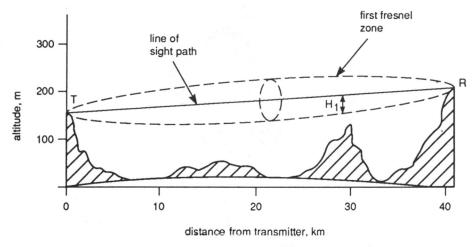

Figure 3.22 Path profile for microwave link showing radius H_1 of first Fresnel zone

planned carefully [45,46,47] to avoid intervening obstacles such as high ground, as shown in Figure 3.22. The clearance between the direct ray and the highest obstacle should be sufficient to ensure that the difference between the path lengths of a reflected wave and the direct wave exceeds half a wavelength. This prevents interference between the waves causing cancellation. This minimum clearance distance, shown as H_1 in Figure 3.22, is the radius of the first Fresnel zone. It can be shown that

$$H_1 = \{\lambda d_1 (1 - d_1 / d)\}^{1/2}$$

where d is the total path length and d_1 is the distance from the transmitter.

Fading occurs on microwave paths because of variations of refractive index due to temperature, humidity and pressure changes [43]. At frequencies above 10 GHz, the wavelength becomes comparable with the size of rain drops. Thus, at these frequencies, additional fading due to absorption is produced by rainfall [48]. Microwave transmission over distances beyond the optical horizon can be obtained by means of the mechanism known as *tropospheric scattering* [43]. It is believed to be caused by local perturbations in the atmosphere producing variations in refractive index. If a powerful transmitter sends a narrow beam to its horizon, it illuminates a region which is within line-of-sight of the receiving station, as shown in Figure 3.23. A high-gain receiving aerial can obtain a signal refracted by the scatter region. The received signal is much less than would be obtained by free-space transmission and it is subject to considerable fading because the scattering is random. It is therefore necessary to use both space and frequency diversity reception [49,50]. Bandwidths of several MHz can be obtained over distances of up to 500 km.

For satellite-communication links [51,52] it is necessary to transmit radio waves beyond the earth's atmosphere. Ideal free-space propagation condi-

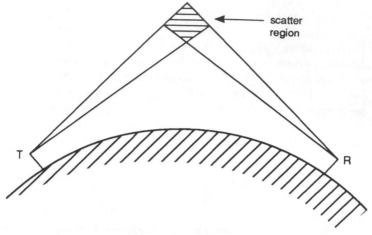

Figure 3.23   Tropospheric scattering

tions are then obtained. To escape through the ionosphere, frequencies well above 30 MHz must be used. However, absorption caused by water vapour occurs in the troposphere at frequencies near 26 GHz. Between these frequencies there is a 'window' which can be used for extra-terrestrial propagation. However, because of the high attenuation of the very long path length (36 000 km) of a satellite link, it is necessary to use antennas with very high gains. This restricts satellite systems to microwave frequencies.

Initially, satellite communication systems were used solely for inter-continental links and employed the 4 GHz and 6 GHz bands, which are also used for terrestrial microwave links. However, satellite systems are now also used for 'domestic' communication and direct broadcasting by satellite (DBS). Crowded conditions in the 4 and 6 GHz bands have led to the use of the 11 GHz and 14 GHz bands. Some use is also being made of frequencies below 4 GHz and above 14 GHz. At frequencies above 26 GHz there are some smaller windows, lying between various absorption bands due to water vapour and oxygen, which would also allow radio waves to pass through the troposphere.

3.11.3 Optical-fibre transmission

Electromagnetic waves can be propagated in a waveguide consisting of a dielectric bounded by another dielectric of lower permittivity instead of by a conductor. The use of such a waveguide, propagating waves of optical wavelengths, for telecommunication transmission was proposed by Kao and Hockham [53] in 1966. The practicability of such systems depends on the availability of materials for the guide having sufficiently low energy losses, and suitable optoelectronic devices for launching light waves into the guide and detecting them after transmission. A large research and development effort over the last 30 years has resulted in optical fibres with sufficiently low attenuation (typically <0.5 dB/km) and suitable solid-state transmitting devices (light-emitting diodes (LEDs) and lasers) and receiving devices (PIN and avalanche photodiodes). As a result, transmission over optical fibres developed from a scientific curiosity to application in commercial systems. Today, most of the world's largest telecommunications operating organisations have systems in service and the major telecommunications equipment suppliers are manufacturing systems.

Transmission over optical fibres has several advantages over transmission by metallic conductors:

(*a*) Since optical-fibre systems operate at much higher frquencies than coaxial-cable systems, they are potentially able to provide much wider bandwidth (e.g. 60 000 GHz).
(*b*) The loss per km of optical fibres is now much lower than that of electrical cables operating at high frequencies. Thus, repeater sections can be much longer and even long-haul systems do not require any intermediate repeaters.
(*c*) The fibres provide electrical isolation between transmitter and receiver.
(*d*) Transmission is unaffected by electrical radiation. (Hence optical fibres can be used in noisy electrical environments.)
(*e*) The light energy is contained within the fibre. Thus, no fibre-to-fibre crosstalk occurs.
(*f*) Optical-fibre cables are smaller and lighter than electrical cables of the same channel capacity.

Disadvantages of optical fibres are:

(*a*) Fibres must have very small diameters (about the size of a human hair). Care must therefore be taken to avoid subjecting them to mechanical stresses. In addition, making efficient connections to optical fibres and splices between them is more difficult than for metallic conductors.
(*b*) Electric power cannot be sent over the fibres. However, this is no drawback when intermediate repeaters are not required.
(*c*) Locating faults on cables is more difficult for optical fibres than for metallic conductors.

An optical-fibre guide consists of a small-diameter cylindrical core of silica or glass surrounded by a cladding of lower refractive index [54,55,56,57]. The usual type is silica-based fibre produced by deposition from the vapour phase (CVD) fibre [54,55].

One form of fibre has a step change in refractive index between the core and cladding, as shown in Figure 3.24*a*. A typical fibre has a silica core of 50 μm diameter surrounded by a glass cladding of 125 μm diameter.

Figure 3.25 shows a light ray R_1 inside a cylindrical core which meets the surface at an angle θ_1. There is total reflection of the ray if $\theta_1 < \theta_c$, where θ_c is the critical angle, given by

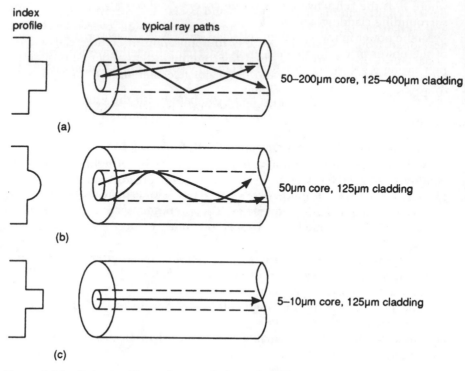

Figure 3.24 *Index profiles and ray paths in optical fibres*

    a Multimode step-index fibre
    b Multimode graded-index fibre
    c Monomode step-index fibre

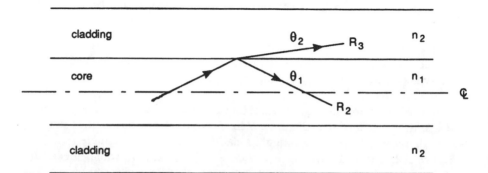

Figure 3.25 *Internal reflection of ray within fibre*

$$\cos \theta_c = n_2/n_1$$

where n_1 is the refractive index of the core, n_2 is the refractive index of the surrounding cladding, and $n_2 < n_1$.

All the power in ray R_1 is then reflected into ray R_2. However, if $\theta_1 > \theta_c$, some of the power in ray R_1 appears in an external refracted wave R_3. Thus, only rays having $\theta_1 < \theta_c$ will survive to travel for any considerable distance along the glass.

If the diameter of the glass is large compared with the wavelength of the light, many different ray paths can be traversed, as shown in Figure 3.24a. This is called *multimode propagation*. The propagation time for a ray at an angle θ_1 to the axis exceeds that of a ray parallel to the axis by a factor $1/\cos \theta_1$. Thus, if the end of the fibre is illuminated by a short pulse of light, the light will arrive at the far end as a pulse whose arrival time spreads between Ln_1/c and $Ln_1/c \cos \theta_c$, where L is the length of the fibre and c is the velocity of light in free space. The duration of the received pulse is thus

$$\Delta t = \frac{Ln_1}{c}\left(\frac{n_1}{n_2} - 1\right)$$

This dispersion of transmitted pulses increases with the length of the fibre; i.e. the bandwidth decreases. This is clearly disadvantageous for the design of wideband systems with long repeater sections. Nevertheless, multimode fibres are useful for short-distance systems of relatively low bandwidth. For example, if $n_1 = 1.5$ and $\Delta n = 0.01$, then $\Delta T/L = 34$ ns/km.

A method of preventing the large amount of dispersion which occurs in the simple multimode fibre is to use a *graded-index fibre*. The refractive index is made to vary continuously across the diameter as shown in Figure 3.24b. Light in the outer regions of the core travels faster than light near its centre. Thus, the difference between the propagation times for steep-angle and shallow-angle rays is much smaller than for a fibre with uniform refractive index. It can be shown that the refractive index n should vary with radial distance r according to the law

$$n = n_1\{1 - K(2r/d)^2\}$$

A ray launched into the fibre at an angle to the axis oscillates sinusoidally with distance, as shown in Figure 3.24b, and the propagation time is almost independent of the angle. In practice, it is difficult to match refractive index to the parabolic law exactly. Fortunately, large reductions in dispersion can be obtained with relatively poor approximations to this law.

Graded-index fibres provide a transmission path of almost uniform delay and have a core of reasonable diameter. Consequently, they were widely used in optical-fibre transmission systems until monomode-fibre cables became practicable.

If the core diameter of the fibre is so small (e.g. 5 or 10 μm) that it is comparable with the wavelength of light, only a single wave mode can propagate, as shown in Figure 3.24c. This is called a *monomode fibre*. It is also desirable for the light to be provided by a coherent source (a laser) in order to obtain the high intensity needed to launch sufficient power into such a small-diameter fibre.

For monomode propagation, the simple ray theory given above is inadequate; it is necesary to use field theory. It can be shown [55] that the necessary condition for monomode propagation is

$$r_1 < 2 \cdot 405 \lambda_0 / 2\pi (n_1^2 - n_2^2)^{1/2}$$

where λ_0 is the wavelength in free space and r_1 is the core radius.

Although there is now no pulse dispersion due to multimode effects, small amounts of dispersion remain because of:

(a) the dependence of n_1 and n_2 on wavelength λ_0, known as material dispersion or chromatic dispersion; and

(b) the dependence of the field distribution on λ_0, known as waveguide dispersion or modal dispersion.

Since practical sources emit light not at a single wavelength but over a range of wavelengths, these variations cause the components of a pulse to travel at different velocities and become dispersed in time. Thus, dispersion can be reduced by using a light source with a narrow spectrum. For a typical silica glass, the material dispersion at 0.85 μm may limit the signal bandwidth to 2.4 GHz for a kilometre length for a source having a spread in wavelength of 1 nm. Thus, for a laser having a spectral spread of 2 nm the bandwidth for 1 km is 1.2 GHz. For a light-emitting diode (LED) with a spectral spread of 33 nm, the bandwidth is only 70 MHz.

Unless the required bandwidth (or bit rate) is so large that dispersion is the limiting factor, the permissible distance between repeaters is limited by the attenuation of the fibre. Figure 3.26 shows the variation of attenuation with wavelength for a typical monomode fibre.

Early systems operated in the infra-red region at wavelengths of about 0.85 μm giving attenuation of 2 or 3 dB/km. This wavelength matched the response of available aluminium–gallium-arsenide devices. However, sources have now been developed to operate in the region of 1.3 μm where the attenuation of very high-purity fibres is less than 0.5 dB/km. This enables systems to operate with repeater spacings of about 40 km. As a result, few repeaters are needed and these can all be situated in buildings. Many inland routes do not need any repeaters and optical-fibre submarine cables are practicable.

The reduction of scattering attenuation with wavelength provides an incentive to develop systems operating at wavelengths of about 1.5 μm to achieve an attenuation of about only 0.2 dB per km. However, the material dispersion per kilometre increases at wavelengths beyond 1.3 μm. Fortunately, the waveguide dispersion is of the opposite sign (i.e. it makes delay increase with wavelength instead of decreasing). Therefore, appropriate

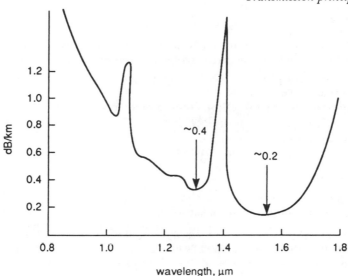

Figure 3.26 Typical attenuation/wavelength characteristics for silica monomode fibre

choice of core diameter and index difference enables the waveguide and material dispersions to cancel at the wavelength of interest [58]. Such dispersion-shifted monomode fibres have already been made.

Since the scattering varies as λ^{-4}, materials with the fundamental absorption peak at much longer wavelengths may be possible [59], which in theory could offer attenuations of the order 0.001 dB/km. Other possibilities [59] include the use of wavelength-division multiplexing, and coherent-detection systems in place of the intensity-modulated transmitters and receivers currently used. The use of 'soliton' waves offers the potential of almost lossless transmission over enormous distances [60]. Thus, although the high performance of current monomode systems appears to indicate a mature technology, rapid development is likely to continue.

3.12 References

1. Bell Telephone Laboratories Staff: 'Transmission systems for communications' (Bell Telephone Laboratories, 1980) 4th edn.
2. HILLS, M.T., and EVANS, B.G.: 'Transmission systems' (Allen and Unwin, 1973)
3. FLOOD, J.E., and COCHRANE, P. (Eds.): 'Transmission systems' (Peter Peregrinus, 1991)
4. SHANKS, P.H.: 'A new echo suppressor for long-distance communications', *Post Off. Electr. Eng. J.*, 1968, **60**, pp. 288–292
5. SONHI, M.M.: 'An adaptive echo canceller', *Bell Syst. Tech. J.*, 1967, **46**, pp. 497–511
6. JOHNSON, J.B.: 'Thermal agitation of electricity in conductors', *Phys. Rev.*, 1928, **32**, pp. 97–109

7. NYQUIST, H.: 'Thermal agitation of electric charge in conductors', *Phys. Rev.,* 1928, **32,** pp. 110–113.
8. BENNETT, W.R.: 'Electrical noise' (McGraw-Hill, 1960)
9. BULL, C.S.: 'Fluctuations of stationary and non-stationary electron currents' (Butterworth, 1966)
10. CCITT Recommendation P.53: 'Psophometers (apparatus for the objective measurement of circuit noise)'
11. COHRAN, W.T., and LAVINSKI, D.A.: 'A new measuring set for message circuit noise', *Bell Syst. Tech. J.,* 1960, **39,** pp. 911–931
12. CARTER, R.O.: 'Theory of syllabic compandors', *Proc. IEE,* 1964, **111,** pp. 503–513
13. WATT-CANTER, D.E., and WHEELER, L.K.: 'The Lincompex system for protection of HF radio telephone circuits', *Post Off. Electr. Eng. J.,* 1966, **59,** pp. 163–167
14. BLACK, H.S.: 'Modulation theory' (Van Nostrand, 1953)
15. TUCKER, D.G.: 'Modulators and frequency changers for amplitude-modulated line and radio systems' (Macdonald, 1953)
16. CCITT Recommendation G.135: 'Error on the reconstituted frequency'
17. BENNETT, W.R. and DAVEY, J.R.: 'Data transmission' (McGraw-Hill, 1965)
18. LUCKY, R.W., SALZ, J., and WELDON, E.J.: 'Principles of data communication' (McGraw-Hill, 1968)
19. ALEXANDER, A.A., GRYB, R.M., and NAST, D.W.: 'Capabilities of the telephone network for data transmission', *Bell Syst. Tech. J.,* 1960, **39,** pp. 431–475
20. WILLIAMS, M.B.: 'The characteristics of telephone circuits in relation to data transmission', *Post Off. Electr. Eng. J.,* 1966, **59,** pp. 151–162
21. CCITT Recommendation V.23: '600/1200 baud modem standardized for use in the general switched telephone network'
22. BREWSTER, R.L. (Ed.): 'Data communication and networks' (Peter Peregrinus, 1986)
23. TAUB, H., and SCHILLING, D.L.: 'Principles of communication systems' (McGraw-Hill, 1986) 2nd ed.
24. CATTERMOLE, K.W.: 'Principles of pulse code modulation' (Iliffe, 1969)
25. NYQUIST, H.: 'Certain topics in telegraph transmission theory', *Trans. Am. Inst. Electr. Eng.,* 1928, **47,** pp. 617–644.
26. CCITT Recommendation G.711: 'Pulse code modulation of voice frequencies'
27. SCHROEDER, M.R.: 'Vocoders: Analysis and synthesis of speech—30 years of applied speech research', *Proc. IEEE,* 1966, **54,** pp. 720–734
28. WALLACE, A.D., HUMPHREY, L.D., and SEXTON, M.J.: 'Analogue-to-digital conversion', *in* FLOOD, J.E., and COCHRANE, P. (Eds.): 'Transmission systems' (Peter Peregrinus, 1991) Chap. 6
29. BULLINGTON, K., and FRASER, J.M.: 'Engineering aspects of TASI', *Bell Syst. Tech. J.,*1959, **38,** pp. 353–364
30. CAMPANELLA, S.J.: 'Digital speech interpolation', *Comsat Tech. Rev.,* 1976, **6,** pp. 127–159
31. BYLANSKI, P., and INGRAM, D.G.W.: 'Digital transmission systems' (Peter Peregrinus, 1987)
32. BYRNE, C.J., KARAFIN, B.J., and ROBINSON, D.B.: 'Systematic jitter in a chain of digital regenerators', *Bell Syst. Tech. J.,* 1963, **43,** pp. 2679–2714
33. KEARSEY, B.N., and McLINTOCK, R.N.: 'Jitter in digital telecommunications networks', *Br. Telecom Eng. J.,* 1984, **3,** pp. 108–116
34. CHIPMAN, R.A.: 'Theory and problems of transmission lines' (McGraw-Hill, 1968)
35. GRIVET, P.: 'The physics of transmission lines at high and very high frequencies' (Academic Press, 1970)
36. JARVIS, R.F.J., and FOGG, G.H.: 'Formulae for the calculation of the theoretical characteristics and design of coaxial cables', *Post Off. Electr. Eng. J.,* 1937, **30,** pp. 138–151
37. JONES, I.O. and ADCOCK, R.C.: 'Group delay in the audio data network', *Post Off. Electr. Eng. J.,* 1971, **64,** pp. 9–15
38. WAIT, J.R.: 'Introduction to antennas and propagation' (Peter Peregrinus, 1986)

39. BULLINGTON, K.: 'Radio propagation fundamentals', *Bell Syst. Tech. J.*, 1957, **36,** pp. 593–626
40. WATT, A.: 'VLF radio engineering' (Pergamon Press, 1967)
41. BETTS, J.A.: 'HF communications' (English Universities Press, 1967)
42. MASLIN, N.: 'HF communications: a systems approach' (Pitman, 1987)
43. HALL, M.P.M.: 'Effects of the troposphere on radio communication' (Peter Peregrinus, 1986)
44. KRAUS, J.D.: 'Antennas' (McGraw-Hill, 1950)
45. CARL, H.: 'Radio relay systems' (Macdonald, 1966)
46. DUMAS, K., and SANDS, L.: 'Microwave system planning' (Hayden 1967)
47. MARTIN-ROYLE, R.D., DUDLEY, L.W., and FEVIN, R.J.: 'A review of the British Post Office microwave radio-relay network', *Post Off. Electr. Eng. J.*, 1977, **69,** pp. 162–168 and 1977, **69,** pp. 225–234
48. MEDHURST, R.G.: 'Rainfall attenuation of centimetre waves: comparison of theory and measurement', *IEE Trans.*, 1965, **AP-13,** pp. 550–563
49. GUNTER, F.: 'Troposcatter scatter communication—past, present and future', *IEEE Spectrum*, 1966, **3,** pp 79–100
50. HILL, S.J.: 'British Telecom transhorizon radio services to offshore oil/gas production platforms', *Br. Telecom Eng.*, 1982, **1,** p. 42–48
51. EVANS, B.G. (Ed.): 'Satellite communication systems (Peter Peregrinus, 1994), 2nd Ed.
52. DALGLEISH, D.I.: 'An introduction to satellite communications (Peter Peregrinus, 1989)
53. KAO, K.C., and HOCKHAM, G.A.: 'Dielectric fibre surface waveguides for optical frequencies', *Proc. IEE*, 1966, **113,** pp. 1151–1158
54. MIDWINTER, J.: 'Optical fibre communication systems' (John Wiley, 1980)
55. CHERIN, A.H.: 'An introduction to optical fibres' (McGraw-Hill, 1983)
56. GOWAR, J.: 'Optical communication systems' (Prentice-Hall, 1984)
57. SENIOR, J.: 'Optical fiber communications: principles and practice' (Prentice-Hall, 1985)
58. WHITE, K.I., and NELSON, B.P.: 'Zero total dispersion in step index monomode fibres at 1.3 and 1.5 μm', *Electron. Lett.*, 1979, **15,** pp. 396–397
59. O'REILLY, J.J.: 'Approaching Fundamental limits to digital optical-fibre communications', *Oxford Surveys in Information Technology*, 1986, **3,** p. 147 (Oxford University Press)
60. DORAN, N.J.: 'Solitons', *IEE Rev.*, 1992, **38,** pp. 291–294

Chapter 4

Transmission systems

J.E. Flood

4.1 Introduction

Transmission systems provide circuits between the nodes of a telecommunication network. Usually, the circuits are required to be *duplex circuits*, i.e. they are required to provide simultaneous communication in each direction. A long distance normally necessitates a four-wire circuit, as described in Section 3.4. This uses a separate channel for transmission in each direction. In general, a complete channel passes through sending equipment at a *terminal station*, a *transmission link*, which may contain *repeaters* at *intermediate stations*, and receiving equipment at another terminal station. In practice, a building which houses any kind of transmission equipment is usually called a repeater station. Present-day transmission systems [1,2,3,4] range in complexity from simple unamplified audio-frequency circuits to satellite-communication systems.

Traditionally, customers' access networks [5] have employed unamplified audio-frequency balanced-pair cables, using the smallest wire gauge consistent with meeting the overall transmission plan of the national network and providing a sufficiently low resistance for loop/disconnect DC signalling. For very long customers' lines, analogue FDM carrier systems have sometimes been used. Frequency-division multiplexing enables several customers to be served by a four-wire circuit. These systems are said to provide *pair gain*. Nowadays, digital multiplexers are being used to provide pair gain. The multiplexers may be connected to the local exchange over metallic pairs or optical fibres.

The advent of *integrated-services digital networks* (ISDN) has also required digital transmission over customers' lines, as described in Chapter 14. For *basic-rate access*, 144 kbit/s are used to provide two B channels at 64 kbit/s and a D channel (for signalling) at 16 kbit/s. For *primary-rate access*, a 24-channel or 30-channel PCM system is used.

Increasingly, optical fibres are being used to provide access for large business customers which require many circuits. The penetration of optical fibres into customers' access networks may be expected to increase as the technology develops [5]. Radio links are sometimes used to provide access to isolated remote customers and radio is, of course, used to provide access for mobile customers in cellular-radio systems. In future, microcellular radio systems may be used as an alternative to cable distribution in access networks.

Junction networks have also, traditionally, used audio cables. Since the number of circuits is less than in customer-access networks, it is economic for these cables to have larger conductors in order to reduce attenuation. Loading has also been used. FDM carrier systems have been used on junction routes requiring very large numbers of circuits. The first application of PCM primary multiplex systems was to increase the number of circuits provided on junction cables. PCM is now employed extensively and higher-order multiplexing is used on routes requiring very large numbers of circuits.

The high cost of audio transmission for long-distance circuits led to the introduction of FDM carrier systems in trunk networks. Initially, four-channel, 12-channel and 24-channel systems were used on open-wire lines and twisted-pair cables. Later, high-capacity systems operating over coaxial cables and microwave radio links have been employed.

The success of PCM primary multiplex systems in junction applications led to the development of higher-order multiplex systems for providing large numbers of circuits over long distances. These digital systems use coaxial cables, microwave radio links and optical-fibre cables. Digital transmission has now made analogue carrier systems obsolete. Digital transmission on optical fibres has been introduced for most applications, ranging from short-haul systems to intercontinental submarine cables.

4.2 Frequency-division multiplexing

4.2.1 The basic group

FDM multichannel carrier systems use the basic 12-channel group described in Section 3.8.1 and shown in Figure 3.12. It produces an assembly of 12 channels in the frequency band 60–108 kHz. For transmission over a balanced-pair cable [6], this group modulates a carrier of 120 kHz to produce a lower sideband in the frequency range 12–60 kHz. The basic group also forms a building block from which larger systems are constructed, as described below.

4.2.2 Higher-order channel asemblies

The CCITT recommended a range of channel assemblies containing from 12 to 10 800 channels [6]. Most countries use systems conforming to these recommendations, although other assemblies are used, most notably in North America. These systems are built on the hierarchical principle, in which the required channel capacity is built up from a number of basic 12-channel groups, using successive stages of single-sideband suppressed-carrier modulation. All speech channels are allotted a nominal 4 kHz bandwidth and all carriers have frequencies which are multiples of 4 kHz.

The next-higher channel assembly after the 12-channel group is the 60-channel *basic supergroup*. In a group-translating equipment, five basic groups each modulate a different carrier frequency to produce a composite signal in the band from 312 kHz to 552 kHz, as shown in Figure 4.1*a*.

Further stages of modulation can be used to build up still larger assemblies of channels. A standard method [6] is to combine five supergroups to form a basic *mastergroup*, which contains 300 channels in the frequency band from 812 to 2044 kHz. A basic *supermastergroup* can be made up of three mastergroups, comprising 900 channels in the frequency band from 8.516 to 12.388 MHz.

There are variations in this procedure. In the USA, a mastergroup consists of 10 supergroups (i.e. 600 channels). The L1 system transmits the 600-channel mastergroup [6] in the frequency band 60 kHz to 2.788 MHz. The L3 system [6] transmits 1860 channels by means of three of these mastergroups in the frequency range 564 kHz to 8.284 MHz, together with an additional supergroup in the range 312–552 kHz.

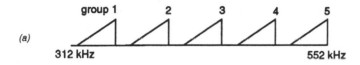

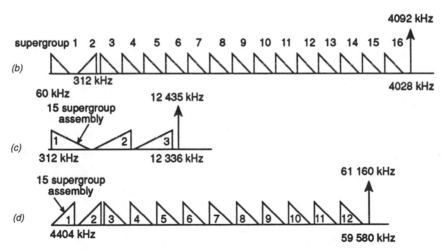

Figure 4.1 *Hierarchy of FDM channel assemblies used in the UK*

    a Basic supergroup
    b Assembly of 15+1 supergroups (as used in 4 MHz system)
    c Assembly of three hypergroups (as used in 12 MHz system)
    Note that supergroup 1 is omitted on later systems, the new unit being known as a 15-supergroup assembly
    d Assembly of 12 hypergroups (as used in 60 MHz system)

In the UK, an assembly of 15 supergroups, known as a basic *hypergroup*, is formed without using mastergroups [6]. This occupies the band from 312 to 4028 kHz. Figure 4.1*b* shows the frequency spectrum used for the 4 MHz system. This uses the 15-supergroup assembly together with an additional supergroup in the frequency range 60–300 kHz. The 12 MHz system uses three hypergroups, arranged as shown in Figure 4.1*c*. The 60 MHz system uses 12 hypergroups, as shown in Figure 4.1*d*.

The channel assemblies recommended by the CCITT to match coaxial-cable transmission systems and radio-relay systems are listed in Table 4.1. The channel assemblies used in North America for coaxial systems are listed in Table 4.2.

4.2.3 Other channel assemblies

Multi-channel voice-frequency telegraph systems [6] were early users of FDM. Double-sideband modulation is commonly employed, with 120 Hz carrier

Table 4.1 Channel assemblies recommended by CCITT for use on coaxial-pair FDM line systems and high-capacity radio-relay links

		Total channel assembly		Recommended for	
Channels	Basic assemblies used	Frequency Band	Bandwidth	Radio systems	Line systems
		kHz	MHz		MHz
300	1 mastergroup	64–1 296	1.232	Yes	1.3
300	5 supergroups	60–1 300	1.240	Yes	1.3
900	1 supermastergroup	316–4 188	3.872	Yes	4.0
960	1 15 SGASSs plus 1 supergroup	60–4 028	3.968	Yes	4.0
1 200	4 mastergroups	312–5 564	5.248	Yes	6.0
1 260	1 15 SGASSs plus 6 supergroups	60–5 636	5.570	Yes	6.0
1 800	2 supermastergroups	316–8 204	7.888	Yes	No
1 800	2 15 SGASSs	312–8 120	7.808	Yes	No
2 700	3 supermastergroups	316–12 388	12.072	Yes	12.0
2 700	3 15 SGASSs	312–12 336	12.024	Yes	12.0
2 700	1 15 SGASSs plus 2 supermastergroups	312–12 388	12.076	Yes	12.05.15
3 600	4 supermastergroups	316–17 004	16.688	No	18.0
3 600	4 15 SGASSs	312–16 612	16.612	No	18.0
10 800	12 supermastergroups	4 332–59 684	55.352	No	60.0
10 800	12 15 SGASSs	4 404–59 580	55.17	No	60.0

Table 4.2 Coaxial-cable systems used in North America

Coaxial-cable system	Number of channels	Frequency band used kHz	Channel assemblies used
L1	600	60–2 788	10 supergroups
L3	1 860	312–8 284	3 mastergroups + 1 supergroup
L4	3 600	564–17 548	6 mastergroups
L5	10 800	3 124–51 532	18 mastergroups (3 jumbogroups)

spacing, to transmit 24 telegraph channels in one telephone channel.

For long submarine systems, the cost of the cables is very high. This makes it desirable to multiplex as many telephone channels on them as is possible. The CCITT therefore recommended the use of a channel spacing of 3 kHz instead of 4 kHz. This has been made possible by designing superior filters, giving a nominal passband from 250 Hz to 3.05 kHz for each channel. In this way, the basic group is increased from 12 channels to 16 channels in the frequency band from 60 kHz to 108 kHz.

Sound-programme channels can be sent over FDM systems by using the bandwidth normally occupied by several telephone channels [6]. The CCITT standardised the following programme channels:

(*a*) 50 Hz to 6.4 kHz, using the bandwidth of two speech channels
(*b*) 50 Hz to 10 kHz, using the bandwidth of three speech channels
(*c*) 50 Hz to 15 kHz, using the bandwidth of six speech channels.

For stereophonic transmission, a pair of 15 kHz circuits, occupying the whole of a group band, is used.

Analogue television signals can be transmitted over high-capacity FDM radio links and over 12 MHz, 18 MHz and 60 MHz line systems. For radio links, the video signal can be applied directly to the baseband input. However, for transmission over line systems, special modulation and demodulation equipment is required to preserve the television waveform. Vestigial-sideband transmission is employed and coherent demodulation is used to eliminate any frequency shift during transmission.

4.2.4 Through-group and through-supergroup filters

To provide flexibility of routing, it is necessary to interconnect groups of telephone channels between different FDM systems. Instead of demodulating the channels to audio frequencies in one system and then translating them again in the other system for retransmission, it is more economical to interconnect the systems at the group or supergroup stage by bandpass filters. These are known as through-group or through-supergroup filters.

4.3 The PCM primary multiplex group

Analogue speech signals may be converted to digital signals by using PCM, as described in Section 3.7.5, and multiplexed onto a common bearer by using time-division multiplexing (TDM), as described in Section 3.8.2. In this way, telephone channels are combined to form an assembly of 24 or 30 channels [7,8,9]. This is known as the *primary multiplex group*. The primary multiplex group is also used as a building block for assembling larger numbers of channels in *higher-order multiplex systems*, as described in Section 4.4.

The block diagram of a typical PCM primary multiplex equipment is shown in Figure 4.2. Since the coder and decoder must each operate in the time slot of one channel, they are made common to all the channels of the system. To use the system in a switched telephone network, provision must be made for transmitting the necessary signalling conditions. As described in Chapter 8, either channel-associated signalling or common-channel signalling may be used. Figure 4.2 shows built-in channel-associated signalling.

In the sending direction, the speech signal from each incoming audio channel passes through an anti-aliasing filter and a sampling gate before being encoded. The sampling gate of each channel is operated by a pulse at a different time TC_n to multiplex their pulse-amplitude-modulated (PAM) pulses onto a common highway to the encoder. The DC-signalling condition on the incoming line is sampled at the same time and the signalling samples are subsequently inserted in the frame of speech samples. A distinctive *frame-alignment signal* is also inserted in the frame to synchronise the receiving equipment at the far end of the transmission link.

In the receiving direction, the incoming signal is regenerated and its timing is extracted to provide clock pulses for the decoder. The frame-alignment signal is extracted and used to synchronise the channel-pulse generator to ensure that the receive gate of each channel operates at the appropriate time RC_n. Thus, the receiving gate of each channel selects the correct PAM pulse train on the common highway from the decoder and this is demodulated by a lowpass filter. Signalling pulses are similarly directed to the correct outgoing circuits.

The digit stream on the transmission path is arranged in frames of 125 μs duration (i.e. the interval corresponding to 8 kHz sampling). Each frame contains one coded speech sample from each channel, together with digits for signalling and synchronisation. Two frame structures are widely used: the European 30-channel system and the 24-channel system (known as DS1) used in North America and Japan. Both have been standardised in CCITT recommendations [10]. Both use 8 kHz sampling and 8-bit samples. However, the 30-channel system uses A-law companding and the 24-channel system uses μ-law companding [11].

The 30-channel frame is shown in Figure 4.3. It is divided into 32 time slots each of eight digits, so the total digit rate is $8 \times 8 \times 32$ kbit/s=2.048 Mbit/s. Time slot 0 is allotted to the *frame-alignment word* (FAW). Time slots 1–15 and

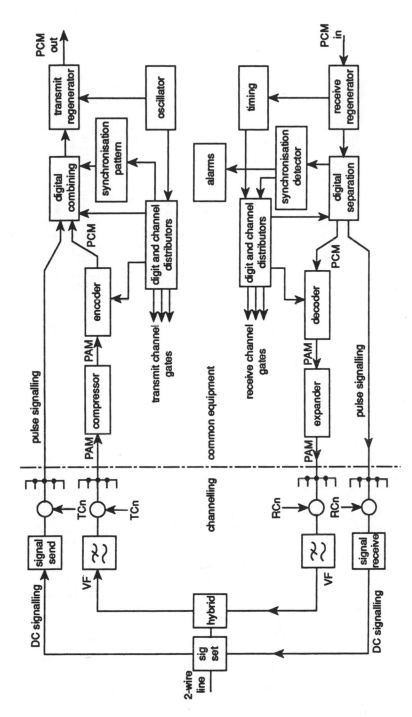

Figure 4.2   Block diagram of primary multiplex equipment

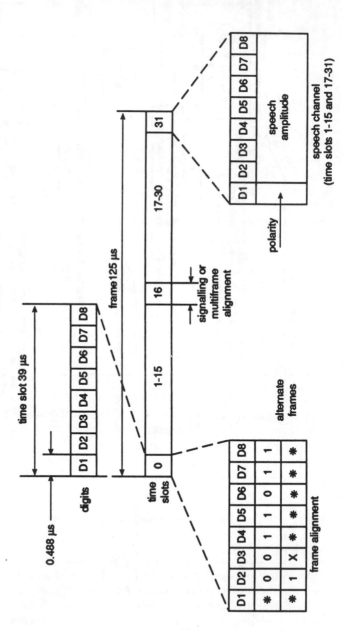

Figure 4.3   30-channel frame format

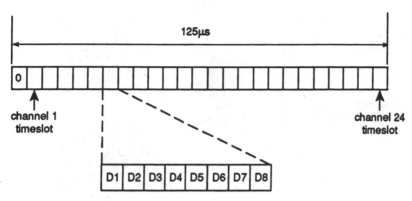

Figure 4.4   24-channel frame format

17–31 are each allotted to a speech channel. Time slot 16 is allotted to signalling. For channel-associated signalling, this may be shared between the speech channels by a process of *multiframing*, as described in Section 8.3.4. In one frame in every 16, each channel is allocated four bits in channel 16. This gives it four signalling channels at 500 bit/s in each direction. Alternatively, the whole of time slot 16 may be used to provide a 64 kbit/s data link for common-channel signalling.

The DS1 frame format used in the 24-channel system is shown in Figure 4.4. The basic frame consists of 193 bits. Of these, 192 bits convey 24 time slots containing the 8-bit speech samples and the additional bit is used for frame alignment. The overall digit rate is thus 8×193 kbit/s = 1.544 Mbit/s.

On odd-numbered frames, the additional bit (the F bit) has the pattern '1,0,1,0,...', which is used for frame alignment. This is thus a distributed frame-alignment signal as opposed to the block-alignment signal used in the 30-channel system. The even frames carry the pattern '01110,...', which defines a 12-frame multiframe. On frames 6 and 12 of the multiframe, bit D8 of each channel time slot is used for signalling for that channel. This process of 'bit stealing' causes a small degradation in quantising distortion, which is none the less considered acceptable. To provide 64 kbit/s common-channel signalling, it is necessary to sacrifice one of the 24 channels. The system then only conveys 23 telephone channels.

4.4 Higher-order digital multiplexing

4.4.1 General

The primary multiplex group of 30 or 24 channels is used as a building block for larger numbers of channels in higher-order multiplex systems. At each level in the hierarchy, several bit streams, known as *tributaries*, are combined by a multiplexer/demultiplexer, known as a *muldex* (often abbreviated to

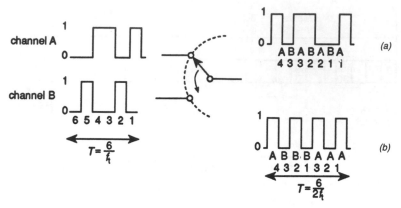

Figure 4.5   Interleaving digital signals

a Bit interleaving
b Word interleaving (word length = 3 digits)

mux). The output from a multiplexer may serve as a tributary to a multiplexer at the next-higher level in the hierarchy, or it may be sent directly over a line or radio link.

If the inputs to a multiplexer are synchronous, i.e. they have the same bit rate and are in phase, they can be interleaved by taking a bit or a group of bits from each in turn. This can be done by a switch that samples each input under the control of the multiplex clock. As shown in Figure 4.5, there are two main methods of interleaving digital signals: *bit interleaving* and *word interleaving* or *byte interleaving*. In bit interleaving, one bit is taken from each tribatary in turn. If there are N input signals, each with a rate of f_i bit/s, then the combined rate will be Nf_i bit/s and each element of the combined signal will have a duration equal to $1/N$ of an input digit. In word interleaving, groups of bits are taken from each tributary in turn and this involves the use of storage at each input to hold the bits waiting to be sampled. Since bit interleaving is simpler, it was chosen for the first-generation hierarchy of higher-order multiplexers. This is known as the *plesiochronous digital hierarchy* (PDH) and it is described in Section 4.4.2. Later, word interleaving was chosen for the second-generation hierarchy. This is known as the *synchronous digital hierarchy* (SDH) and it is described in Section 4.4.3.

4.4.2 The plesiochronous digital hierarchy

In a transmission network which has not been designed for synchronous operation, the tributaries entering a digital multiplex will not generally be exactly synchronous. Although they have the same nominal bit rate, they commonly originate from different crystal oscillators and can vary within the clock tolerance. They are said to be *plesiochronous*. The first generation of higher-order digital multiplex systems was designed for this situation. It forms

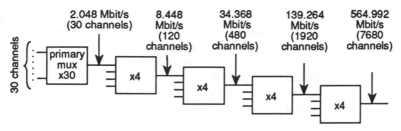

Figure 4.6   European plesiochronous digital hierarchy

the plesiochronous hierarchy (PDH) [12].

There are three incompatible sets of standards for plesiochronous digital multiplexing, centred on Europe, North America and Japan. The European standards are based on the 30-channel primary multiplex and the north American and Japanese standards on the 24-channel primary multiplex. The European and North American hierarchies are shown in Figures 4.6 and 4.7 and they are summarised in Table 4.3. Transoceanic digital transmission has led to the introduction of further muldexes to bridge between the different standards of the three 'islands'. These are also included in Table 4.3. The corresponding CCITT recommendations are listed in Table 4.4

These systems all use bit interleaving. The frame length is the same as for the primary multiplex, i.e. 125 μs, since this is determined by the basic channel sampling rate of 8 kHz. However, as shown in Figures 4.6 and 4.7, when *N* tributaries are combined, the digit rate of the higher-order frame is more than *N* times the digit rate of the tributary frames. This is because it is

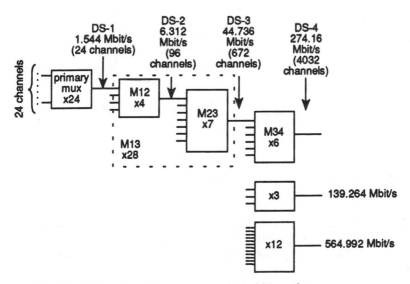

Figure 4.7   North American plesiochronous digital hierarchy

Table 4.3 Digital PDH 'islands' and interworking between them

Island	PCM used	High-order hierarchy
A (Europe):	A-law 30 channel:	2048—8448—34 368—139 264— (564 992) kbit/s
B (Japan):	μ-law 24 channel:	1544/3152—6312—32 064— 97 728 kbit/s
C (North America):	μ-law 24 channel:	1544—6312—44 736–139 264(new)/(274 176) (old)/–(564 992) kbit/s
Between A & B/C:	A-law 30 channel:	2048—6312–44 736–139 264 kbit/s
Between B & C:	μ-law 24 channel:	1544–6312–44 736–139 264 kbit/s

1 International links are to be at the 139 264 kbit/s level, corresponding with 1920 telephone channels for the A-law hierarchy, and 2016 channels for μ-law.
2 Rates in brackets are not CCITT standards.
3 In North America, 1544 = 'DS1', 6312 = 'DS2', 44 736 = 'DS3', 139 264 = 'DS4'.
4 The UK originally used a 1536 kbit/s 24-channel system, then adopted 2048 kbit/s when it became a CCITT standard. Also the next level above 8448 kbit/s was originally 120 Mbit/s in the UK and was replaced by 34 and 140 Mbit/s when they were adopted by CCITT.

Table 4.4 CCITT recommendations for European and north American high-order muldexes

Europe Bit rate	Recommendation	North America Bit rate	Recommendation
Mbit/s		Mbit/s	
2–8	G742	1.5–6	G743
8–34	G751	2–6	G747
34–140	G751	6–32 or 45	G752
140–565	G954 Annex B (see note 2)	45–140	G755

1 These are for positive justification. Positive/zero/negative-justification muldexes for plesiochronous networks are also recommended by CCITT (e.g. G745) but are not widely used.
2 This is not a CCITT recommendation, but CCITT gives it for information.

necessary to add extra 'overhead' digits for two reasons.

The first reason is frame alignment. A higher-order demultiplexer must recognise the start of each frame in order to route subsequent received digits to the correct outgoing tributaries, just as a primary demultiplexer must route received digits to the correct outgoing channels. The same tehcnique is employed. A unique code is sent as a frame-alignment word (FAW), which is recognised by the demultiplexer and used to maintain its operation in synchronism with the incoming signal. The European hierarchy uses a bunched FAW at the start of each frame, but the other hierarchies use distributed FAWs.

The second reason for adding extra digits to the frame is to perform the process known as *justification* [12]. This process is to enable the multiplexer and demultiplexer to maintain correct operation, although the input signals of the tributaries entering the multiplexer may drift relative to each other. In *positive justification*, the transmitted digit rate per tributary is slightly higher than the nominal input rate. If an input tributary is slower, a dummy digit (i.e. a *justification digit*) is added to maintain the correct output digit rate. If the input tributary speeds up, no justification digit is added. These justification digits must be removed by the demultiplexer, in order to send the correct sequence of signal digits to the output tributary. Consequently, further additional digits, called *justification service digits*, must be added to the frame for the multiplexer to signal to the demultiplexer whether a justification digit has been used for each tributary. In *negative justification*, instead of dummy digits being inserted when the digit rate of a tributary is too slow, a data digit is occasionally removed when a tributary is too fast and is transmitted in a spare time slot. Another option is to use both positive and negative justification in the same multiplexer. This is called *positive/zero/negative justification*. The European PDH uses only positive justification.

The term 'justification' originated in the printing trade. Since different lines of print on a page contain unequal numbers of letters, the printer inserts additional spaces to ensure that all the lines on a page are of equal length. Word processors can also perform justification.

When bit interleaving is used, bits for a particular channel occur in different bytes of a higher-order frame. To separate one channel from the aggregate bit stream, a total demultiplexing process is required. This results in the 'multiplexing mountain' shown in Figure 4.8. The new synchronous digital hierarchy, described in Section 4.4.3, employs byte interleaving. This enables *drop and insert* or *add/drop muldexes* to insert or remove lower-order assemblies, down to a primary group, with relative ease.

4.4.3 The synchronous digital hierarchy

Networks are becoming fully digital, operating synchronously, using high-capacity optical-fibre transmission systems and time-division switching. It is advantageous for the multiplexers used in these networks to be compatible

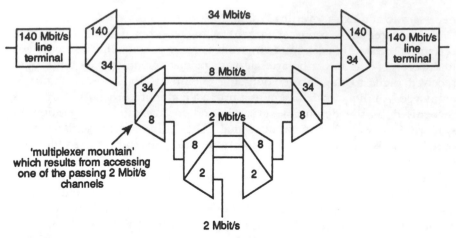

34 Mbit/s

8 Mbit/s

2 Mbit/s

'multiplexer mountain'
which results from accessing
one of the passing 2 Mbit/s
channels

2 Mbit/s

Figure 4.8   The PDH multiplex mountain

with the switches used at the network nodes, i.e. they should be synchronous rather than plesiochronous. In 1990, the CCITT defined a new multiplex hierarchy, known as the *synchronous digital hierarchy* (SDH) [13,14,15]. In the USA, this is called the *synchronous optical network* (SONET), since the muldexes use optical interfaces. The SDH uses a digit rate of 155.52 Mbit/s and multiples of this by factors of $4n$, e.g. 622.08 Mbit/s and 2488.32 Mbit/s. Any of the existing CCITT plesiochronous rates can be multiplexed into the SDH common transport rate of 155.52 Mbit/s. The SDH also includes management channels, which have a standard format for network-management messages [16,17].

The basic SDH signal, called the *synchronous transport module at level 1* (STM-1) is shown in Figure 4.9*a*. This has nine equal segments, with 'overhead' bytes at the start of each. The remaining bytes contain a mixture of traffic and overheads, depending on the type of traffic carried. The total length is 2430 bytes, with each overhead using nine bytes. Thus, the overall bit rate is 155 520 kbit/s, which is usually called '155 Mbit/s'.

This frame is usually represented as nine rows and 270 columns of 8-bit bytes, as shown in Figure 4.9*b*. The first nine columns are for *section overheads* (SOH), such as frame alignment, error monitoring and data. The remaining 261 columns comprise the *payload*, into which a variety of signals can be mapped.

Each tributary to the multiplex has its own payload area, known as a *tributary unit* (TU). In North America, a TU is called a *virtual tributary* (VT). Each column contains nine bytes (one from each row), with each byte having 64 kbit/s capacity. Three columns (i.e. 27 bytes) can hold a 1.5 Mbit/s PCM signal, with 24 channels and some overheads. Four columns (i.e. 36 bytes) can hold a 2 Mbit/s PCM signal with 32 time slots. The STM-1 frame can also hold payloads at the European rates of 8, 34 and 140 Mbit/s and the North

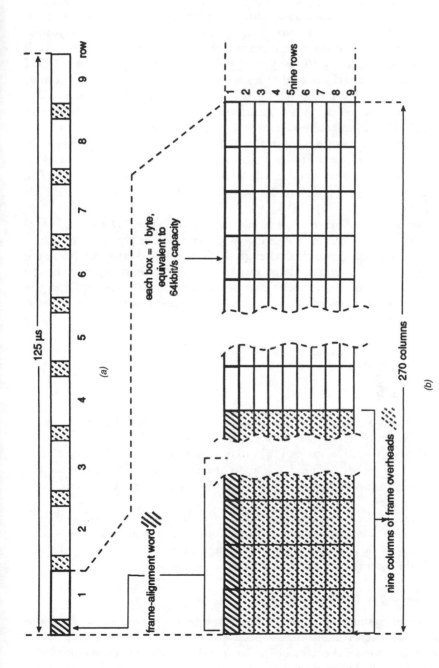

Figure 4.9 SDH: frame structure of STM-1

  a Outline frame structure
  b Frame structure shown in rows and columns

American rates of 6, 45 and 140 Mbit/s.

In the multiplexing process, the bytes from a tributary are assembled into a *container*, and a *path overhead* is added to form a *virtual container* (VC). In North America, the VC is known as a *virtual-tributary synchronous-payload envelope*. The VC travels through the network as a complete package until it is demultiplexed. Since the VC may not be fully synchronised with the STM-1 frame, its start point is indicated by a *pointer*. The VC together with its pointer constitute the TU; thus, it is the TU that is locked to the STM-1 frame. The pointers occupy fixed places in the frame and their numerical values show where the VCs start to enable demultiplexing to be done.

Because the SDH provides interfaces for network-management messages in a standard format, it can lead to a managed transmission-bearer network in which transport capacity can be allocated flexibly to various services. The network can be reconfigured under software control from remote terminals [16,17].

The ability of the SDH to provide add/drop multiplexers can lead to novel network structures. Figure 4.10 shows four remote switching units (RSU) connected to a principal local exchange (PLE) in a ring configuration. There are two alternative routes between each pair of exchanges and the synchronous multiplexers (SMX) can be arranged to reroute traffic in the event of a failure, without any higher-level network-management intervention.

4.5 Line systems

4.5.1 General

A line system consists of terminal equipment, cable and, where necessary, intermediate repeaters. In addition, the system provides operations and maintenance (O&M) features [18] such as performance monitors, alarms and other supervisory signals. Usually, an order wire (OW) is also provided to allow communication between staff working on the system without interrupting service.

Since attenuation increases with distance, a long line is divided into sections, each with its repeater. Since four-wire transmission is used, a repeater normally has an amplifier (in an analogue system) or a regenerator (in a digital system) for each direction of transmission.

4.5.2 Analogue systems

FDM carrier systems were originally developed for open-wire lines and balanced-pair cables. Later, high-capacity systems were made possible by the introduction of coaxial cables [18]. Two cable sizes have been standardised. Large-bore cable has a diameter of 9.5 mm and small-bore cable has a diameter of 4.4 mm. Their attenuations at 10°C are:

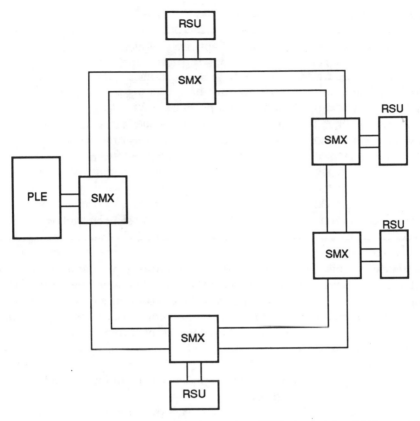

Figure 4.10 *Local-exchange consolidation using SDH ring with add/drop multiplexers*

SMX= synchronous add/drop mux
PLE = principal local exchange
RSU = remote switching unit

(*a*) large-bore cable: 2.355 ($\sqrt{f}$) + 0.006 *f* decibels per kilometre
(*b*) small-bore cable: 5.32 ($\sqrt{f}$) decibels per kilometre
where *f* is the frequency in megahertz.

Their temperature coefficient of attenuation is 0.2% per deg C. Systems which use these cables are listed in Table 4.5.

Intermediate repeaters are usually located in sealed cases in cable manholes, or on poles for aerial cables. A block diagram of an intermediate repeater is shown in Figure 4.11. DC power is fed to intermediate repeaters over the cable pair carrying the traffic channels [18]. A constant direct current is passed along the cable pair between the terminal stations: thus, the repeaters are in series. Separation of the DC power and the HF signals is achieved by the power-separation filters shown in Figure 4.11. Protection of

Table 4.5 FDM systems used on coaxial cables

Bandwidth	Channel capacity	Cable	Repeater spacing	Notes
MHz		mm	km	
4	960	9.5	9.1	
12	2700	9.5	4.55	Also TV transmission
18	3600	9.5	4.55	Also TV transmission
60	10 800	9.5	1.5	Also TV transmission
4	960	4.4	4	
12	2700	4.4	2	
18	3600	4.4	2	
40	7200	4.4	1	Not standardised by CCITT

the repeater against surges is provided by means of zener diodes [18].

The rising loss/frequency characteristic of the line is compensated both by shaping the gain of the amplifier and by means of passive equaliser networks. By splitting these between input and output, good return losses and further attenuation of voltage surges on the cable are obtained. To avoid having to adjust the gain and equalisation to suit repeater sections of different lengths, a *line build-out network* is included. This makes the characteristics of shorter lengths of line the same as those of a repeater section of maximum length, for which the repeater is designed.

The temperature of a buried cable may vary over the year by ±10 deg C and the loss of a 2 km repeater section of small-bore coaxial cable then changes by ±0.12 dB at 300 kHz and ±0.8 dB at 12.5 MHz. After several repeater sections this variation could accrue to an unacceptable amount. Gain regulation is therefore provided to compensate for this effect [18]. A pilot frequency is injected into the signal at the sending terminal station and this operates automatic gain regulators at some intermediate repeaters. The pilot is picked

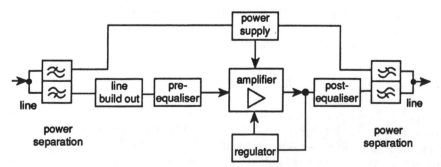

Figure 4.11   Block diagram of intermediate repeater for analogue line system (protection arrangements not shown)

off by a crystal filter, amplified, rectified and smoothed. The direct voltage is compared with a reference voltage and the difference controls the current in a thermistor which varies the gain of the amplifier. To compensate for residual errors, additional pilots are usually transmitted within each group or supergroup and these control flat-gain networks at the receive channelling equipment to compensate for the small drift likely to occur across a limited bandwidth.

Regulation ensures that the signal level at every repeater is within the required limits, i.e. not sufficiently high to cause overloading or sufficiently low to cause deterioration of signal-to-noise ratio. The amplifier has a large amount of negative feedback, both to ensure gain stability and to reduce intermodulation distortion (which causes interchannel crosstalk). The gain required is determined by the attenuation of the section of cable. Since the cable attenuation increases with frequency, the permissible length of a repeater section decreases as the bandwidth of the system increases. The repeater spacings used in standard coaxial-cable systems are shown in Table 4.5.

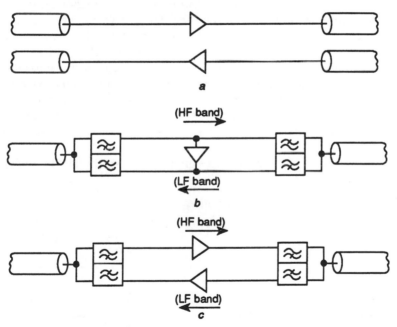

Figure 4.12   Repeater configurations

    a Two-cable-core repeater
    b Single-cable-core repeater with common amplifier
    c Single-cable-core repeater with separate amplifiers

Landline systems with buried cables invariably use four-wire repeaters, as shown in Figure 4.12*a*. However, the single-amplifier configuration shown in

Figure 4.12*b* has been used when an aerially suspended cable uses a single core for both directions of transmission. Different frequency bands are used for the two directions and these are separated by the directional filters shown. Submarine-cable systems also use a single cable core for transmission in both directions. Systems operating up to 5 MHz use the configuration of Figure 4.12*b* for their submerged repeaters. However, systems of greater capacity use separate amplifiers for the two directions of transmission, as shown in Figure 4.12*c*. This halves the bandwidth required for each amplifier.

4.5.3 Digital systems

The first application of PCM systems operating at 1.5 Mbit/s and 2 Mbit/s was to enable balanced-pair cables previously used for audio transmission to provide more circuits. These cables were originally installed with loading coils every 1.83 km (2000 yards). The parameters of the PCM systems were therefore chosen to enable the regenerators to replace loading coils in the same locations. The systems are vulnerable to near-end crosstalk [18]. Therefore, separate cables are often used for the two directions of transmission. If PCM transmission is required in both directions within one cable, only a proportion of pairs can be used. The pairs for the two directions are chosen on opposite sides of the cable with the intervening pairs acting as a screen between them. New cables have been installed which have a transverse metallic screen to separate the two directions of transmission. In North America and Japan, some cables with lower capacitance have been installed to enable systems to operate at 6 Mbit/s and carry 96 channels [18]. Subsequently, coaxial-cable systems were developed to convey higher-order multiplex channel assemblies. Examples of digital landline systems are shown in Table 4.6.

A digital transmission system requires greater bandwidth than an analogue

Table 4.6 Digital systems used on copper cables

Bit rate	Channel capacity	Line code	Cable	Repeater spacing	Example country
Mbit/s				km	
1.5/2	24/30	AMI/HDB3	Audio quads	1.8	US (T1) Europe
3	48	AMI/PR	Audio quads*	1.8	US(T1C)/US(T1D)
6	96	B6ZS	Low capacitance pairs	1.8	US(T2)
8	120	HDB3	0.7/2.4 mm coax.	4	Italy
34	480	MS43	0.7/2.4 mm coax.	2	Italy
140	1920	6B-4T	1.2/44 mm coax.	2	UK
140	1920	MS43	2.6/9.5 mm coax.	4.5	FRG

*In the AMI case, separate cables are needed for the two directions of transmission

system. Thus, there is more attenuation and delay distortion and more noise, but these can be accommodated because of regeneration [9,19]. Consequently, the repeater spacings achievable for digital systems are very similar to those used for corresponding analogue systems. Digital-line systems for coaxial cables have been designed for operation on either 4.4 mm or 9.5 mm cable at the same repeater spacings as used for FDM systems [18]. Thus, a 140 Mbit/s digital system can replace a 12 MHz analogue system and convey 1920 voice channels.

A repeater for a digital-line system [19] must perform the same functions as the analogue repeater shown in Figure 4.11. However, as shown in Figure 4.13, it must also perform the following additional functions:

(*a*) timing extraction;
(*b*) threshold detection; and
(*c*) output-pulse generation.

Another important difference is that, in a digital system, the equalisation need not be as accurate as, and the noise level can be higher than, in an analogue system. This is because distortion and noise are removed by each regenerator, whereas they accumulate, link by link, in an analogue system.

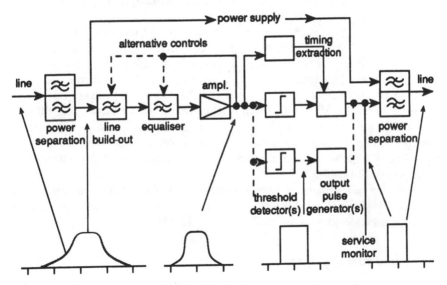

Figure 4.13   *Block diagram of digital repeater*

The simple binary code is unsuitable for the transmission of a digital signal over a line system, because power separation at repeaters loses the DC component of the signal. If the signal contains a long run of identical digits, the waveform will 'droop' and so shift in relation to the decision threshold of regenerators and cause errors. The necessary timing information can also be lost. To overcome this, a binary-coded signal is converted to a different code, called a *line code*, before transmission.

One of the simplest line codes is the *bipolar* or *alternate-mark-inversion* (AMI) code. This is a three-level code. Binary '0' is represented by zero voltage and '1's are represented alternately by positive and negative pulses. Since each mark ('+' or '−') cancels any imbalance due to the previous one, this removes DC and low-frequency energy. Moreover, any single error in transmission will cause a violation of the alternate-mark inversion and so can be detected. A long sequence of '0's still results in a period of zero voltage and so can cause loss of timing. This can be overcome by including a *scrambler* at the sending end to produce a virtually random sequence of '0's and '1's and descrambling the bit stream at the receiving end [19]. AMI is the line code most extensively used in 1.5 Mbit/s PCM systems, e.g. in the North American T1 system.

Modified forms of AMI code have been designed to guarantee timing information by inserting extra '1's in the coded signal when more than a certain number of consecutive '0's occur. These are called *high-density bipolar* (HDB) codes or *bipolar zero-substituted* (BZS) codes [19]. As an example, in the HDB3 code, after a succession of three spaces the next space is replaced by a mark. In the B6ZS code, six consecutive zeros are substituted by the sequence 'mark, zero, violation, mark, zero, violation'.

The AMI, HDB and BZS codes are inefficient in that they use a ternary symbol to transmit each binary digit. A more efficient line code is the 4B-3T code [19], which uses three ternary symbols to transmit four binary symbols. This reduces the transmission rate on the line by a factor of 3/4. Consequently, transmitting a 2 Mbit/s PCM signal requires no more bandwidth than a 1.5 Mbit/s signal in AMI code and so the same cable fill and the same repeater spacings can be used. A further saving in bandwidth can be obtained by using longer code words. For example, the 6B-4T code uses four ternary digits to send six binary digits. This has been used for 140 Mbit/s systems on coaxial cables.

Use has also been made of *partial-response codes* (PR) to reduce bandwidth [19]. An example is the duobinary code, which adds the binary information in adjacent digit periods to give a three-level signal.

4.6 Optical-fibre systems

The major change in transmission technology in recent years has been the introduction of optical-fibre systems [20,21,22]. These have been developed to operate at all the bit rates used on metallic-line systems and at the still higher rate of 565 Mbit/s. Optical-fibre cables are smaller and lighter than copper cables, they provide greater bandwidth and lower attenuation, they are immune to electromagnetic interference and they are even cheaper. Initially, multimode fibres operating at a wavelength of 0.85 μm were used. Subsequently, the bandwidth limitation (due to dispersion) in multimode fibres was overcome by the development of monomode fibres. Monomode-fibre systems, operating at a wavelength of 1.3 μm or 1.5 μm, have been installed in most modern trunk networks.

Optical-fibre systems use semiconductor light sources as transmitters [20,23], i.e. light-emitting diodes (LEDs) and laser diodes (LDs). Currently available devices can operate in the range of wavelengths from 650 nm to 1550 nm, with the key operating windows centred on 850, 1300 and 1500 nm. The LED has a relatively broad spectrum (tens of nanometres) and a carrier lifetime of about 1 ns. Thus, it is used for systems of limited bandwidth over moderate distances. The laser diode is used for high-data-rate and long-distance applications.

Receivers [20,23] use avalanche photodiodes (APDs) and PIN diodes. The advantage of the APD is its gain. However, the PIN diode has a simpler structure and is therefore cheaper. It is also insensitive to temperature variations and therefore does not need temperature stabilisation.

Repeaters have normally used an electronic regenerator placed between an opto-electronic receiver and transmitter. However, optical amplifiers have now become available. These use erbium-doped fibres, pumped by light of a different wavelength [24], as shown in Figure 4.14.

Optical-fibre systems use 'on—off' modulation of their light sources; thus ternary codes cannot be transmitted. However, because of the flat attenuation/frequency characteristics over a very wide bandwidth, there is no need to use multilevel line codes to conserve bandwidth. Nevertheless, there may still be a need to use a redundant code to provide error monitoring and to avoid a DC shift and possible loss of timing as a result of a long succession of '0's being transmitted. One code employed is the *coded-mark-inversion code* (CMI), which is related to the binary and AMI codes as follows:

binary-code character	0	1
AMI:	0	− 1 and +1 alternately
CMI:	01	00 and 11 alternately

The CMI code was recommended by the CCITT as the interface between

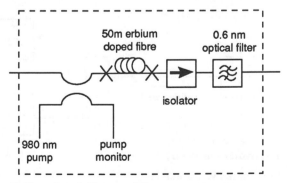

Figure 4.14   Erbium-doped-fibre amplifier

equipments at 140 Mbit/s [19]. Long-haul systems commonly use the 5B–6B code, in which groups of five binary symbols are encoded into six-bit words [19]. In SDH transmission, adequate DC balance and timing content are provided by scrambling and other line-coding functions are performed by digits in the overheads [19].

Optical-fibre transmission systems normally send a single bit stream in one direction on each fibre. However, the use of optical couplers enables *wavelength-division multiplexing* (WDM) to be used [20]. Independent bit streams can be transmitted on a single fibre by using different wavelengths, either in a single direction (as shown in Figure 4.15*a*) or in opposite directions (as shown in Figure 4.15*b*).

An outstanding feature of single-mode optical fibres is that their low attenuation permits long distances between repeaters. The repeater spacing of 140 Mbit/s systems is typically 50 km. Thus, the majority of junction systems require no intermediate repeaters. In densely populated countries, such as the UK, accommodation for repeaters can be found in existing exchanges and repeater stations. In less-populated areas, power may be fed by a metallic pair from the nearest building or local power sources, such as solar cells, may be used.

Optical fibres are now being used for submarine cable systems. The first transoceanic cable (TAT-8) was opened in 1988 between the USA and Europe

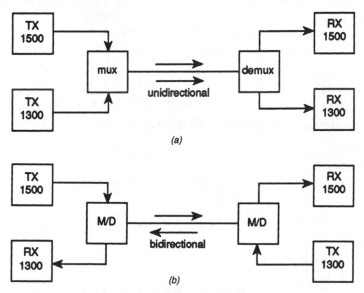

(a)

(b)

Figure 4.15   Wavelength-division multiplexing

    a Unidirectional WDM system
    b Bidirectional WDM system

[25]. It is 9360 km in length and uses a bit rate of 295.6 Mbit/s to convey 3780 channels per fibre pair. Use of DSI provides 40 000 circuits, which effectively doubled the traffic capacity available on cables across the North Atlantic. Several other transoceanic systems have been installed subsequently. TAT-12 is now in operation and TAT-13 is planned.

Optical fibre has now also penetrated customer-access networks. Optical fibres are being used to provide access for large customers which require many circuits. Other customers may be served by multiplexers or concentrators connected to the main exchange by optical fibre. A possible future architecture for access networks is called *telephony on a passive optical network* (TPON) [5,26]. A number of customers' premises are connected to a single fibre by passive optical splitters. The system uses either time-division multiple access (TDMA) or wavelength-division multiple access (WDMA).

Continuing research is likely to result in further dramatic advances in optical-fibre transmission systems [27]. Demonstrations have been given of the transmission of bit rates up to 20 Gbit/s over distances up to 500 km. Dispersion-shifted monomode fibre can give an attenuation of only 0.2 dB per km. Wavelength-division multiplexing is re-introducing FDM, but at optical frequencies. Modulation and signal-processing methods used in radio technology (e.g. coherent demodulation and heterodyne detection) have been used to provide increased receiver sensitivity. Finally, *soliton* waves, arising from interaction between dispersion and nonlinearity in a fibre, have the potential of travelling over distances of the order of tens of thousands of kilometres without regeneration [28].

4.7 Microwave radio-link systems

4.7.1 General

A microwave radio link [1, 2, 29] is an alternative to the cable transmission systems described in Sections 4.5 and 4.6. Its relative advantages and disadvantages may be summarised as follows.

Advantages:
(a) Capital cost is usually lower.
(b) Installation is easier and quicker, since no work is required along the route between repeaters.
(c) As a result of (b), problems are not caused by the route crossing difficult terrain between repeaters.
(d) The system is immune to interruptions caused by road works or other events which damage cables.

Disadvantages:
(e) Operation is restricted to line-of-sight distances.
(f) Repeater stations may need to be sited in remote locations, where access by road is difficult.

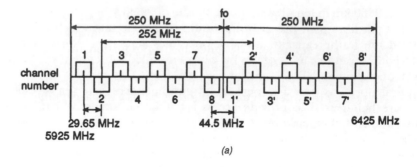

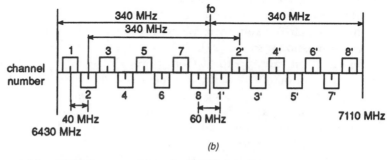

Figure 4.16   CCIR frequency plans for 6 GHz bands

        a Lower 6 GHz band (*Recommendation 383*)
        b Upper 6 GHz band (*Recommendation 384*)

(*g*) Provision of adequate power supplies may be difficult in remote locations.

(*h*) Adverse weather conditions can cause fading of the signals, so measures must be taken to combat this.

Microwave-radio systems predominate in the trunk networks of countries where the terrain is difficult. Other countries use cable systems on some routes and radio on others, as appropriate. In some countries, both may be used on the same route, to obtain security of operation if either system breaks down.

Although narrow-beam antennas are used, available frequency spectrum is the limiting factor on the number of channels which can be carried on a given route, so it must be treated as a scarce resource. Channel spacings and frequency assignments have been agreed internationally. Bands at 4 GHz, 6 GHz and 11 GHz have been used in trunk systems and higher-frequency bands of 19, 23 and 29 GHz are being used for short links. The use of millimetric bands above 30 GHz is being planned.

The frequency plans specified by the CCIR for the two 6 GHz bands are shown in Figure 4.16. At any station, all the transmit channels are in one block and the receive channels are in the other block. At any intermediate repeater, a frequency shift is made to the received signals before they are retransmitted to minimise interference between the two directions of transmission and so ensure stability. The equipment designer has to apportion the filtering to minimise adjacent-channel interference, while maintaining an acceptable channel performance. The system planner has to site the stations and antennas to restrict the interference from other paths to an acceptable level.

4.7.2 Analogue systems

The application which gave the initial impetus to microwave radio-relay systems was the transmission of television signals, whose wide bandwidth was difficult to carry on cable systems. Systems have subsequently been installed with capacity for 60, 960, 1800, 2700 and 3600 telephone channels.

A transmitter contains equipment to perform the following functions [29]:

(*a*) Baseband processing converts the broadband input signal to one suitable for the modulator.
(*b*) A modulator converts the information into a modulated carrier for transmission over the radio path.
(*c*) Radio-frequency equipment provides the means of transmitting the modulated signal. In addition to power amplification, it may combine the signals of several broadband channels for transmission on a single antenna.

Equipment at the receiver performs the inverse functions.

The block diagram of a repeater is shown in Figure 4.17*a*. The input signal of a particular broadband channel is separated from others by a circulator and a bandpass filter. It is then downconverted to the intermediate frequency (IF), where it is amplified. The IF signal then passes to the upconverter to shift to the frequency to be transmitted. Finally, the signal is amplified by the power amplifier, combined with the signals of other channels and retransmitted.

Initially, frequency modulation (FM) was universally adopted. This enabled the necessary linearity to be obtained in the modulation and demodulation process and the modulated signal was not distorted by the limiting amplifiers (e.g. travelling-wave tubes) in the output of the transmitter. More recently, pressure for increased capacity in the available frequency bands has resulted in the use of single-sideband (SSB) modulation for 6000 channels on a single bearer.

4.7.3 Digital systems

Radio-relay systems have been designed to convey digital signals at the standard rates used in the plesiochronous digital hierarchies described in Section 4.4.2. In Europe, 34 Mbit/s and 140 Mbit/s are common and in North America, multiples of 45 Mbit/s.

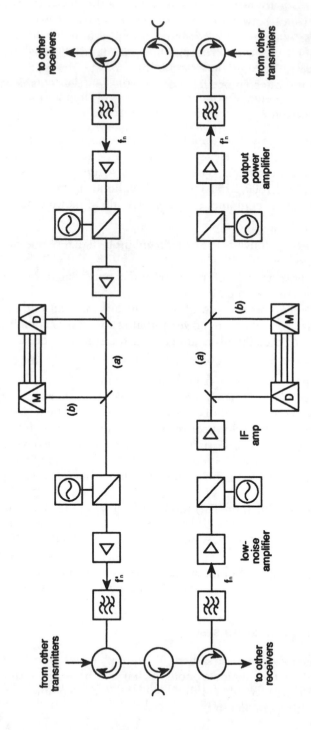

Figure 4.17 Block diagram of repeater for microwave radio link

   a For analogue system
   b For digital system

In a digital system, a transmitter and receiver must perform the same functions as in an analogue system. However, as shown in Figure 4.17*b*, digital systems, unlike analogue systems, demodulate the signal to baseband at each repeater in order to perform regeneration. Adaptive equalisation is also employed to minimise intersymbol interference.

For digital transmission, a single-frequency carrier can be modulated in amplitude, phase or frequency. To obtain efficient use of the radio spectrum, multilevel modulation schemes are employed. Four-phase and eight-phase PSK are used. 16-phase PSK is not used, because a better separation between 16 signal states is obtained by using quadrature amplitude modulation (16 QAM). This produces 16 states by means of three different amplitudes and 12 different phases, as shown in Figure 3.9*c*. The use of 16 QAM can fit 140 Mbit/s within 40 MHz channel spacings, but not within a 30 MHz spacing; this requires 64 QAM. A comparison between the performances of different modulation methods is given in Table 4.7.

In future, higher-capacity systems, based on 155.52 Mbit/s and multiples of it, will be required to transmit signals of the synchronous digital hierarchy (SDH). This need may be met by using multicarrier 256 QAM or possibly trellis-code modulation.

Table 4.7 Spectrum efficiency of modulation schemes for microwave radio links

Modulation	Effective eye opening	Performance at 10^{-5} BER		Copolar channel spacing for 140 Mbit/s	Spectrum efficiency
		Ideal systems	Practical systems		
		CNR dB	CNR dB	MHz	bit/s Hz
QPSK	0.88	13.8	15.5	134	2.2*
RBQPSK	0.62	17.0	18.4	90	3.1†
16 QAM	0.32	22.7	25.1	80	3.5*
64 QAM	0.14	29.5	30.5	59.3	4.7*

* Systems using interleaved crosspolarised channels
† Systems using cofrequency crosspolarised channels

4.8 Satellite communication systems

4.8.1 General

A satellite in a circular orbit above the equator 36 000 km from the earth appears stationary to users on earth. Three transponders in such geostationary satellites can provide global coverage of the earth's surface apart from the polar regions. Such satellite links can provide a service which is independent of distance and thus can be economical for fixed long-distance links, mobile communications and broadcasting [30,31,32]. Some of the major applications of satellite communications and their frequency bands are listed in Table 4.8.

The Intelsat system developed using 6 GHz uplinks and 4 GHz downlinks with earth stations using 32 m diameter antennas. Further expansion occurred with the introduction of smaller earth stations (11 m diameter). Domestic networks (e.g. the Canadian Telsat system) resulted in further growth of satellite networks and a mobile service was established for ships. During the 1980s, satellite communication was made accessible to individual business users by the introduction of smaller earth stations (using 3 m antennas operating at 11 GHz and 14 GHz) and satellites with spot beams. Now very small-aperture terminals (VSATs) [33] are available with 1.2 m antennas for two-way communication and 60 cm antennas for receive-only service. Television direct-broadcasting-satellite (DBS) services require only a 30 cm-diameter dish for a receiving terminal. As a result of this growth, congestion has now arisen in the C and Ku frequency bands, and bands at 20 GHz and 30 GHz have been allocated.

Table 4.8 Major application areas and frequency bands for satellite communication

Types of service	Frequency band uplink/downlink	Band designation	Allocated bandwidth
	GHz		GHz
Fixed satellite service (FSS)	6/4	C band	1.0
	14/12 (or 11)	Ku band	1.1
	30/20	K band	3.5
Broadcasting satellite	14/12	DBS band	0.8
service (BSS)	18/12	DBS band	0.8
Mobile satellite service (MSS)	1.6/1.5	L band	34/29 MHz

4.8.2 Access methods

Multiple-access methods enable a satellite transponder to be shared by many ground stations within its *footprint*, i.e. the coverage area of its antenna beam. The following methods are currently employed:

(*a*) frequency-division multiple access (FDMA),
(*b*) time-division multiple access (TDMA), and
(*c*) code-division multiple access (CDMA).

Frequency-division multiple access (FDMA) was the first method to be used.

The major international systems used for telephony, e.g. Intelsat, transmit FDM groups similar to those used in terrestrial systems, with blocks of channels allocated to each country. Usually, a block of 972 channels (12 kHz to 4.028 MHz) occupies a transponder of 36 MHz bandwidth, using frequency modulation. For domestic satellite systems, providing small numbers of channels for a large number of earth stations, *single-carrier-per-channel* (SCPC) access is used [30]. Each channel may carry an analogue voice channel (using FM) or a 64 kbit/s digital channel (using PSK).

Time-division multiple access (TDMA) is based on time sharing a transponder between different earth stations [30]. Each station transmits a short-duration signal according to a *burst plan*. This ensures that different bursts reach the transponder in assigned slots within a TDMA frame without any overlap. Each earth station is allotted one or more bursts per frame, depending on the required capacity. In the Intelsat and Eutelsat systems, the transmission rate is 120 Mbit/s and the frame period is 2 ms. Thus, a transponder can accommodate 1800 channels.

In code-division multiple access (CDMA), each user's signal is spread over the bandwidth of the transponder, using a spread-spectrum technique [34,35]. A family of orthogonal pseudonoise codes is used and each receiver must maintain the same code as the corresponding transmitter. The main application of CDMA has been in military systems. However, it has recently been used in commercial microterminals, mainly to combat interference from adjacent satellites and other sources.

4.8.3 Future applications

Satellite-communication systems were introduced to provide large numbers of telephone channels across oceans. However, their inherent large propagation delay (of about 0.25 s) makes them less suitable than submarine cables for this application and optical-fibre submarine cables are able to convey such large volumes of traffic at much lower cost.

In future, satellite links are likely to be used in applications where they are uniquely suitable. They will still be used for lower-capacity routes, where links to many different destinations can share a common transponder. In addition, applications will increase in the areas of broadcasting, mobile communications (to ships and aircraft), business services for individual companies and personal communications. Digital techniques will be used extensively and the trend will be to smaller earth stations but more powerful satellites having regenerative transponders with on-board signal processing and scanning spot-beam antennas.

4.9 Conclusion

Digital transmission has made analogue carrier systems obsolete and optical fibres are superseding metallic cables. Digital transmission on optical fibres has now been introduced for most applications, ranging from short-haul systems to intercontinental submarine-cable systems and optical fibres have already begun to penetrate customers' access networks. Continuing research may be expected to produce further advances in optical-fibre systems and extend their applications still further.

Microwave radio links can now transport high-order digital-multiplex-channel assemblies. They will continue to provide an alternative to cable systems, particularly across difficult terrain. Radio is essential to provide access for mobile users in cellular-radio systems. It is also used to provide access to customers in remote locations and microcellular systems may be employed as an alternative to cable distribution in urban areas.

Satellite-communication systems are now more expensive than optical-fibre submarine cables for conveying large amounts of traffic across oceans. However, they will still be economic for low-capacity routes, where links to different destinations can share a common satellite transponder. Applications will also increase in broadcasting, mobile communications and providing private systems for businesses.

4.10 References

1. Bell Laboratories Staff: 'Transmission systems for communications', Bell Telephone Laboratories, 1980, 4th Ed.
2. HILLS, M.T., and EVANS, B.G.: 'Transmission systems' (Allen and Unwin, 1973)
3. FREEMAN, R.L.: 'Telecommunications transmission handbook', (Wiley, 1981), 2nd Edn.
4. FLOOD, J.E. and COCHRANE, P. (Eds.): 'Transmission systems' (Peter Peregrinus, 1991)
5. ADAMS, P.F., ROSHER, P.A., and COCHRANE, P.: 'Customer access' in FLOOD, J.E., and COCHRANE, P. (Eds.): 'Transmission systems' (Peter Peregrinus, 1991), chap. 15.
6. KINGDOM, D.J.: 'Frequency-division multiplexing' in FLOOD, J.E., and COCHRANE, P. (Eds.): 'Transmission systems' (Peter Peregrinus, 1991), chap. 5.
7. WALLACE, A.D., HUMPHREY, L.D., and SEXTON, M.J.: 'Analogue to digital conversion and the primary multiplex group' in FLOOD, J.E. and COCHRANE, P. (Eds.): 'Transmission systems' (Peter Peregrinus, 1991), chap. 6.
8. CATTERMOLE, K.W.: 'Principles of pulse code modulation' (Iliffe, 1969)
9. BYLANSKI, P. and INGRAM, D.G.W.: 'Digital transmission systems' (Peter Peregrinus, 1980), 2nd ed.
10. CCITT Recommendation G.704: 'Functional characteristics of interfaces associated with network nodes'
11. CCITT Recommendation G.711: 'Pulse code modulation of voice frequencies'
12. FERGUSON, S.P.: 'Plesiochronous higher-order digital multiplexing' in FLOOD, J.E., and COCHRANE, P. (Eds.): 'Transmission systems' (Peter Peregrinus, 1991), chap. 8.
13. CCITT Recommendation G.707: 'Synchronous digital hierarchy bit rates'
14. CCITT Recommendation G.708: 'Network node interface for the synchronous digital hierarchy'
15. CCITT Recommendation G.709: 'Synchronous multiplexing structure'
16. SEXTON, M., and REID, A.: 'Transmission networking: SONET and the synchronous digital hierarchy' (Artech House, 1992)
17. SEXTON, M.J., and FERGUSON, S.P.: 'Synchronous higher-order digital multiplexing' in FLOOD, J.E., and COCHRANE, P. (Eds.): 'Transmission systems' (Peter Peregrinus, 1991), chap. 9.

18. HOWARD, P.J., and CATCHPOLE, R.J.: 'Line systems' in FLOOD, J.E., and COCHRANE, P. (Eds.): 'Transmission systems' (Peter Peregrinus, 1991),chap. 10.
19. DORWARD, R.M.: 'Digital transmission principles' in FLOOD, J.E., and COCHRANE, P. (Eds.): 'Transmission systems' (Peter Peregrinus, 1991), chap. 7.
20. BICKERS, L.: 'Optical-fibre transmission systems' in FLOOD, J.E., and COCHRANE, P. (Eds.): 'Transmission systems' (Peter Peregrinus, 1991), chap. 11.
21. HOWES, M.J., and MORGAN, D.V.: 'Optical fibre communication' (Wiley, 1980).
22. SENIOR, J.: 'Optical fibre communication' (Prentice Hall, 1985)
23. WILSON, J., and HAWKES, J.F.B.: 'Optoelectronics: an introduction' (Prentice Hall, 1983)
24. DESURVIRE, E.: 'Erbium-doped fibre amplifiers' (Wiley, 1994)
25. SMITH, R.L., and WHITTINGTON, R.: 'TAT-8: an overview', *Br. Telecom Eng. J.* 1986, **5**, pp. 148–152.
26. OAKLEY, K.A., TAYLOR, C.G., and STERN, J.R.: 'Passive fibre local loop for telephony with broadband upgrade'. *Br. Telecom Eng. J.* 1989, **7**, pp. 233–236
27. COCHRANE, P.: 'Future trends' in FLOOD, J.E., and COCHRANE, P. (Eds.): 'Transmission systems' (Peter Peregrinus, 1991), chap. 17.
28. DORAN, N.J.: 'Solitons: permanent waves', *IEE Rev.* 1992, **38**, pp. 291–294.
29. DE BELIN, M.J.: 'Microwave radio links' in FLOOD, J.E. and COCHRANE, P. (Eds.): 'Transmission systems' (Peter Peregrinus, 1991), chap. 12.
30. NOURI, M.: 'Satellite communication' in FLOOD, J.E., and COCHRANE, P. (Eds.): 'Transmission systems' (Peter Peregrinus, 1991), chap. 13.
31. DALGLEISH, D.I.: 'An introduction to satellite communications' (Peter Peregrinus, 1989)
32. EVANS, B.G. (Ed.) 'Satellite communication systems' (Peter Peregrinus, 1991), 2nd ed.
33. EVERETT, J. (Ed.): 'VSATs: very small aperture terminals' (Peter Peregrinus, 1992)
34. FLOOD, J.E.: 'Principles of multiplex communication' in SKWIRZYNSKI, J.K. (Ed.): 'New directions in signal processing' (Noordhoff, 1975), pp. 271–287
35. SKAUG, R. and HJELMSTAD, J.F.: 'Spread spectrum communication' (Peter Peregrinus, 1985)

Chapter 5
Switching principles
S.F. Smith

5.1 Introduction

Preceding Chapters have described the nature of telecommunication networks and identified the need for switching at the nodes of a network. The users of communication terminals, such as telephones and data terminals, make calls to other users over a common network. Such networks may be public or private, but all have the characteristic that communication paths are provided to users only when they require them, usually on demand but sometimes after a delay. To provide permanent paths for all possible communicating groups of users would be impractical and uneconomic, so the networks contain exchanges (central offices) at their nodes to set up and break down paths as required.

This Chapter is concerned with the general principles applying to the design of exchanges and some of the practical implementations of them. Later Chapters will deal with these topics in greater depth.

5.2 General principles

In its simplest form, the required communication path consists of a metallic connection between two terminations. The fundamental requirement of an exchange can then be seen as physically interconnecting any two out of the total population of terminations such as customers' telephone instruments or data terminals.

We can classify switching systems according to their connection technique as follows:

(*a*) *Circuit switching*
This consists of establishing for each call a connection which is then held throughout the call for the exclusive use of that call.

(*b*) *Packet switching*
This consists of temporarily establishing the connection at intervals during the call to transmit a discrete packet of data each time.

In circuit switching, if a route is selected for a call but all the circuits on the route are busy, the call is lost. Routes are dimensioned so that the probability of this happening is small. In packet switching, when outgoing routes are busy, packets are stored in a queue until circuits become free and they are then retransmitted. Thus, packets are not lost during busy periods; they are delayed. A circuit switch is therefore an example of a *lost-call system* and a packet switch is an example of a *queueing system*.

Each of these techniques can be implemented using *time-division switching* or *space-division switching* or a combination of the two.

Time division consists of assigning regular discrete time slots cyclically to each of several calls so that they share a physical connection (i.e. the calls are separated in time).

Space division consists of providing each call with its own physical connection (i.e. the calls are separated in space).

The transmission mode in any of these cases may in principle be either analogue or digital, but only certain combinations have practical applications. This Chapter is concerned with analogue space-division circuit switching. However, many of the principles introduced here also apply to exchanges employing other techniques. Time-division switching is discussed in Chapter 6 and packet switching in Chapter 13.

More than one connection or call will be required at any one time, of course, but not all terminations will be wanting service at the same time. Practical systems are designed to provide a number of links which is less than the number of terminations. Thus, at times of peak traffic, a customer may be unable to make a call because there is no free link. The number of links provided is determined on a statistical basis to make the proportion of lost calls very small. Figure 5.1 shows such an arrangement based on a matrix of *crosspoints*. Each crosspoint consists of some form of on/off switch, such as a manually-operated key, an electromechanical relay or a transistor.

The simple example in Figure 5.1 can be regarded as a single-node network. All connections pass through one single exchange. This is clearly unreasonable on a global scale and practical networks have many nodes and consequently many exchanges. Each exchange therefore has connections to other exchanges. Clearly, the exchange must be able to handle calls between

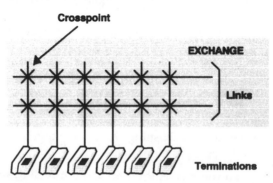

Figure 5.1   Switching principle

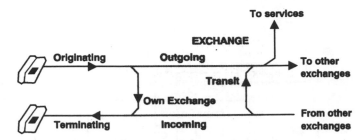

Figure 5.2 Call routing in a telephone exchange

its own customers' terminations and between them and the external connections (junctions). In large networks it is also usual to switch calls between other exchanges. Transit (tandem) exchanges exist solely for this purpose and have no customer-line terminations of their own.

The various types of call routing, as shown in Figure 5.2, are:

(*a*) own-exchange calls (between customers on the same exchange);
(*b*) outgoing junction calls (to other exchanges or various services);
(*c*) incoming junction calls (from customers on other exchanges); and
(*d*) transit calls (incoming from one exchange and outgoing to another).

Traffic is not usually divided evenly over these types of call. Only a small proportion (5–15%) of the traffic is own-exchange traffic, with the rest divided between incoming and outgoing traffic. Services such as emergency, speaking clock etc. are usually centralised at one exchange in an area, so the outgoing traffic at other exchanges normally exceeds the incoming traffic. The proportion of transit traffic varies from zero at small rural exchanges to 100% at exchanges provided solely for this purpose (i.e. trunk, tandem or transit exchanges).

5.3 Exchange structure

The simple matrix in Figure 5.1 concentrates originating calls from six terminations down to two links and expands the terminating calls to the same six terminations. The concentration and expansion functions in this case are combined in one switch, but in some systems they are separated. To extend the principle to a larger number of terminations, we could obviously use a larger switch. However, any type of switch will have size limitations and it is more usual to use several switches, each serving its own group of terminations.

Figure 5.3 shows an exchange with each customer's line connected to a *concentrator* and a separate *expander.* The number of these units that are installed increases with the size of the exchange. The separate switches are interconnected through yet another switching stage, which does not provide

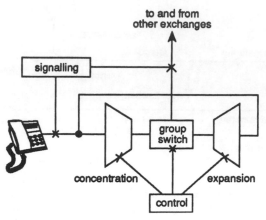

Figure 5.3   Elements of a complete exchange

concentration or expansion. This *group-switching stage* or *distributor* can also be used to make connections to or from other exchanges. It is therefore also known as a *route switch*.

To complete the exchange, two other functional blocks are needed. The call-connection and disconnection instructions have to pass between the terminations and the exchange and also between exchanges for inter-exchange (junction) calls. These messages are conveyed by the exchange *signalling systems*, which may transmit the relevant data over the same paths as the calls they control (channel associated signalling) or they may use separate links for the purpose (common channel signalling), as described in Chapter 8. Also needed is a *control system* for interpreting these instructions to enable appropriate paths to be set through the switches. In early systems, control was performed manually by telephone operators. In electromechanical switching systems, it is performed by relay circuits and modern systems use electronic control.

The operation and design of exchange switching systems can be conventionally separated into these three separate functions: switching, signalling and control. However, these are interdependent; they interact to enable customers to obtain the services they need. The functions required of a switching system to provide these services are called *facilities*. In practical terms, the choice of switching technology is very dependent on the transmission methods used in the network concerned and this, in turn, influences the choice of signalling and control methods.

5.4 The manual exchange

The first telephone exchanges were operated by hand. Instructions were passed verbally between the customer and the operator. Similar principles were used on early telegraph exchanges. As the network grew and operators were expected to handle increasing volumes of traffic, the signalling and control elements were developed which formed the basis of the first automatic systems.

In its most common form, the manual switchboard [1,3] has customers' terminations connected to jacks (sockets). A link consists of a pair of plugs on a flexible cord which can be used to interconnect any pair of jacks. Signalling apparatus consists of calling indicators (e.g. relays and lamps) associated with the jacks and clearing indicators associated with the cords and plugs. Call-routing instructions are passed verbally.

The great strength of the system is that the control is human and therefore intelligent; thus, the variety of connections that can be made is virtually unlimited. Services, such as advice of duration and charge, transfer of calls when absent, wake-up calls etc., which are so complex to provide automatically, present no problem at all on manual exchanges.

The weakness of the manual exchange, which has led to its almost complete disappearance, was essentially its slowness. Unless operators could set up calls faster, it would have been impossible to find enough operators to handle the volume of traffic on the modern network.

5.5 The step-by-step system

The first successful 'automatic' system was introduced by Almon B. Strowger in 1892. In the Strowger system [2,3], ratchet-driven switches, controlled directly by pulses from the dial of the calling telephone, are used to simulate the plug-and-cord switchboard. As shown in Figure 5.4a, wipers (moving contacts) are connected by short flexible cords and perform a similar function to the switchboard plugs. The place of the jack field is taken by a bank of fixed contacts and the wipers are moved one step from each contact

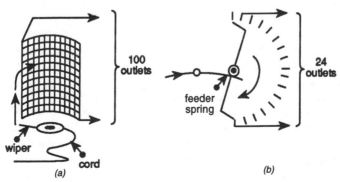

Figure 5.4   Strowger selectors

      a Two-motion selector
      b Uniselector

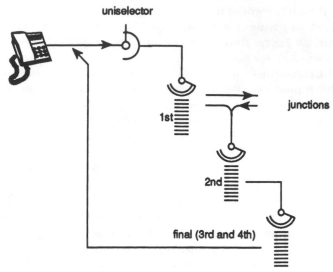

Figure 5.5   Strowger step-by-step exchange

to the next for each pulse, by means of a ratchet-and-pawl mechanism.

At first, Strowger's exchange provided every customer with their own 100-outlet *two-motion selector.* The expense of providing one large selector for each customer is now avoided by using a stage of smaller cheaper *uniselectors,* as shown in Figure 5.4*b,* to connect the callers to the selectors as and when they require them. Extension of the exchange beyond 100 lines is by adding further stages of selectors, as shown in Figure 5.5. The setting up of a call progresses through the exchange stage by stage. Each selector connects the call to a free selector in the next stage in time for the train of dial pulses to set that selector. The wipers step vertically to the required level under dial-pulse control and then rotate automatically to select a free outlet. This is *step-by-step* operation and needs only simple control circuits which can be provided economically at each selector.

The concentration function shown in Figure 5.3 is provided by the uniselectors. In this example there are then two stages of group selectors to provide the distribution function. The final two digits of the number are used to position the wipers of the *final selector* or *connector* to the contacts connected to the required line, thus providing the expansion function.

The final selector also contains the circuit element known as the *transmission bridge* [4], which provides current to the calling and called lines from the central battery at the exchange. It is here that the 'answer' and 'clear' conditions are detected to control charging (e.g. by metering) and to release the connection. These functions are called the *supervision* of the call.

In the step-by-step system, each switching stage required in a connection uses one dialled digit. Thus, to reach a particular customer, more digits must be dialled for a junction call than for an own-exchange call. Further digits are required if the call is routed through a tandem exchange. A large city often has hundreds of exchanges with overlapping service areas. Then, it would be unacceptable for callers to have to consult a different list of dialling codes according to which exchange the telephone they happen to be using is connected. To allow the same codes to be used anywhere in the area, *director equipment* [2] is provided which translates the code dialled by the caller into the actual routing digits required.

Even with the addition of the director, the step-by-step system lacks flexibility. The convention of ten finger holes in a dial means that the switching network branches in a decadic fashion. Each switch requires a decadic ability and is mechanically designed accordingly. If, however, traffic is unevenly distributed throughout the exchange numbering range, some parts of this capability will be under used. The relative slowness of the switches also makes it impractical to attempt rerouting for call transfer or to bypass congestion in the network. The mechanical complexity of the switches themselves presents reliability problems and requires expensive maintenance.

5.6 Register control

The director, as used in the Strowger system, is an example of *register control.* Some of the disadvantages of the Strowger switches have been overcome with other designs of mechanical switch [2], but these are generally unsuitable for direct control from the customer's dial. Register control provides a way of using such switches.

The most successful register-controlled systems are based on the *crossbar switch* [2,3,5] which is mechanically simpler than the Strowger switch and can be designed for greater reliability and smaller size. Translation facilities like those provided by the director system, but not limited by considerations of mechanical-switch sizes, are readily obtained because there is no inherent relationship between number allocations and particular switch outlets.

A crossbar switch [6] is shown in Figure 5.6. It has a rectangular array of contact sets forming its crosspoints. Thus, operating a crosspoint connects a terminal associated with its column to a link associated with its row. The crosspoint is operated by energising a 'select' electromagnet corresponding to the row and a 'hold' electromagnet corresponding to the column. These cause horizontal and vertical bars to pivot and close the contacts at the point where they cross. Typical mechanism contain 200 or more crosspoints per switch.

Crossbar switches are generally used in combinations to form a group-selection (distribution) stage and a combined concentration/expansion stage, as shown in Figure 5.7. (In practice, each of these stages itself usually contains two stages of crossbar switches in order to be able to provide a large number of connections [2,7].) Additional crossbar switches may be used for establishing other connections necessary for the operation of the exchange,

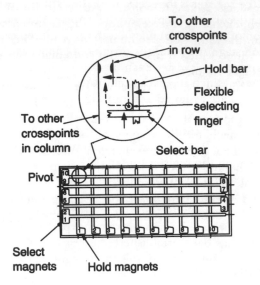

Figure 5.6   Crossbar switch

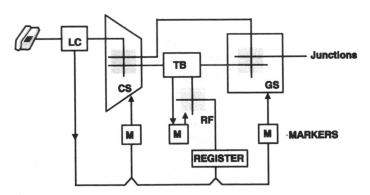

Figure 5.7   Crossbar exchange

CS = concentrator/expander stage
GS = group-selector stage
LC = customer's line circuit
 M = marker
RF = register finder
TB = transmission bridge

e.g. switch RF, which connects registers as required to callers via the concentration stage (CS).

The function of the register is to receive the caller's instructions in the form of dialled digits and convert them into the appropriate signals for route selection. Operation of the relevant magnets in the required crossbar switches in the concentration stage (CS) and the group-selection stage (GS) is carried out by apparatus known as a *marker*. Each marker is directly associated with a set of crossbar switches which it serves exclusively. The registers are available in a common pool and are assigned as required to incoming calls.

The customer's line circuit (LC) detects the calling condition and signals to the relevant marker which causes CS to connect the calling line to a free transmission bridge (TB). This, in turn, co-operates with another marker to set a connection through the register finder (RF) to seize a free register.

When a register has received the dialled digits and analysed them to determine the required connection, it seizes a marker to set the appropriate switches in the group-selector stage. For an own-exchange call, a marker also sets the required switches in the concentrator/expander of the called line. Once the connection has been established, the register and the markers are released. Supervision of the connection is then provided by the transmission bridge for the duration of the call.

Since a register is only used during dialling and setting up the connection, the number required is much less than the number of connections that can exist (which corresponds to the number of transmission bridges). A marker is held only for the short time taken to set a crossbar switch. Consequently, even a large and busy exchange needs only a few markers.

Unlike the Strowger system, the number of switching stages need not be related to the exchange numbering scheme. Moreover, the number of the terminal to which a customer's line is connected on the concentrator/expander, which is known as its *equipment number*, need not be the same as the customer's directory number. The use of markers enables *directory-number-to-equipment-number translation* to be performed.

5.7 Central control

The use of registers represents one application of *common control*, where one control circuit (e.g. a marker or a register) is shared by several switches.

Greater centralisation of the control, by introducing *central processors*, offers further operational advantages. For example, when numbering changes are made during the life of the exchange, it may be possible to avoid the need to change a large number of circuits such as registers. It is also possible to employ more effective search patterns to find free paths, thus making economies possible in switch provision. For example, in a register system, the register chosen initially for the call does not necessarily have an unrestricted choice of switch paths for the forward connection of the call. It may have access only to some of the switches at the next stage, and it cannot govern the choice of free switches beyond that. Although suitable free paths may exist, the exchange may be unable to allocate them and the customer may fail to get the call. This situation is known as *blocking* [7].

In any group of switches under one control, it is only possible to set one call at a time. To extend the control across a whole exchange needs faster switches and a control speed to match. Although some central control has been used with crossbar switches, it has been more widely applied to reed-relay exchanges in the public network and later to fully-electronic exchanges. There are systems using solid-state analogue crosspoints such as thyristors, but these are confined mainly to PABX applications where crosstalk, transmission and overvoltage requirements are often less onerous than on public systems.

A *reed relay* [3,5] consists of sealed contact units inside an operating coil. They are interconnected to form switch matrices, similar to Figure 5.1, in which each crosspoint has its own reed relay. Being sealed, and having no external moving parts, they are faster and more reliable than crossbar switches. Switching systems using reed-relay switches controlled by electronic central processors have been developed in several countries [8].

A typical arrangement for a reed-relay exchange with central control is shown in Figure 5.8. This is the TXE4 system [3,6,9], which serves about one-third of all telephone lines in the UK network.

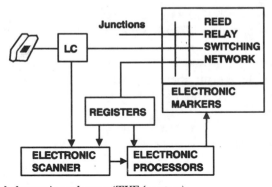

Figure 5.8   Reed-electronic exchange (TXE4 system)

LC = customer's line circuit

An electronic scanner continuously updates a record of the state of all switch terminations. This record is accessible to all processors on a 'read-only' basis. In response to this information, the electronic processors employ the markers to set paths as required in the switching network. A connection is first established between a calling line and a register-interface circuit, which passes received dialled data to the processor for evaluation and action. Paths through the switching network are set and disconnected as required by the markers under instruction from the processors. The speed of operation is such that several re-attempts can be made if the first attempt at connection is unsuccessful.

This system shows a return to the concepts of the simple manual exchange. All terminations on the exchange are connected to the same set of switches and the whole operation of setting up a path is controlled by a processor, which corresponds to a manual-exchange operator and can observe directly the state of the whole exchange. Several processors are provided per exchange and they work as independently of one another as do the operators on a multiple-position switchboard, sharing only common directory information and operating rules. Provided that the processors are made powerful enough, systems organised in this way have again begun to acquire the power and flexibility associated with an operator-controlled manual exchange.

5.8 Stored-program control

One advantage of central control by processors is that facility changes may be effected by changes to a few processors instead of many registers. This advantage is further increased if the processors are effectively digital computers and based on 'software' programs [5]. This method of operation is known as *stored program control* (SPC) and the exchange shown in Figure 5.8 has a large part of its logic contained in a stored program which, for security reasons, is held on programmable read-only memory rather than a more flexible, but volatile, storage medium. In fact, it is possible, in the limit, to convert into processor software almost all of the logical functions of the exchange, including signalling.

The advantages of SPC are that the variety of hardware may be minimised, so easing production and stocking problems, and that facility changes may be readily accomplished. Disadvantages are that software development can be exceedingly complex and that real-time problems may occur. Stored-program control is described more fully in Chapter 7.

5.9 Distributed control

In the Strowger system, control elements are associated with each selector so that control is distributed with no centralisation. Communication between control elements is minimal, instructions being received from the customers at the relatively slow rate of ten pulses per second.

Although centralisation of the control brought many benefits, as we have seen, it also has its problems. Exchanges grow to meet traffic demand during their life and their processing-power requirements grow with them. This is not easy to achieve economically if all the processing power resides in one central processor. Also, it is usual to provide at least two central processors to guard against failure, yet doubling the capacity in this way is wasteful. Various architectures [7,10] have, therefore, been devised which retain the essential

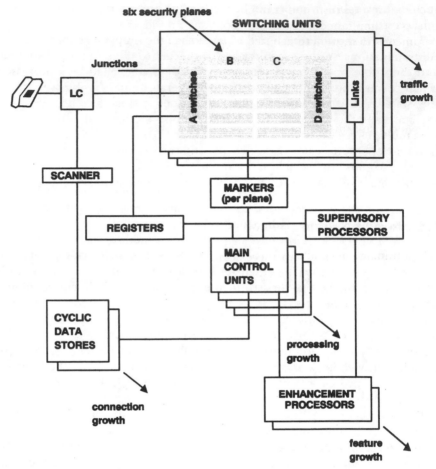

Figure 5.9   Example of distributed control (TXE4 system)

functional centralisation of processing while allowing decentralisation of the processors themselves. This facilitates growth and permits a more economical replication strategy.

As an example, the solution adopted in the TXE4 system is shown in Figure 5.9. The cyclic stores contain data relating to line and circuit provision. Associated with the cyclic store is scanning equipment which continuously updates a state-of-line store. The number of cyclic-store units can be extended to provide additional line capacity during the life of the exchange up to a maximum of 40 000 lines. The main control units process the setting up of calls. Their number can be extended as required to handle growth in the calling rate up to a maximum of some 500 000 BHCA (busy-hour call attempts) depending on the complexity of the call routing. The switching units contain reed-relay matrices, which can be extended as required to provide for growth in the traffic to be carried up to a maximum of 10 000 erlangs (simultaneous calls) of both-way traffic. Each of the principal parameters determining the size of an exchange is, therefore, provided for separately.

5.10 Feature evolution

The newer fully-digital time-division switching systems described in Chapter 6 are more economical to provide in a 'green field' situation. However, existing analogue space-division circuit switches represent considerable invested capital and reed-relay systems, in particular, can provide just as good a service for ordinary telephone users. It has been necessary over the years, however, to enhance the features provided to include such things as itemised billing, centralised maintenance and common-channel signalling.

In the TXE4 system, this has been achieved by adding specific functional processors interfacing with the existing exchange control elements [11], as shown in Figure 5.10. This enhancement process is still in progress and is likely to continue for the life of the exchanges, which is expected to extend well into the 21st century.

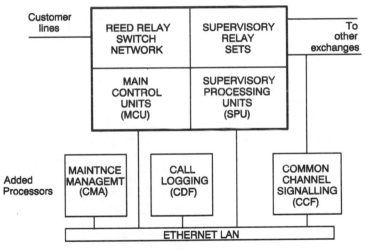

Figure 5.10   Enhancement of TXE4 system

5.11 References

1. ATKINSON, J.: 'Telephony' (Pitman, 1970), Vol. 1 (reprint)
2. ATKINSON, J.: 'Telephony' (Pitman, 1972), Vol. 2 (reprint)

3. SMITH, S.F.: 'Telephony and telegraphy' (Oxford, 1978), 3rd Ed.
4. GRIFFITHS, J.M. (Ed.): 'Local telecommunications' (Peter Peregrinus, 1983)
5. BRILEY, B.E.: 'Telephone switching' (Addison Wesley, 1983)
6. SMITH, S.F.: Chapter 56 *in* MAZDA, F.F. (Ed.): 'Electronic engineer's reference book' (Butterworth, 1988)
7. FLOOD, J.E.: 'Telecommunications switching, traffic and networks' (Prentice Hall, 1995)
8. JOEL, A. E. (Ed.): 'Electronic switching: central office systems of the world' (IEEE Press, 1976)
9. SMITH, S.F., and WATERS, D.B.: 'Chapter 5.9 *in* HOLDSWORTH, B., and MARTIN, G.R. (Eds.): 'Digital systems reference book' (Butterworth Heinemann, 1991)
10. HUGHES, C.J.: 'Switching, state of the art', *Br. Telecom Technol. J.*, 1986, **4**, (1, 2, 5)
11. SMITH, S.F., and BRYAN, D.G.: Chapter 31 *in* MAZDA, F.F. (Ed.): 'Communication engineer's reference book' (Butterworth Heinemann, 1992)

Chapter 6

Digital time-division switching systems

N.R. Winch

6.1 Introduction

Digital time-division switching systems [1] are now well established and the installation of digital switching networks is progressing at a rapid rate around the world. However, it is worth remembering that the introduction of electronics to telephone switching was an activity which began over 30 years ago. In reaching this present state of the art, the line of development has undergone several changes of direction. It includes a family of exchanges implemented in various countries [2] which utilise reed-relays as the crosspoint elements and are described as 'electronic'; however, this description is only really applicable to their control elements.

The main problem was always to develop a switching mechanism which utilised electronics to its greatest ability rather than simply design the switches as functional models of, for example, Strowger selectors. Since FDM had been successfully implemented within transmission systems, the possibility was considered of using similar techniques within switching systems. Although some work was done, it was not a realistically cost-effective technique. The speed of electronics lends itself to time-division-multiplexed (TDM) solutions, so that elements of the switch are time shared by many users. This has led to the development of TDM switching systems.

An early example of time division was installed at Highgate Wood exchange in North London [3] in the early 1960s. This used pulse-amplitude modulation (PAM). Its architecture, shown in a simplified form in Figure 6.1, consisted essentially of a main communication highway to which the subscribers involved in calls were connected for a short period 10 000 times per second, allowing a transfer of the instantaneous analogue voltage between one subscriber and the other. It is worth noting that two highway accesses are required per call, to provide the two directions of speech; thus a 2/4 wire interface is needed at each subscriber line circuit. Because only an extremely short sample time could be used, the energy transferred was low, thus requiring amplification.

The subscriber interface was a very expensive part of the exchange, and since this is replicated thousands of times it is essential that such costs are minimised. At the time of this development, discrete components were being

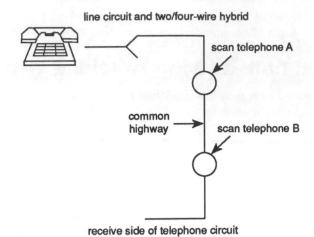

Figure 6.1   Time-division-multiplex switching using pulse-amplitude modulation (PAM)

used; LSI was a long way off! Thus, TDM switching of analogue signals was dropped. It was followed by the development of reed-relay exchanges mentioned above.

During the early 1960s transmission systems were being developed using PCM coding and TDM principles and the TDM format meant that the information received over the junction circuits was suited to a form of TDM switching, thus permitting the first steps towards an integrated digital network [4]. If space-division switching is employed in a junction or trunk network which uses PCM transmission, it is necessary to instal terminal equipment at every tandem exchange to demultiplex circuits before switching and to multiplex them again after switching. If TDM switching of the PCM signals is employed, the cost of the terminal equipment is eliminated. Moreover, tandem exchanges have no customers' lines. Thus, the problem mentioned above of the cost of the customer line circuit does not arise.

A field trial was therefore carried out at Empress exchange [5] in West London in 1968. Although it only carried a few 24-channel PCM circuits and utilised a very elementary call control, it did prove the viability of PCM switching. The first commercial digital trunk exchange was the Bell No. 4 ESS system [6], which was first installed in 1976.

PCM is based on digital encoding of successive samples of voice-frequency analogue waveforms. The samples are converted into binary codes representing the instantaneous height of the waveform and transmitted as digits to line. By time-sharing the encoding and decoding mechanisms across several analogue circuits and multiplexing the codes to line, PCM became a cost-effective transmission system, eliminating many forms of distortion found in analogue transmission systems.

Within PCM systems the digitally encoded 'word' consists of eight bits, and a channel is sampled at 8 kHz, thus providing a channel data rate of 64 kbit/s. In telephony, a form of companded coding is used, in accordance with μ-law or A-law standards.

At the primary rate, PCM systems convey either 24 or 30 channels. 24-channel systems are in use generally in the USA and Japan, while 30-channel systems were developed for Europe as second-generation systems. The 30-channel system actually uses 32 time slots, one of which is used for framing purposes, and one for signalling. It thus signals to line at 2048 kbit/s, encoding the signal according to the A-law standard, as described in Chapter 4. This system will be assumed where relevant in the rest of this Chapter.

6.2 Digital TDM switching networks

6.2.1 Time and space switching

Within an analogue environment the switching function of a tandem exchange may be readily understood in terms of spatial connection of an incoming to an outgoing circuit. When the incoming circuit carries TDM channels then, in addition to spatial switching between the incoming and outgoing system, there is a need to match the relevant time slots between the two. This is achieved by delaying, or storing, the incoming word from the selected channel, until the moment at which it is required by the outgoing channel. Such a mechanism is known as *time switching*.

Thus, a tandem exchange requires both space switching and time switching. Space switching is needed to transfer PCM samples from the line on which they are received to the line on which they are sent out. Time switching is needed because, in general, a connection is required between PCM channels which occupy different time slots. Various systems have used switching networks having different combinations of time and space switching. A three-stage network may have central time switches, with the first and third stages each consisting of a space switch. This is a space–time–space (S–T–S) network. Alternatively, the central stage may be a space switch, with the first and third stages using time switches. This is a time–space–time (T–S–T) network.

6.2.2 Time switch

A simple incoming time switch is shown in Figure 6.2. It consists of a speech store which is a random-access memory (RAM) with 32 addresses, each storing 8 bits. It can thus store one complete frame of the incoming PCM stream, being cyclically written to under the control of a time-slot counter.

Associated with the speech store is a connection store. This contains as many locations as there are internal exchange time slots. The words in this identify which location in the speech store should be addressed to read information out for passing across the exchange.

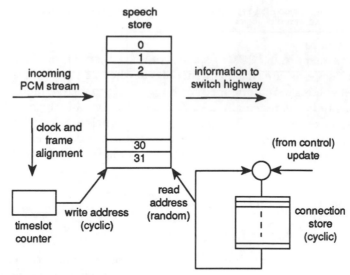

Figure 6.2   Incoming time switch

A corresponding outgoing time switch reverses these roles, receiving information across the exchange for storing under control of the connection store and subsequently reading the information cyclically to an outgoing PCM system.

6.2.3 Space switch

The function of a space switch is shown in Figure 6.3. It consists of a matrix of crosspoints which are simple electronic gates, and the operation of these is also controlled by a set of connection stores. One connection store is provided per vertical and the outputs are decoded and used to address individual crosspoints on the vertical.

6.2.4 S–T–S switching

An S–T–S switching network is shown in Figure 6.4. This shows the basic design of the Empress field trial [5]. The technology then meant that time switching was expensive. LSI had not arrived and electronic delay elements were used for that function. Each delay element acted as a preset time switch.

To establish a connection between time slot A of incoming PCM system X, and time slot B of outgoing PCM system Y, a link L is selected which provides a delay $(B–A)$ and has time-slot A free at its input. The connection store of the incoming space switch is set to close the crosspoint connecting incoming trunk X to link L in time slot A. The connection store of the outgoing space switch is set to close the crosspoint connecting link L to outgoing trunk Y in time slot B.

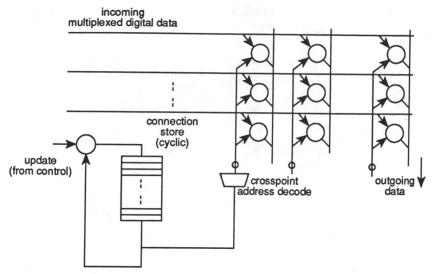

Figure 6.3   Space switch

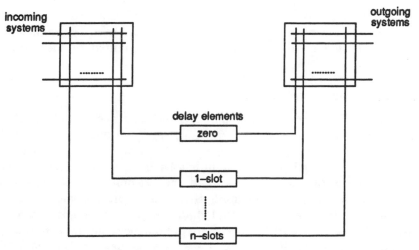

Figure 6.4   Principle of S–T–S switch

6.2.5 T–S–T switching

With the advent of LSI technology the cheapening cost of storage made time switching much more cost effective, and developments have generally followed the T–S–T architecture shown in Figure 6.5.

The principle of operation can be understood by recognising that the

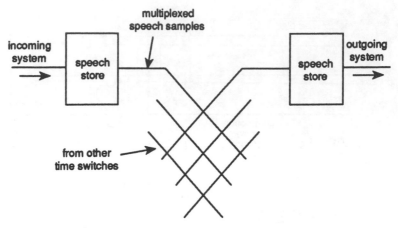

Figure 6.5   Principle of T–S–T switch

incoming time switch is written into cyclically every 125 μs, while the outgoing time switch is read out cyclically every 125 μs. The function of the connection stores is then to set up a given 'exchange slot', whereby information is read from the incoming time switch to the associated inlet of the space switch; simultaneously the required crosspoint is closed on the space switch and the information passed to the outgoing time switch. The connection stores thus cycle every 125 μs, ensuring that each sample received from a given incoming channel is passed across the exchange and transmitted to the required outgoing channel.

6.2.6 Both-way connections

Most electronic devices are one way, and the above descriptions of the switches show how the multiplexed information is transferred across the exchange from the incoming to outgoing side. However, since a telephone circuit has both transmit and receive functions, it is necessary to set up both directions across the switch. Thus, both the 'incoming' (call-originating) circuit and 'outgoing' (destination) circuit have an appearance at the incoming side for received channel data and at the outgoing side for transmitted channel data. All PCM systems are symmetrical with regard to the switching function; the differentiation between circuit types is a matter for call-control software.

In an S–T–S configuration the delay between incoming channel A and outgoing channel B is derived by introducing a delay of $(B–A)$ time slots. However, in the reverse direction the incoming channel B can only be transferred to outgoing channel A by delaying into the next frame, since channel A has already occurred in the present frame. The asymmetrical aspects of the time switching for a 2-way S–T–S connection is shown in Figure 6.6.

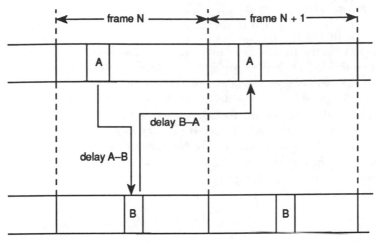

Figure 6.6   Asymmetry of delays across S–T–S switch

In comparison, the switching between incoming and outgoing time switches within a T–S–T configuration is a purely internal function of the switching elements. Thus, the times of transfer of the two directions could, if desired, be random. However, while this could result in a minimal both-way delay, at the other extreme the both-way delay across the switch could be maximised to four frame times, in addition to delays caused by aligners (see Section 6.3). Within T–S–T switches, the preferred mechanism is therefore to establish an internal transfer slot for one direction of transmission and to arrange for the transfer in the reverse direction to utilise the slot exactly half a frame later.

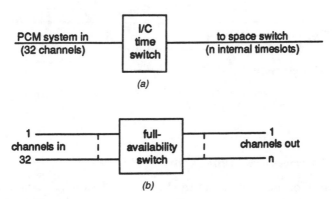

Figure 6.7   Traffic equivalence of time switch

a Time switch
b Space-division equivalent

6.2.7 Traffic through TDM switches

Calculation of the traffic capacity of TDM switches is straightforward. It is possible to map the functions of such switches into equivalent classical space-divided switches, and then to apply traffic theory in the usual way [7].

Figure 6.7b shows the equivalent of the time switch described in Section 6.2.2. The 32 (30 speech) channels from the PCM system are the equivalent of 32 individual circuit inlets, and the n internal time slots for transfer via the space switch are the equivalent of n individual circuit outlets. Similarly, the TDM version of a space switch may be considered as a replicated set of individual switches, as shown in Figure 6.8b. The trunking through a T–S–T switch is then as shown in Figure 6.9.

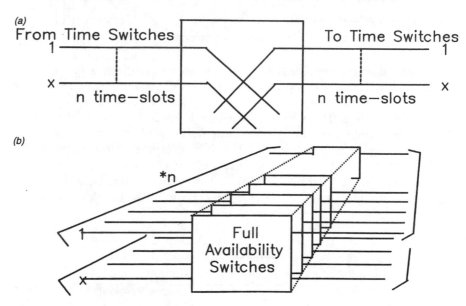

Figure 6.8   Traffic equivalence of space switch

> a Time-divided space switch
> b Space-division equivalent

6.2.8 Improvement in traffic capacity

Consideration of the trunking of the basic T–S–T switch shown in Figure 6.5 and its equivalent in Figure 6.9 shows that the number of possible paths between an incoming and an outgoing circuit is $n/2$ both-way connections, where n is the number of time slots. If the switch design uses only a few (say

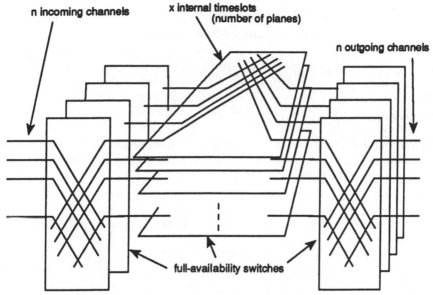

Figure 6.9   Equivalent space-division trunking for a T–S–T switch

$n=32$) time slots, in line with the capacity of a single-PCM time switch, then the resulting availability of 16 is typical of many space-division switches, the blocking of which was subject to considerable attention by traffic engineers over the years.

The problem of increasing link availability in a space-division switch is one of cabling. In a TDM switch, it is a matter of speed. If the number of time slots is increased, the link availability is increased accordingly.

In the digital route switch implemented in System X, 32 PCM systems are terminated on a single time switch [8], which thus has a nominal capacity for $32 \times 32 = 1024$ channels. The exchange itself switches 1024 slots per frame. Thus, the link availability for a both-way connection is 512. A value of this magnitude makes the exchange virtually nonblocking. It thus makes the switching engineer's job one of basically sizing routes between exchanges to achieve the necessary grades of service.

It will be appreciated that 32 systems of 2 Mbit/s each gives a nominal data rate of 64 Mbit/s through the time switch and via the inlets to the space switch. This very high rate is achieved by passing the eight bits of each PCM word over a parallel bus. In addition, the space switch may be segmented into two identical elements, one switching odd timeslots and the other even timeslots [8]. This then reduces the data rate on each wire to 4 Mbit/s.

The System X space switch is constructed around a basic 32×32 matrix, and nine elements of this are capable of being linked into a 96×96 matrix. Thus, the switch can handle 96 time switches of 32 PCM systems, each carrying 30 speech circuits. Allowing for the both-way requirements of the switch, and an average 0.8 erlangs circuit loading, the capacity of the switch is theoretically

$$96 \times 32 \times 30 \times \tfrac{1}{2} \times 0.8 = 37 \text{ k erlangs}$$

In practice, because the switch is required to switch other functions in addition to established calls, this figure is not achieved, but such a switch is capable of in excess of 20 k erlangs.

6.2.9 Security against failure

The failure of a crosspoint or selector within an electromechanical exchange is of little consequence, since the configuration of switch elements allows traffic to bypass the failure with little effect on the grade of service. Within a TDM switch, however, a single failure may affect many circuits, and could be quite catastrophic. The time switch described above carries 32 PCM systems. Failure here could therefore affect $32 \times 30 = 960$ circuits, or about 750 erlangs of traffic.

Within System X, the T–S–T switch is secured by duplication [8]. The PCM termination unit is designed to give simultaneous access to two identical switches. It is, of course, at this stage that the PCM bipolar coding is removed, and thus the transmission-error-detection mechanism is lost. To check the data across the switch, a parity bit is generated; thus a 9-bit highway is used. When the data are read from the outgoing time switches, they are checked for parity and also for consistency between the two versions. Failure of one of the switching networks means that, normally, the sample from the other is used; thus the data remain intact.

The T–S–T switch, using a space switch which is segmented for speed and duplicated for security against failure, is shown in Figure 6.10. Such digital switching is highly cost effective when compared with analogue switches. Even

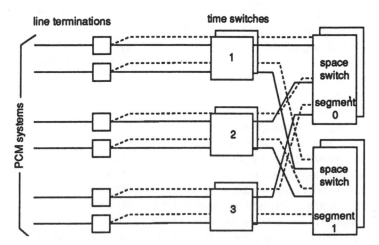

Figure 6.10   Duplicated and segmented T–S–T switch

fairly early designs of switch and control show approximately 10:1 space savings when compared with crossbar systems.

6.3 Network synchronisation

When PCM systems were introduced for transmission, they were stand-alone systems; thus the accuracy of the clocks which they used was not critical. Indeed, the structure of second-order multiplexers is such that the accuracy of the primary tributaries may be within 50 parts in 10^6.

When such systems are to be switched, however, it is important that the clocks are extremely accurate. Otherwise, information provided by a fast-running source may fail to be transferred to a slow-running sink; alternatively, information from a slow source may be read twice into a faster sink. Either of these occurrences is known as *slip*, and it causes degradation of a connection. It is important to understand the impact of slip on the data being transferred.

The loss or repetition of an occasional PCM speech sample will, when decoded, cause a small discontinuity in the output waveform, and will normally be insignificant. Analysis shows that only about 1 in 25 such events would be perceived by a listener.

Loss or repetition of bits over a data link can be much more serious. Normally data must be received in an exact form to retain the meaning. It is possible to allow occasional errors where a sequence of data is used continuously to update a real-time display (e.g. in radar), since the error will be updated by new information within a few seconds. However, a high error rate would result in such spurious outputs as to make the system unusable. Where data are transferred in the form of files for storage or manipulation, precautions must be taken to ensure that errors are detected and corrected by suitable algorithms. Normally, HDLC techniques at layer 2 of the OSI model are a means of ensuring correct receipt of data. However, even this can fail in the case of high error rates, owing to the queuing of messages caused by the need to retransmit earlier ones.

Thus, while it is possible to operate a network which is not perfectly synchronised, there is clearly a need for a standard of performance to be defined. The level looked for in public networks is for 99.95% error-free seconds due to slip. To meet this requirement calls for a tolerance of exchange clocks of 3 parts in 10^7. Additionally, the ITU-T has established a standard for performance between countries of not more than 1 slip in 70 days, thus requiring clock accuracies of 1 part in 10^{11}.

In the UK, the System X network has been designed with a caesium atomic clock of this accuracy at the top of a synchronisation hierarchy, as shown in Figure 2.12, and for selected traffic-carrying PCM links to be used to control the clocks at lower-level exchanges (unilateral control) [9]. To protect against loss of master links, exchanges at the same level may also utilise bilateral links, so that isolation of an area can allow the clocks to achieve a mutually agreed rate.

The decision as to whether a local-exchange clock is running fast or slow compared with the master clock from the higher level is done by comparison of the framing signals of the 'go' and 'return' paths of a selected PCM link. Absolute phase is unimportant, but phase change could indicate that the clock rates are different. It is important, however, to eliminate effects due to seasonal changes in the electrical length of the links. This shows up as a phase change of identical sign at both ends, whereas clock differences show up as phase changes of opposite sign at each end of the link.

To establish the actual situation, a signalling channel is required between the ends of the links and System X achieves this by using bit 5 of time-slot zero in alternate frames in the framing pattern [10]. This bit is extracted (at 4 kbit/s) and is structured into its own frame pattern, giving a 32-bit word including phase-change information, a speed-up/slow-down indicator and various housekeeping digits including check bits.

The speed-up/slow-down information is ignored at the master end of the link. At the slave end, it is compared with its own decision on phase shift. If both have concluded the same, this is caused by seasonal drift and is ignored. If the two ends have identified that one end is slowing and the other gaining, then the slave end responds by activating the varactor control of its oscillator to correct the drift.

While this complexity is justified within the main trunk network, it is not necessary at the bottom of the hierarchy and a simpler procedure is used. Here, the local exchanges are arranged to derive their synchronous clocks directly from selected PCM systems. In the event of failure of these links, the exchanges can still run on their own internal clocks.

Phase drift which occurs due to seasonal fluctuations still needs to be absorbed by the switches. The most extreme example of this form of drift is that due to the daily cycling of geostationary satellites, which can cause a change in path length equivalent to several milliseconds. However, terrestrial systems only require to accommodate fractions of a millisecond, and this is done by feeding the input data via a two-frame elastic store. The average delay caused by this is one frame in each direction of transfer. Only if the phase changes to half a frame is the readout point shifted by 180° to cause loss or repetition of data.

6.4 Local exchanges

6.4.1 Concentration

The description of PCM switching has so far dealt with tandem or trunk switching, whereby the exchanges have been required to connect between primary multiplex PCM systems, i.e. route switching. However, as described in Section 5.3, the route switch has a role in local exchanges, where it is used to provide connectivity between bundles of customers' traffic and the various interexchange routes. In addition, there is the requirement to provide concentration of many low-occupancy customer lines to a smaller number of efficiently utilised links.

During the early PCM switching developments, the coding function was only cost effective when done with concentrated traffic, as with PCM-transmission systems. The concentration was therefore done in the analogue domain, using reed-switching technology as shown in Figure 6.11.

Only in recent years has the use of LSI allowed single-channel codecs to become a cost-effective alternative and for the complex line-interface card to reduce sufficiently in cost for digital concentration to be used [11]. The result is shown in Figure 6.12.

The principle of digital concentration may be understood by considering a time switch which has a store location for each telephone which it serves, and from which it receives PCM information. The output of the switch is limited by the number of time slots available. By the use of additional internal time slots it is possible for the concentrating switch to handle signalling information which is also received from the lines, and to look for new calling requests [12].

Digital concentrators, as shown in Figure 6.12, can be located in the same building as the route switch in order to form a complete local main exchange. However, control signals between the concentrators and the central processors of the exchange can be transmitted (in both directions) over time slot 16 of the connecting links. This enables concentrators to be located remotely from the main exchange and be connected to it only by their PCM links over copper-cable pairs or optical fibres. As a result the size of an exchange area is no longer limited by the resistance and attenuation of direct lines between customers' premises and the exchange. Digital local exchanges can therefore

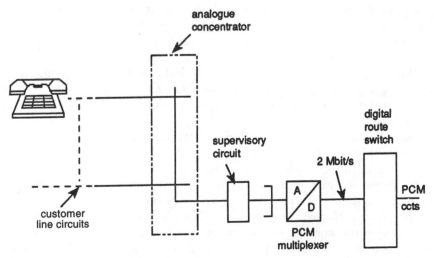

Figure 6.11   Local exchange with analogue concentrator

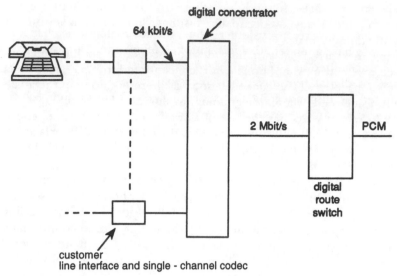

Figure 6.12   Local exchange with digital concentrator

serve many more customers than analogue electromechanical exchanges and may have a capacity of the order of 50 000 lines rather than only 10 000 lines.

6.4.2 Digital telephones

A single-channel codec having been realised as a cost-effective entity, the next stage was to consider its removal from the exchange to the telephone. This gave rise to the problem of the development of suitable transmission mechanisms which would enable the transfer of 64 kbit/s coded speech samples over customers' two-wire lines and provide a suitable means of conveying signalling.

The first approach was to add two extra bits to the PCM word, using one as an 8 kbit/s signalling channel, with the other as a truncated data channel, also at 8 kbit/s. To utilise existing customer cables, a burst-mode (or time-compression duplex) method was used to enable the two-way digital transmission to take place over a single cable pair [13]. Latterly, the reducing cost of echo cancellation, again due to LSI technology, has enabled the existing pairs to be utilised on a continuous-transmission basis [13].

An important feature of the digital telephone is the independence of the signalling and speech channels. The nature of the well-established analogue telephone was that since signals (i.e. loop/disconnect and ringing) and speech used the same pair of wires, they could not take place concurrently. This limitation is removed by the use of separate signalling bits when digital telephones are employed.

6.5 ISDN

To transmit data over a customer's line via an analogue exchange, it is necessary to have a modem in addition to the telephone. Moreover, data can only be transmitted at a relatively slow speed (e.g. 1.2 kbit/s or 4.8 kbit/s). The use of digital transmission over customers' lines enables data to be transmitted by disconnecting the codec instead of connecting a modem. Moreover, data can be transmitted at 64 kbit/s.

This opens the way to using customers' lines and local exchanges for the provision of many different services. For example, these can include high-speed data and facsimile and slow-scan television in addition to telephony. The development of an integrated digital network (IDN) based on PCM transmission and switching thus leads to the subsequent development of an integrated services digital network (ISDN). This is described in Chapter 14.

6.6 Digital switching systems

Several digital telephone-exchange systems have been developed on the basis of the architecture shown in Figure 6.12, with a large central route switch interconnecting incoming junctions, outgoing junctions and peripheral customers' concentrators. Tandem and trunk exchanges omit the concentrators [14]. Such systems are now operating in many countries [15].

Examples of these systems include the AXE-10 System [16] (developed in Sweden), the DMS-10 System [17] (Canada), the E-12 System [18] (France), the No. 5 ESS System [19] (USA), the EWS System [20] (Germany), the NEAX System [21] (Japan) and System X [22,23] (UK). All these systems use stored-program control, which is discussed in Chapter 7.

The development of cheap single-channel codecs has also led to the widespread introduction of digital private branch exchanges (PABXs). A small PABX needs only a single bus, to which are connected the codecs of all extension lines and exchange lines, as shown in Figure 6.1. A connection is made by setting the connection store to operate the codecs of the calling and called lines in the same time slot. However, a different time slot is needed for each direction of transmission, as described in Section 6.2.6. A single 8-bit parallel bus operating at 2 Mbit/s can handle 128 simultaneous calls and can thus serve several hundred lines [24].

The introduction of inexpensive microprocessors has made stored-program control economic, even for the smallest PABXs. This enables these systems to provide their extension users with a very wide range of facilities.

6.7 Conclusion

TDM switching, after the early PAM work, finally found its realisation through PCM. The change of technology over the past 10–15 years has accelerated the conversion of networks at an ever increasing pace. Digital switching, having been based on voice requirements during its initial phases, is now meeting the requirements of ISDN.

However, life never stands still! No sooner have networks become established based on 64 kbit/s fixed channels, than the transmission engineers have begun to develop systems with greater flexibility of bandwidth. The switching of these variable-bandwidth signals is considered in Chapter 15. Consequently, the fixed TDM switching of the last decade must be viewed as just another stage in network evolution.

6.8 References

1. REDMILL, F.J., and VALDAR, A.R.: 'SPC digital telephone exchanges' (Peter Peregrinus, 1990)
2. JOEL, A.E. (Ed.): 'Electronic switching: central office systems of the world' (IEEE Press, 1976)
3. HARRIS, L.R.F., MANN, V.E., and WARD, P.W.: 'The Highgate Wood experimental electronic telephone exchange system', *Proc. IEE*, 1960, **107,** Part B, Suppl. 20, pp. 70–80
4. DUERDOTH, W.T.: 'Possibility of an integrated PCM switching and transmission network'. *International conference on Electronic Switching*, Paris, 1966 (Editions Chiron, 1966)
5. CHAPMAN, K.J. and HUGHES, C.J.: 'Field trial of an experimental pulse-code modulation tandem exchange', *Post Off. Electr. Eng. J.*, 1968, **61,** pp. 186–195.
6. VAUGHAN, H.E.: 'Introduction to No. 4 ESS'. *International Switching Symposium Record*, 1972, pp. 12–25
7. FLOOD, J.E.: 'Telecommunications switching, traffic and networks' (Prentice Hall, 1995)
8. RISBRIDGER, J.N.A.: 'System X subsystems. Part II: The digital switch subsystem', *Post Off. Electr. Eng. J.*, 1980, **73,** pp. 19–26
9. BOULTER, R.A. and BUNN, W.: 'Network synchronisation', *Post Off. Electr. Eng. J.*, 1977, **70,** pp. 21–28.
10. BOULTER, R.A.: 'System X subsystems. Part 2: The network synchronisation subsystem', *Post Off. Electr. Eng. J.*, 1980, **73,** pp. 88–91
11. OLIVER, G.P.: 'Architecture of System X, Part 3: Local exchanges', *Post Off. Electr. Eng. J.*, 1980, **73** pp. 27–34
12. FOXTON, M.C.: 'System X subsystems, Part 5: The subscribers' switching subsystem', *Post Off. Electr. Eng. J.*, 1981, **73,** pp. 216–222
13. ADAMS, P.F., ROSHER, P.A. and COCHRANE, P.: 'Customer access' *in* FLOOD, J.E., and COCHRANE, P. (Eds.): 'Transmission Systems' (Peter Peregrinus, 1991), chap. 15
14. VANNER, N.J.: 'Architecture of System X. Part 2: 'The digital trunk exchange', *Post Off. Electr. Eng. J.*, 1979, **72,** pp. 142–148
15. JOEL, A.E. (Ed.): 'Electronic switching: digital central office systems of the world' (IEEE Press, 1981)
16. MEURLING, J.: 'Presentation of the AXE-10 switching system', *Ericsson Rev.*, 1976, **53,** pp. 54–59
17. WATKINSON, B.G. and VOSS, B.E.: 'The DMS-10 digital switching system'. IEEE *National Telecommunication Conference Record*, 1977, pp. 07.2.1–07.2.4
18. VIARD, F., JACOB, J.P., and FRITZ, P.: 'A new range of CIT-Alcatel time-switching telephone exchanges', *Commut. Electron.*, 1979, (1), pp. 87–102
19. DANIELSON, W.E., and MACURDY, W.B.: 'Digital local switching trends in the Bell System network'. *International Switching Symposium*, Paris, 1979, pp. 53–61
20. SUCKFULL, H.: 'Architecture of a new line of digital switches'. *International Switching Symposium*, Paris, 1979, pp. 221–228

21. SUEYOSHI, H., SHIMASAKI, N., and KITAMURA, A.: 'A versatile digital switching system for central offices: NEAX 61'. *International Switching Symposium*, Paris, 1979, pp. 688–695
22. JONES, W.G.T., KIRTLAND, J.P., and MOORE, W.: 'Principles of System X', *Post Off. Electr. Eng. J.*, 1979, **72**, pp. 75–80
23. TIPPLER, J.: 'Architecture of System X. Part 1: an introduction to the System X family', *Post Off. Electr. Eng. J.*, 1979, **72**, pp. 138–141
24. RAYFIELD, D.A.T., and GRACIE, S.J.: 'The Plessey PDX: a new digital PABX'. *International Conference on Private Electronic Switching Systems*, London, 1979, *IEE Conf. Pub. 163*, pp. 211–214

Chapter 7
Stored-program control
M.B. Kelly

7.1 Introduction

Stored-program control (SPC) is a term used to describe computer-controlled telecommunication systems. When SPC was originally conceived, computers were still large and expensive machines and their programs were generally written in low-level languages to achieve run-time efficiencies. Integrated circuits had not yet appeared on the scene and the idea of using computers for controlling a telephone exchange, rather than as number-crunching data processors, was a novel one.

Today, of course, VLSI, microprocessors and dense memory chips are commonplace hardware building blocks, and software engineering is emerging as a recognised discipline from the various software crises through which computing has passed in the intervening period. SPC remains as a slightly anachronistic term from the early days, but the application of computers and software-based control to telecommunications is at the vanguard of present-day concerns in information technology.

Around the turn of the 20th century, the first automated telephone switching systems used the patterns of interconnection between step-by-step switches to effect call routing [1]. From the perspective of modern systems, exchanges of this sort used Strowger selectors as very simple 'distributed controllers' with their interconnecting wires representing the controlling program and the streams of dialled digit pulses providing the input data for each call (see Figure 7.1). It was soon recognised that hard-wired control of this kind produced major constraints on the ability of the system to respond readily to change and that it precluded any form of overall decision making during call handling within the exchange. The idea of centralising control over the whole of an exchange (see Figure 7.2) thus became something of a 'holy grail' for switching, and was first attempted in crossbar electro-mechanical systems [2,3] and later realised in semi-electronic systems such as TXE4 [4]. However, these systems still incorporated a significant amount of hardware-based control. For example, TXE4 originally used wires threaded selectively through ferrite cores for storage, although these were replaced by integrated circuits in TXE4A. It was not until the advent of digital switching,

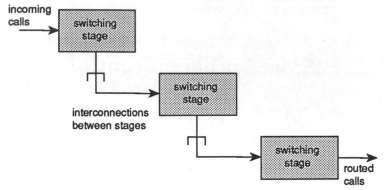

Figure 7.1   Principle of step-by-step control

with its synergy between digital computer processing and digital data transmission, that SPC came into its own [5].

In the late 1990s, the shrinking costs and constantly increasing capabilities of microelectronics are bringing this process full circle, with the opportunity to distribute intelligence within telecommunication networks to street-cabinet equipment such as multiplexers and concentrators, and even to the customer terminals themselves [6]. There is therefore a shifting balance between the concentration of network control in central switching nodes and in peripheral network elements, but the common thread linking these alternative architectures is the use of software and computer technology to implement the control. The original concept of SPC as a function of central switching nodes has thus become generalised to include the entire hierarchy of telecommunications-network elements, and has led to a growing importance for distributed control and its associated management throughout these networks.

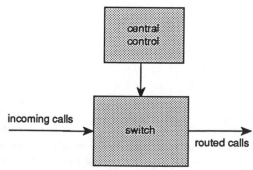

Figure 7.2   Principle of centralised control

7.2 Characteristics of SPC systems

SPC systems in telecommunication networks share many features with other computer-controlled systems, particularly those involved with real-time applications such as process control. However, telecommunications presents the following combination of requirements which is not commonly faced in other fields:

(*a*) *Very high reliability and availability*
SPC systems are required to provide service for 24 hours a day, year in, year out. During a lifetime which may span two or more decades, they are expected to be out of action for at most a few hours.

(*b*) *Very large peripheral connectivity*
Telephone networks support many millions of customers and a single telephone exchange can have tens of thousands of customer terminals.

(*c*) *High degree of event-driven concurrency*
SPC systems make extensive use of time-multiplexed operation in all areas of the network, to maximise the cost-effectiveness of equipment. A large telephone exchange, for example, must handle hundreds of processing-intensive call setups simultaneously and thousands of calls may be in speech at the same time.

(*d*) *Close timing tolerances*
External signalling interactions have demanding time constraints and call timing for charging purposes must be extremely accurate.

(*e*) *Extensive networked operation*
Telecommunication systems cannot be considered as isolated units but must fit within the local, national and international networks of which they are a part.

(*f*) *Compatibility with existing systems*
New systems must interwork with those already in place within the overall network, where it is not practical or economic to replace or upgrade the older systems.

(*g*) *Evolutionary potential*
New systems must anticipate future network modifications, by providing a straightforward means for their upgrading to accommodate new services or methods of operation.

(*h*) *Powerful management support*
As telecommunication networks become more sophisticated and complex, there is an increasing requirement for correspondingly sophisticated management systems to support the network operator in controlling the network and in planning for network growth and development.

SPC systems must be designed to operate within an environment defined by these pressures, and all share a common approach intended to address the issues concerned [7]. This approach is based on the premise that standard features can be implemented in general-purpose hardware elements, while features specific to particular applications or areas of the network should be implemented in software. This approach allows customised operation to be provided without the need for physical changes and it facilitates the evolution of the telecommunication network. At the same time, hardware- and software-design techniques are used which provide a high system reliability and which support the necessary processing within the system to meet the needs of the relevant network applications.

7.3 Implementations of SPC

7.3.1 System architecture

When the first-generation SPC systems were designed, about 30 years ago, the high cost of digital computers enforced the use of a single controlling processor (with a standby), as shown in Figure 7.2. Special-purpose processors were used to provide a reliable base on which the software could run [8,9]. Because of the high cost of central processors, SPC systems were only economic for large exchanges (e.g. for 10 000 lines or more).

Later, the falling cost of integrated circuits and the introduction of standard microprocessors made it economic to use SPC for even the smallest exchanges, including PABXs. It also enabled the number of processors in a large system to be increased. This led to more distributed forms of control [10]. As shown in Figure 7.3, separate processors are now often used as follows:

(a) to control separate parts of the switching network, e.g. customers' concentrators; and

(b) to relieve the central processor of specific functions, e.g. common-channel signalling.

As a result, the functions of the central processor are reduced to:

(a) co-ordinating the actions of 'local' processors in order to set up complete connections; and

(b) performing overall maintenance and administration functions.

Thus, the architecture of Figure 7.3 is characteristic of several modern systems, such as AXE-10, DMS-10, E-12, EWS, NEAX and System X.

More recently, distributed control has spread from the exchange into the customers' access network as a result of the introduction of remote concentrators and multiplexers [11]. These are controlled by their own microprocessors.

7.3.2 Fault tolerance

The use of centralised control emphasises the need for the control system to be resilient to faults, since telecommunication systems must be able to provide

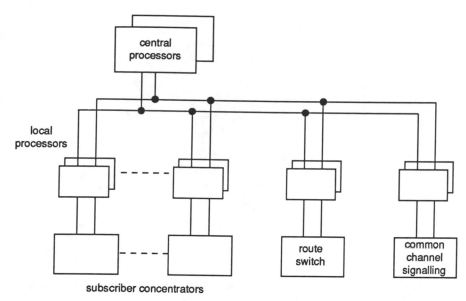

Figure 7.3   Distributed stored-program control

service for close to 100% of the time, and accommodating this need is a major problem for the system designers. The allowed downtime (or unavailability) can be of the order of hours per year, or even less in some cases, and such stringent demands cannot usually be met directly by the intrinsic reliability of the systems, subsystems and components involved. Instead, the design must incorporate *fault tolerance*, i.e. the ability to continue to function even when faults exist [12]. Fault tolerance implies the existence of some form of redundancy and centralised control systems include redundant processing hardware as part of their fault-tolerant design. For most hardware, the dominant form of expected fault is associated with the manufacture of the unit, i.e. the well known 'bathtub' curve describing wear out [13]. With software, as will be seen later, the important fault causes are related to design errors.

The basic principles of fault tolerant hardware are straightforward. As shown in Figure 7.4, two or more similar units, each of which is capable of performing the same function, are provided in the subsystem or system involved. The units can then be used to carry out the same task, and thus provide a means of each checking the other's correct functioning. Alternatively, they can carry out different tasks, with one or more healthy units taking over the tasks of a failed unit if this is necessary.

To be fully effective in achieving this, a fault-tolerant system must be able to carry out the following:

(*i*) detect and locate faults in units when these occur;
(*ii*) transfer work in progress from a failed unit to one or more of the

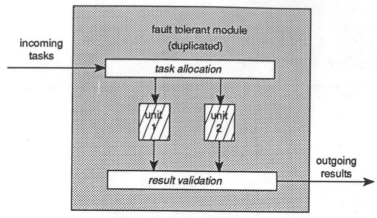

Figure 7.4   Fault-tolerant operation

remaining units, so that system operation can be maintained during recovery from the fault concerned;

(*iii*) isolate failed units to avoid any continuing disruption to system operation;

(*iv*) repartition the allocation of new tasks amongst the remaining units, to maintain continuing system operation following unit failure; and

(*v*) restore system redundancy after repair or replacement of failed units.

7.3.3 Worker–standby systems

In worker–standby systems, the system tasks are carried out by 'worker' units, each of which has associated with it one or more 'standby' units able to take over responsibility for its tasks should it become faulty. The commonest case is a duplicated system as shown in Figure 7.5. This involves a single worker unit and a single standby unit [14], although more complex arrangements

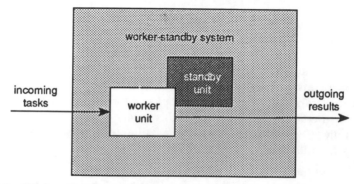

Figure 7.5   Worker–standby operation

are possible, as described later. The state of readiness of a standby unit to take up the tasks of a worker unit which becomes faulty can vary between systems.

It is possible that only a 'cold' standby is provided, so that tasks in progress are lost and the standby unit takes over the worker role from a predefined initialised state. This may involve actually powering-up the standby unit, possibly under manual control. However, service can be restored in such cases without the necessity to diagnose and correct the fault which has affected the previous worker unit. This is not a common strategy in practical SPC systems, because of the disruption to service which results while the standby unit comes into full operation.

'Hot' standby systems, on the other hand, ensure that a standby unit is fully up-to-date on the state of the tasks carried out by the relevant worker unit; thus, the standby is able to take over these tasks without interruption to service should the worker fail. This is a common technique and many SPC systems employ it, including AXE-10, DMS-10 and DMS-100 and NEAX-61. The shadowing of worker operation by a standby unit can be effected by the worker constantly communicating its status to the standby, or by the standby unit actually carrying out the tasks in parallel with the worker (as in AXE-10, for example). The latter approach is more usual, as it provides an inbuilt checking mechanism by comparison of worker and standby-unit status; any discrepancy indicates that one or other unit has developed a fault. However, resolution of which is faulty may not be straightforward. One solution is to operate a triplicated, rather than duplicated, redundancy scheme so that a single faulty unit can be identified by a 'two-out-of-three' voting system, as shown in Figure 7.6. In such cases, individual units are not usually identified as worker or standby, since all are contributing equally to fault-free system operation. This approach can be generalised to larger groups of units for critical applications. A special case of these arrangements is where all units operate in synchronism, so that precisely the same events are occurring in all

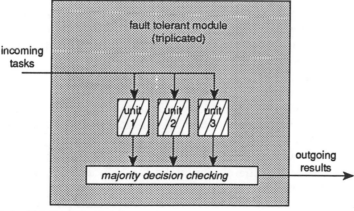

Figure 7.6   Triplicated redundancy

units concurrently, and this is the mode used in triplicated hardware units with majority-decision gating between critical areas of the logic. This is not a common technique for processor-based systems themselves. However, it can still be seen in important areas of dedicated logic circuitry, for example in the switch core of the Digital Switching Subsystem in System X.

Between the extremes of hot and cold standby, a range of 'warm' standby systems is possible. In these, standby units are able to take over from a failed worker unit with some knowledge of the state of work in progress, but that knowledge is out-of-date to a greater or lesser extent according to the 'warmth' of the standby concerned. In such systems, particular tasks may have to be aborted because the new worker unit is unable to deal with the discrepancies between its view of the task status and the real position reached by the previous worker unit before failure. For example, a common basis for warm standby of SPC processors in digital exchanges is to ensure that calls which have reached the speech phase can be maintained if a fault develops, but to accept that some of those still in the process of being set up may be lost. The System X Processor Utility, for example, uses a warm-standby approach.

7.3.4 Load-sharing systems

In load-sharing systems, each unit is responsible for part of the total system operation, unless a fault is detected in one of the units. Then one or more of the remaining units takes over responsibility for the affected area. Fault detection in load-sharing systems depends on some additional checking and validation, since the units are each dealing with different tasks and so cannot compare results. This method has the advantage that each unit is directly contributing to the effective power of the system in the absence of faults [15]. Load-sharing can also offer a very attractive mechanism for growth during the lifetime of a system, by addition of new load-sharing units and repartitioning of system tasks across the larger hardware configuration.

There are many ways in which processing can be shared among the units involved. Units can be responsible for some subset of the physical resources handled by the overall system, so that each individual unit carries out all processing involving that subset, as shown in Figure 7.7. This approach is most obviously used where the system includes remote elements, such as concentrators or multiplexers placed throughout a local exchange area to serve adjacent customers. Even entire exchanges can be replicated where the traffic load or number of connections exceeds the capacity of a single system. In the latter case, telephone numbers themselves can provide a basis for the sharing algorithm applied to the overall demand (e.g. all customers with numbers starting with a particular digit can be connected to one exchange unit, while those with a different initial digit are connected to another).

Alternatively, units can be functionally dedicated, so that overall system operation involves passing through a succession of such units each handling an aspect of the processing involved, as shown in Figure 7.8. A common

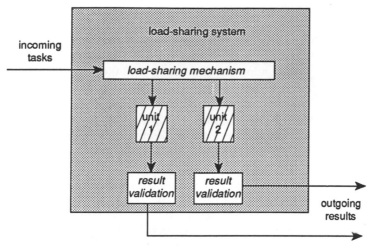

Figure 7.7 Load-sharing operation

instance of this approach is the use of peripheral processors to remove some of the processing load from central computer systems: concentrators and multiplexers are again examples.

Yet another approach is not to partition the responsibilities of hardware units at all, but instead to treat these units as a general-purpose processing pool used by the system software. In such cases, it is the software which provides for load-sharing and this involves similar techniques to those described above for hardware partitioning. Where this is done, the hardware is configured as a multi-unit computer, using methods which correspond with

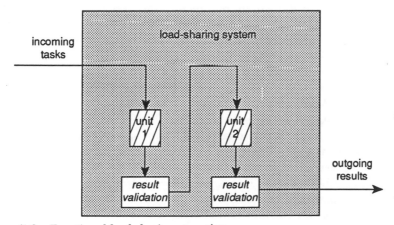

Figure 7.8 Functional load-sharing operation

those adopted more generally to increase the available power of computer systems.

A common solution to fault tolerance in SPC systems of this kind is to provide for '1-in-N sparing' of the units, so that a full system consists of $N+1$ units, of which any N are adequate for system operation. Thus, failure of any unit does not prejudice the system. This can be generalised to 'M-in-N sparing' where a number (M) of spare units are included for even greater fault tolerance. System X, for example [16], which originally was based on a centralised multi-processor system using M-in-N sparing, evolved to use a number of co-operating computer clusters each of which employs internal 1-in-N sparing.

7.3.5 Distributed control

All these load-sharing strategies can occur individually. However, it is also possible for combinations of these to be used, and for mixed worker–standby and load-sharing designs to be employed. In particular, technical advances have made the concept of distributed computing networks more attractive and these often involve mixtures of the schemes described. The TXE4/A Enhancements system [17] which provides common-channel signalling and itemised call logging for TXE4 and TXE4A exchanges, includes a number of functionally-dedicated processing modules based on commercial single-board computers (SBCs) which interwork over a duplicated Ethernet local-area network. Each processing module contains a number of duplicated SBCs, with each duplicated pair handling a subset of the processing load of that module in a load-sharing arrangement. Each of the SBCs within a duplicated pair load shares half of that processing subset and also provides a hot standby for the other SBC in the pair.

A particular form of distributed control involves a hierarchy of control elements (such as the remote concentrators and multiplexers already mentioned) in which the system architecture incorporates layers of processing to provide each hierarchically-superior layer with a simplified view of the tasks being carried out below it, as shown in Figure 7.9. This is a very powerful technique; indeed it can form the basis for an overall system-design approach [18]. It has become especially attractive as microelectronics has made available cheap and plentiful hardware controllers which can form the building blocks of such a control hierarchy. System 12, for example [19], uses a hierarchical distributed control philosophy.

The development of distributed control within and directly associated with network nodes has led to the spread of this approach throughout networks generally. This has brought with it an increased emphasis on sophisticated communication between the intelligent network elements which now exist, and on the importance of co-ordinated network management to gain full benefit from the interacting capabilities of the individual SPC systems involved.

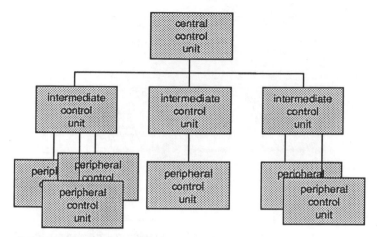

Figure 7.9   Hierarchical control

7.4 Software aspects of SPC

7.4.1 Fault tolerance

Software within SPC systems must provide a highly reliable service, just as must the supporting hardware computer system. However, software differs significantly from hardware in its lack of an inherent failure mechanism by which software can 'wear out' [20]. Thus, software is subject only to failure through design faults leading to erroneous operation, and these can manifest themselves in a variety of ways.

Software used in telecommunication networks can be expected to have undergone extensive proving and verification before its introduction to service. Thus, most software faults will arise in little-used areas of program operation, or will only occur when unusual system conditions arise. Great care is taken during development, integration and testing of software to weed out program bugs which may affect its behaviour. However, the absence of such bugs cannot be definitely proved, since the possible combinations of system state and input conditions which may arise in service are astronomical in number.

As a result, fault tolerance may be needed in software as well as in hardware, and analogous solutions have been developed to those discussed for hardware [21]. In particular, *N-version programming* [22] is the software equivalent of triplicated (or replicated) hardware systems which compare outputs using a 2-out-of-3 (or *M*-out-of-*N*) voting system. In software, a decision algorithm provides the voting mechanism, by comparing the outputs of several versions of a program and proceeding with the result agreed by a majority, as shown in Figure 7.10. Unlike hardware, the software versions cannot be identical (since the system is guarding against design flaws rather than component failures), so independently designed programs must be employed. This is

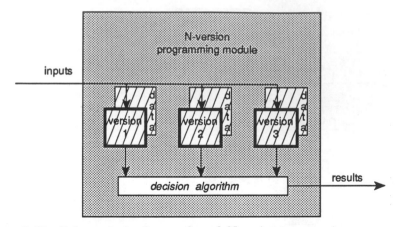

Figure 7.10   Software fault tolerance through N-version programming

expensive, as it multiplies the software development costs. There is also a processing and storage penalty, since it is conventional to execute the software versions in parallel, using independent data. However, there is then no delay in overall system operation, except for that due to differences in the execution times of the software versions involved. Because of its high cost this approach tends to be reserved for safety-critical applications. (It is used, for example, by NASA for the controlling software in space vehicles.) It is not yet applied widely in telecommunications, although the advent of parallel-processing architectures in computer systems could provide an ideal base upon which to develop such systems in the future [23].

A more common approach is the use of acceptance tests to validate software operation and provide a pass/fail result on the outcome. In their simplest form, such tests can be represented by defensive programming checks, which monitor the inputs and outputs of program modules to verify that these are within predefined bounds of reasonableness. These checks can be supported by the underlying operating system software or even, in some cases, by the computer hardware itself. Common examples are operating system or hardware traps for attempted memory accesses outside the allowed area of operation of the program concerned, or for programs stuck in endless processing loops. Where errors are detected by such means, the system response is normally to restore the system (including its stored data) to some previous consistent condition, often by reloading system-state information which had been safely stored before execution of the failed program. This process is known as *backwards error recovery* and the restoration of a previous system state is often termed *rollback*.

It is also possible that, despite the software fault, the system may still be able to proceed to a new consistent state. In fact, this may be unavoidable if events have been triggered in the 'real world' outside which prevent complete reversion to the conditions which existed before the error occurred. For example, if a telephone call has been misrouted because of a software error, the calling customer may already have realised this and have hung up. In such cases, the system must recognise that the outside world cannot now be rolled back and must take this into account in recovering the situation. For telecommunications, there is not usually a serious problem; the misrouted customer above is probably already redialling to overcome the problem, without the telephone network having to intervene. However, in other fields, the difficulties can be real; a software bug which caused a live cruise missile to go astray clearly requires corrective action before normal operation can safely be resumed.

Thus, where erroneous external events have occurred as the result of faults, some form of *forwards error recovery* may be necessary to bring the system (and the real world) to a consistent and acceptable condition before proceeding further. This involves recognising and taking account of any irreversible real-world events which have occurred, by changing the system state to make it both internally consistent and consistent with these existing real-world conditions. In some cases, it may also be necessary to modify the external situation to reach an acceptable state from which to proceed. In the cruise-missile example above, an extreme form of forwards error recovery might be required, involving destruction of the rogue device. A less dramatic example of forwards error recovery in a telecommunication network might be a broadcast message from, say, a network element undergoing this recovery, to inform other parts of the network that an error has occurred. This would allow the recipients to recognise that previous communications from the affected area are suspect and would lead to discarding of any erroneous or incomplete messages received from this area immediately prior to recovery.

The use of acceptance tests can be generalised in software through the use of *recovery blocks* [24]. These involve alternative versions of a software module, just as with *N*-version programming, but normally only one of these modules is executed, as shown in Figure 7.11. If the results pass the relevant acceptance test, then the system proceeds without using the alternative version or

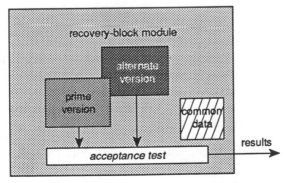

Figure 7.11   Software fault tolerance through recovery blocks

versions. However, if the acceptance test fails, then the first alternative is executed to establish whether it can deliver an acceptable result; if not, the process continues until all the alternatives have executed and been tested. Again, it is necessary to pay the additional development cost of designing several versions of a program. However, if these are executed sequentially in the fashion described, there is no run-time penalty unless errors actually occur. Of course, if the primary version of a program fails, then the system is delayed while one or more alternatives are tried; thus the effects of such delays must be taken into account within the system. It is conventional for only a single version of the state information to be used for all software versions in a recovery-block scheme. Thus, backwards (or forwards) error recovery must be used after an acceptance test fails to·allow the next version to run with a consistent system state.

7.4.2 Error-free software

These approaches to software fault tolerance are based on recognising and correcting in operation the effects of software design errors. There has also been a growing interest in software-engineering techniques focused on producing error-free software [25]. These techniques are based on developing a formal specification of the program requirements which can be mathematically linked to the software design and implementation, with the intention of allowing formal proving of the correctness of the design against its requirements [26,27]. There are formidable problems in transferring these techniques to the industrial arena, and in providing adequate automated support environments in which to develop the software. However, it seems clear that such techniques will become more important as experience in their use grows, particularly for safety-critical applications and areas (such as telecommunications) where the problems caused by system failure can be acute. It is therefore to be expected that future SPC systems may incorporate program ideas developed using formal specification and design methods, although this will not eliminate the need for in-built fault tolerance.

7.4.3 Applications software

Software in SPC systems is required to operate in an event-driven environment in which a large number of individual customers may require service concurrently. The level of demand varies significantly over time in response to the nature of business and domestic use of telecommunications, and the software must be able to cope with this. Part of this demand variation can be catered for where the underlying hardware provides load sharing. However, the software must also be able to service a number of similar tasks simultaneously and to keep track of the status and progress of each as the appropriate input and output signals are received and generated.

It is common practice to meet this need with a combination of operating system and application program features. Each application program deals with a subset of overall system operation, and the complete suite of application programs runs within the software environment provided by the operating system. The operating system, in turn, provides a means of sharing the processing resources of the computer hardware among the application programs, which then compete for processor time and access to these resources to progress tasks on which they are engaged. The application programs maintain data on the state of each task being handled and relinquish their use of the processing resources while they are waiting for any information from other programs or from the outside world, thus allowing other programs to continue in the meantime, as shown in Figure 7.12.

In the real-time environment of telecommunication systems, application-

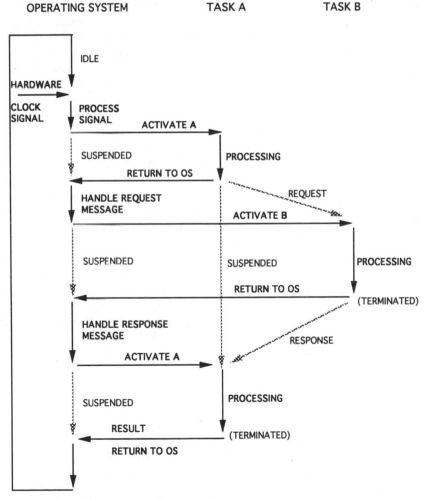

Figure 7.12   Multiprogramming operation

software design uses methods and tools oriented towards the communications needs of the programs involved. These include the use of such techniques as *finite state automata* and *table-driven software* [29], to describe program behaviour in terms of transitions between a limited number of states, and methods of structuring overall system operation to create layered designs in which only selective details need be considered at each stage. Academic interest in such applications, such as the formal specification method CSP [30], offers ways in which these approaches may develop in the future.

7.5 Networks and SPC

The increasing sophistication of telecommunication networks is bringing with it an increasing need for equally sophisticated management and control, to enable the network operating company to exploit the capabilities and flexibility of the network, and to provide customers with the range of services and the responsiveness which they are coming to expect.

More and more, the effectiveness of telecommunication networks and services is being conditioned by their management capabilities [31] and any opportunity to treat networks as a set of independent, or at most loosely coupled, switching nodes is fast disappearing [32]. Network operation and development require access to network facilities, regardless of their location and structure, in an integrated and coherent form. A key aspect of network evolution in this environment is the use of Open Systems Interconnection (OSI) standards for management communications between network elements [33]. As the network and the technology used within it develop, there are inevitable modifications as new services and ways of working are introduced and the boundaries between existing network elements shift due to differences in the rates of change in each area. The use of OSI standards as a common basis for the interfaces between network elements helps to insulate each area of the network from changes in progress elsewhere and allows each network element a degree of independent evolution. There has been rapid progress in the last few years in the use of OSI communication as the basis for network management and in the development of object-oriented information models to be used across these communications facilities [34]. Network management is the subject of Chapter 18.

7.6 Conclusions

Stored-program control has developed from its beginnings in the era of electromechanical switching to take full advantage of the hardware and software technologies now available for the construction of powerful systems able to operate reliably even when their components become faulty. As telecommunication services become more diverse and the management of networks and services more sophisticated, SPC must adapt to cope with the demands of complex network management and the interaction of a wide variety of network elements, each of which is undergoing constantly accelerating development and change in response to the market requirements now faced.

This environment represents an exciting and demanding challenge to the designers of SPC systems and networks, a challenge which is among the most exacting of those faced by those working in the field of information technology. This Chapter gives only an outline of the issues involved. Nevertheless, it is hoped that enough of the concerns and possibilities of present-day telecommunication-systems design has been covered to give an insight into why this field remains as fresh today as when the idea of stored-program control was first conceived.

7.7 References

1. CHAPUIS, R.J.: '100 years of telephone switching' (North-Holland Publishing, 1982)
2. KORN, F.A. and FERGUSON, J.G.: 'The No. 5 crossbar dial telephone switching system', *Trans. Am. Inst. Elect. Eng.* **69** (Pt 1), 1950, pp. 244–254
3. ELLSTAM, S.: 'L. M. Ericsson's crossbar systems', *Ericsson Rev.* **43**, No. 4, 1966, pp. 153–162
4. MAY, C.A.: 'Electronic exchanges: the steps leading to TXE4, *Post Off. Electr. Eng. J.*, 1972, **65**, pp. 134–139
5. McDONALD, J. (Ed.): 'Fundamentals of digital switching' (Plenum Press, 1983)
6. FISHER, D.G., and BAUER, W.: 'Multiplexing with intelligence' *Telecommunications* (international edition), **22**, (2), 1988, pp. 73–74, 79
7. JOEL, A. E.: 'Realization of the advantages of stored program control', *International Switching Symposium*, Munich, 1974
8. EKLUND, M., LARSON, C.G., and SORME, K.: 'AXE10—system description', *Ericsson Rev.* 1976, **53**, (2), pp. 70–89
9. VANNER, N.J.: 'Architecture of System X — the digital trunk exchange', *Post Off. Electr. Eng. J.*, 1979, *Brit. Telecom. Tech. J.* **72**, pp. 142–148
10. HUGHES, C.J.: 'Switching — state of the art', 1986, **4**, (i), pp. 5–19; (2), pp. 5–17
11. CLARK, M.P.: 'Networks and telecommunications, design and operation' (Wiley, 1991)
12. CUAN, W.G.: 'Fault tolerance: tutorial and implementations', *Computer,* November 1989, **22**, (12); Supplement, Winter 1989, pp. 30–39
13. O'CONNOR, P.D.T.: 'Practical reliability engineering' (Heyden and Son, 1981)
14. HAYWARD, W.S. (Ed.): 'The 5ESS switching system', *AT&T Tech. J.* 1985, **64**, pp. 1305–1559
15. KOBUS, S. *et al.*; 'Central control philosophy for the Metaconta L switching system', *Elect. Commun.*, 1972, **47**, pp. 159–163
16. TROUGHTON, D.G., *et al.*: 'System X: the processor utility', *Br. Telecom. Eng. J.*, 1985, **3**(4), pp. 226–240
17. BOWDEN, F.G., and KELLY, M.B.: 'The application of local area networks to the control architecture of a telephone exchange'. *International Switching Symposium*, Florence, 1984
18. PARNAS, D.L.: 'On the criteria to be used in decomposing systems into modules', *ACM Commun.* 1972, **15** (12), pp. 1053–1058
19. BONAMI, J., COTTON, J.M., and DENENBERG, J.N.: 'ITT 1240 digital exchange: architecture', *Electr. Commun.*, 1981, **56** (2,3), pp. 126–134
20. KOPETZ, H.: 'Software reliability' (Macmillan, 1979)
21. ANDERSON, T., and LEE, P.A.: 'Fault tolerance: principles and practice' (Prentice-Hall International, 1981)

22. AVIZIENIS, A.: 'The N-version approach to fault tolerant software', *IEEE Trans.*, 1985, **SE–11** pp. 1491–1501
23. HOCKNEY, R. W., and JESSHOPE, C.R.: 'Parallel computers 2' (IOP Publishing, 1988)
24. RANDELL, B.: 'System structure for software fault tolerance', *IEEE Trans.*, 1975, **SE–1**, pp. 220–232
25. MILI, A.: 'An introduction to formal program verification' (Van Norstrand Reinhold, 1985)
26. BJORNER, D., and JONES, C.B.: 'Formal specifications and software development' (Prentice-Hall, 1982)
27. JONES, C.B.: 'Software development: a rigorous approach' (Prentice Hall, 1980)
28. ALLWORTH, S.T., and ZOBEL, R.N.: 'Introduction to real-time software design', (Macmillan Education, 1987), 2nd Ed., pp. 33–39
29. HEXT, J.: 'Programming structures, vol. 1, machines and programs' (Prentice Hall, 1990), pp. 105–128
30. HOARE, C.A.R.: 'Communicating sequential processes', *ACM Commun.* 1978, **21** (8), pp. 666–677
31. ASHMAN, C.: 'Protecting your investment in networks: the value of network management'. *Proceedings of the international conference on Network Management*. London, June 1988 (Blenheim Online Publications, 1988), pp. 19–33
32. FISHER, D.: 'Architectural concepts for managed transmission networks', *Proceedings of the International Conference on Network Management*, (Blenheim Online Publications 1988), pp. 73–83
33. MARSDEN, B.W.: 'Communication network protocols' (Chartwell–Bratt, 1986)
34. CCITT Recommendation M.3010: 'Principles for telecommunications management network', 1992

Chapter 8

Signalling

E. S. Grundy

8.1 Introduction

In a telecommunication network, signalling systems are as essential as switching systems and transmission systems. Signalling systems must obviously be compatible with the switching systems in a network. They must be able to transmit all the signals required to operate the switches. They must also be compatible with the transmission systems in the network in order to reach the exchanges which they control. Thus, the design of signalling systems is directly influenced by both switching and transmission requirements and the evolution of signalling has followed developments in switching and transmission [1].

For an own-exchange call, it is necessary to send signals in each direction both between the caller and the exchange and between the exchange and the called customer. In a multilink connection, signals must also be sent in both directions between exchanges. Signals sent in the direction away from the caller and towards the called line are known as *forward signals* and those sent in the opposite direction are known as *backward signals*.

Transmitted signals may be either *continuous signals* or *pulse signals*. An example of a continuous signal is the DC off-hook signal on a customer's line. A pulse signal may be either a single pulse or a coded group of pulses. An example of the latter is a decimal digit sent by loop/disconnect pulsing. Signals may also be either *unacknowledged signals* or *acknowledged signals*. Address digits sent by customers are normally unacknowledged. When an acknowledgement signal is returned, it confirms receipt of the signal that was sent. Acknowledgements may be continuous or pulse signals. If pulse signalling is used, a signal may be repeated until it is acknowledged. When continuous signalling is used, a signal is sent until the acknowledgement is received and the acknowledgement signal persists until the original signal has been removed. This is called *compelled signalling* and it is the most reliable method. However, when a circuit has a long propagation time, compelled signalling is slow. Four propagation times elapse before the sending equipment has detected the end of the acknowledgement and can send another signal. Fully compelled signalling is therefore not used over satellite circuits.

Traditionally, exchanges have sent signals over the same circuits in the network as the connections which they control. This is known as *channel-associated signalling*. For a simple telephone call, only the following basic signals are required between exchanges:

(*a*) call request or seize (forward);
(*b*) address signal (forward);
(*c*) answer (backward); and
(*d*) clear signals (forward and backward).

The introduction of stored-program control (SPC) enabled customers to be provided with a wider range of services than those available with electro-mechanical switching systems. These services require more signals to be transmitted between exchanges than have previously been provided. Since the signals are generated by the central processor in one exchange and sent to the processor of another exchange, they can be transmitted directly between the processors over a separate data channel. This is known as *common-channel signalling* (CCS) and is described in Sections 8.4 and 8.5.

Common-channel signalling is now widely used in public telecommunication networks, both nationally and internationally. It is also used in private networks for signalling between digital PBXs, as described in Section 8.7. In an integrated-services digital network, common-channel signalling is also used over customers' lines, as described in Section 8.6.

8.2 Customer-line signalling

In a local telephone network, loop/disconnect signalling is used for sending customers' call-request (seize) and clear signals to the exchange. Since there is a minimum line current which the exchange can detect, there is a maximum permissible line resistance. This limits the maximum length of line and the size of the area served by the exchange. (In addition, the length of lines is also limited by the permissible attenuation at voice frequencies.) Customers' lines on electromechanical exchanges have been limited to about 1000 Ω resistance. However, over 2000 Ω can be achieved with modern digital switching systems.

When dial telephones are used, customers send address information by loop/disconnect pulsing. For each digit, the dial breaks the circuit to send a train of up to 10 disconnect pulses at approximately 10 pulse/s. The exchange is able to detect the end of each pulse train because the minimum pause between digits (e.g. 400 ms to 500 ms) results in a loop state significantly longer than the 'makes' during pulsing (e.g. 33 ms).

A relay circuit to receive dial pulses [2] was required in every selector in a Strowger exchange. However, the introduction of registers reduced the number of dial-pulse receivers needed, so these could be more complex. This led to the introduction of push-button telephones, which send voice-frequency pulses and thus provide faster address signalling.

Hz \ Hz	1209	1336	1477	1633
697	Digit 1	2	3	Spare
770	4	5	6	Spare
852	7	8	9	Spare
941	*	0	#	Spare

Figure 8.1   Dual-tone frequency coding used by pushbutton telephone sets

A push-button (keyphone or touchtone) telephone uses *dual-tone multi-frequency signalling* (DTMF). It sends each digit by means of a combination of two frequencies, one from each of two groups of four frequencies [1], as shown in Figure 8.1. This is done to reduce the risk of *signal imitation.* Since each digit uses two frequencies, and these are not harmonically related, there is much less chance of the combination being produced by speech or room noise picked up by the telephone transmitter than if a single frequency were used.

In addition to the digits '1' to '0', the telephone keypad has buttons with the symbols '*' and '#'. These are used by SPC exchanges for services which are under the control of customers (for example, to enable a customer to divert incoming calls to another telephone).

A call between two customers on the same exchange requires a number of actions to be performed in response to signals. These signals are shown in Figure 8.2. It will be seen from this diagram that there is a 'handshake' protocol. Every signal should result in a response, thus verifying correct operation. For an interexchange call, this sequence of signals must also be sent over the circuits between the exchanges.

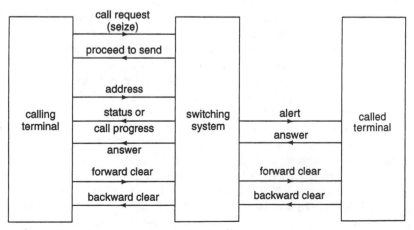

Figure 8.2   Signals transmitted for a local call

8.3 Channel-associated interexchange signalling

8.3.1 Audio-frequency circuits

A two-wire junction circuit can transmit loop/disconnect signals, as used on customers' lines. On an amplified four-wire circuit, centre taps of line transformers can be connected to provide a two-wire phantom circuit for signalling.

In customer-line signalling, the signals to and from the caller and those to and from the called customer are sent over different lines, as shown in Figure 8.2. However, for a junction call, the forward seize and clear signals and the backward answer and clear signals must be sent between the exchanges over the same line. This is accomplished as follows. The originating exchange uses the loop and disconnect states to send its forward seize and clear signals and to send dial pulses. The terminating exchange sends back its answer signal by interchanging the battery and earth connections to the line, thus reversing the direction of the line current. The originating exchange recognises this reversal by means of a polarised relay [1,3]. The terminating exchange sends its backward clear signal by removing the polarity reversal.

The repertoire of different signals may be increased by using pulse signals in addition to continuous signals. When signals are exchanged between registers, it is necessary to delay sending from the originating exchange until a register has been connected at the receiving exchange. This can be achieved by not sending digits forward until a backward proceed-to-send signal has been received.

When address information is sent between exchanges by loop/disconnect pulsing over a long line, distortion of the pulses can cause errors in the received information [1,2]. As shown in Figure 8.3, the capacitance of the line causes a slow decay of the pulse waveform, so the output break from the receiving relay is shorter than that originally sent. On a multilink connection, when all the switches are directly controlled by the originating exchange (as in a network of step-by-step exchanges), the distortion is cumulative. However, when register–senders are used, the pulses can be regenerated at each exchange and the conditions are thus less onerous.

The nonsymmetrical waveform shown in Figure 8.3 occurs with loop/disconnect pulsing because the sending impedance of the circuit is almost zero in the loop state and infinite in the disconnect state. The pulse distortion can be reduced, and the signalling range increased, by using a waveform which is symmetrical. Long-distance direct-current (LDDC) signalling systems were therefore designed which obtain symmetrical waveforms by using *double-current working*. Pulses are sent by polarity reversals instead of by makes and breaks [1,2,4]. The sending impedance is thereby kept constant and the duration of the received pulse (between the crossing points of a zero-voltage reference) is equal to that of the transmitted pulse, as shown in Figure 8.4. The pulses are received by a polarised relay. The signalling range is increased

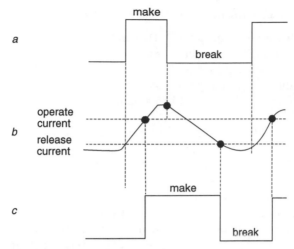

Figure 8.3 Distortion of loop/disconnect pulses caused by capacitance of a long line

    a Input pulse at sending end
    b Received current waveform
    c Output pulse from receiving relay

by the sensitivity of the polarised relay as well as by the reduction in pulse distortion.

Long-distance DC signalling is largely of historical interest, since most long-distance circuits are now provided in high-capacity multiplex transmission systems. The signalling methods used with multiplex systems are described in Sections 8.3.2–8.3.4.

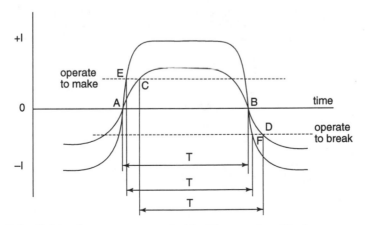

Figure 8.4 Pulsing by current reversals (double-current working)

8.3.2 Outband signalling in FDM carrier systems

In frequency-division-multiplex (FDM) systems, the carriers are spaced at intervals of 4 kHz and the speech band is from 300 Hz to 3.4 kHz. By using channel filters with a sharp cutoff, it is possible to insert a narrowband signalling channel above the speech band (i.e. between 3.4 kHz and 4 kHz). Signal frequencies of 3.7 kHz and 3.85 kHz have been used. This is known as *outband signalling.*

An outband-signalling system is shown in Figure 8.5. A DC signal on the input lead M at one terminal causes the signal frequency to be sent over the transmission channel. This is detected at the other terminal to give a corresponding DC signal on the output lead E. If the repeater station containing the FDM channelling equipment is adjacent to the switching equipment, it is simpler for the latter to send and receive signals over separate E and M wires than to extract them from and re-insert them into the speech circuit. E and M wire signalling has been widely used in North America [4], not only for outband signalling but also for DC- and inband-signalling systems.

To use outband signalling successfully in a network, all routes must use FDM systems with built-in outband signalling. In practice, however, a route may contain a section with audio-frequency transmission or a carrier system

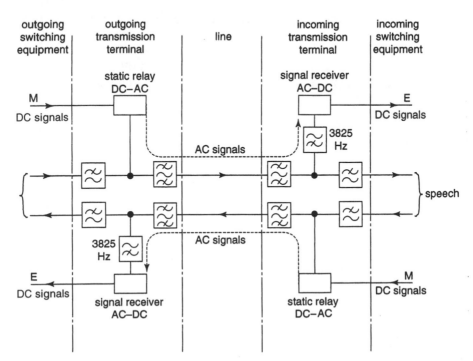

Figure 8.5   Outband signalling for FDM carrier system

not equipped with outband signalling. It is then necessary to use *inband signalling*, as described in Section 8.3.3.

8.3.3 Inband (VF) signalling

Systems which transmit signals within the baseband of FDM systems are known as *inband signalling systems* or *voice-frequency (VF) signalling systems*. They have the important advantages that they are independent of the transmission systems used and they will function over any circuit which provides satisfactory speech transmission. A VF signalling system is shown in Figure 8.6.

Since voice-frequency signals are used, there is a possibility of signal imitation, i.e. the receiver may be operated because the signal frequency is present in transmitted speech. This is obviously undesirable; for example, it could clear down a connection before the users had finished their conversation. The following measures are taken [1] to make the probability of signal imitation almost negligible:

(*a*) A signal frequency is chosen at which the energy in speech is low (i.e. above 2 kHz). For example 2280 Hz is used in the UK [1] and 2600 Hz in north America [5].
(*b*) The durations of signals are made longer than the period for which the speech frequency is likely to persist in speech (e.g. ≥ 50 ms).
(*c*) Use is made of the fact that the signal frequency is unlikely to be produced in speech without other frequencies also being present.

In addition, some systems (2VF systems) have used a combination of two signal frequencies to improve their integrity further [1].

A block diagram of a VF receiver is shown in Figure 8.7. In order to make use of measure (*c*), the receiver contains a signal circuit with a bandpass filter

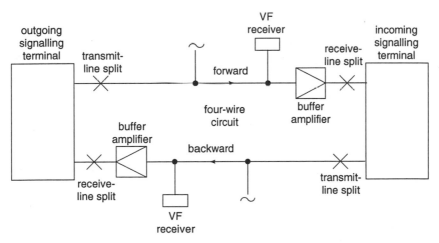

Figure 8.6   Voice-frequency (VF) signalling system

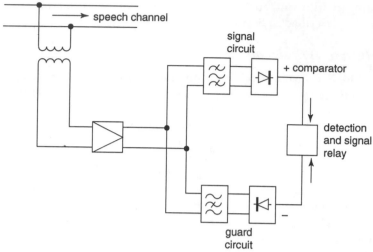

Figure 8.7   Block diagram of voice-frequency receiver

to accept the signal frequency and a guard circuit with a bandstop filter to accept all other frequencies and reject the signal frequency. The outputs of both circuits are rectified and compared. If the output from the signal circuit exceeds that from the guard circuit, the receiver operates; if the output from the guard circuit exceeds that from the signal circuit, the receiver gives no output signal.

To avoid interfering with speech, VF signals must not be transmitted while a conversation is in progress. To avoid this, two signalling methods have been employed:

(*a*) tone-on-idle signalling; and
(*b*) pulse signalling.

Tone-on-idle signalling has been extensively used in North America [5]. Pulse-type VF-signalling systems have been used by many operating organisations. An example is the British AC 9 system [1].

8.3.4 PCM systems

PCM primary multiplexers were designed from the outset to incorporate signalling. The DC signals associated with the audio-frequency baseband circuits in each direction are sampled and the signal samples are transmitted within the frame of PCM channels. It is therefore unnecessary to use VF signalling.

The 2 Mbit/s system has 32 8-bit time slots, but it only provides 30 speech channels. Time slot zero is used for frame alignment and time slot 16 is used

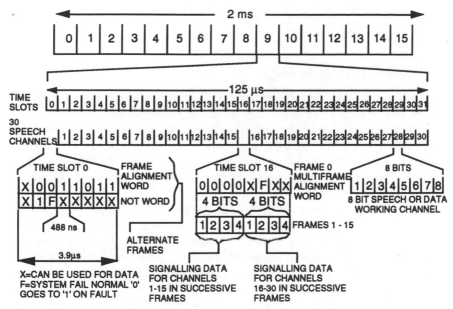

Figure 8.8   Use of multiframe for signalling in 30-channel PCM system

for signalling, as shown in Figure 8.8. The eight bits of channel 16 are shared between the 30 channels by a process of *multiframing*. As shown in Figure 8.8, 16 successive appearances of channel 16 form a multiframe of 8-bit time slots. The first contains a multiframe alignment signal and each of the subsequent 15 time slots contains four bits for each of two channels. Thus, every speech circuit can have, in each direction, up to four independent signalling channels at 500 bit/s. This enables a much larger number of signals to be exchanged than is possible with the DC signalling methods described in Section 8.3.1.

The North American 1.5 Mbit/s PCM system has a frame of 193 bits. This contains 24 8-bit time slots, all of which are used for speech, and the 193rd bit in each frame is used for the frame-alignment signal. In every sixth frame, the eighth bit of each channel is used for signalling instead of for speech. This 'bit stealing' has been found to cause a negligible increase in quantisation distortion. A 12-frame multiframe is used. Thus, associated with each speech channel are two independent signalling channels at 650 bit/s or a single signalling channel at 1.3 kbit/s. The 193rd bit of the PCM frame is used on alternate frames for multiframe alignment instead of for frame alignment.

The methods described above provide channel-associated signalling, but PCM systems can also be used for common-channel signalling. Multiframing is then not required. In the 30-channel system, time slot 16 is used to provide a common signalling channel at 64 kbit/s. In the 24-channel system, the 193rd bit in alternate frames can be used to provide a common signalling channel at 4 kbit/s. However, if a 64 kbit/s channel is required, it is necessary to sacrifice one of the speech channels (i.e. channel 24).

8.3.5 Inter-register signalling

For a multilink connection in a network of register-controlled exchanges, a register in the originating exchange receives address information from the calling customer and sends out routing digits. Each succeeding register both receives and sends out routing digits, until the terminating exchange is reached. This sequence of operations introduces post-dialling delay. To minimise this delay, a more rapid method of sending routine information than loop/disconnect pulsing is needed and inband *multifrequency (MF) signalling systems* have been developed for this purpose [1,6,7].

An inter-register signalling system cannot be used for seize, answer and clear signals. No register is connected when an incoming seize signal is received, since it is this signal which initiates the connection to a register. The register is released after it has set up a connection through its exchange and sent out routing digits; therefore, it cannot receive answer and clear signals. Consequently, *line signalling*, using one of the methods described in Sections 8.3.1–8.3.4, is required in addition to interregister signalling.

Either *en-bloc* or *overlap signalling* may be used. In en-bloc signalling, the complete address information is transferred from one register to the next as a single string of digits. Thus, no signal is sent out until the complete address information has been received and analysed. In overlap signalling, digits are sent out as soon as possible. Thus, some digits may be sent before the complete address has been received and signalling may take place simultaneously on two links (i.e. the signals overlap). This enables subsequent registers to start digit analysis earlier than is possible with en-bloc signalling and it reduces post-dialling delay.

Either *link-by-link signalling* or *end-to-end signalling* may be employed. In link-by-link signalling, information is exchanged only between adjacent registers in a multilink connection, as shown by Figure 8.9*a*. This has the advantage that signals only suffer the transmission impairments of a single link. Also, different signalling systems may be used on different links. Thus, if a network is being modernised, all registers do not need to be modified simultaneously. However, each transit register must receive, store and retransmit the complete address information. Consequently, link-by-link working has the disadvantage that the holding times of registers and the post-dialling delay are long, particularly if backward signalling is used in addition to forward signalling.

In end-to-end signalling, the originating register controls the setting up of a connection until it reaches its final destination, as shown in Figure 8.9*b*. Each transit register receives only the address information required to select the outgoing route to the next exchange in the connection. Having performed its task, it is released and the originating register signals to the next register. End-to-end signalling has the disadvantage that all registers must be compatible with the originating register. Also, signals suffer the transmission impairments of several links in tandem. However, each transit register needs to receive and retransmit only part of the address information, so register holding times and the post-dialling delay are shorter than for link-by-link working. For this reason, end-to-end working has been used in most networks.

The arrangements used for multifrequency (MF) inter-register signalling are shown in Figure 8.10. Each digit is sent by a combination of two frequencies out of six (2/6 MF). Since two frequencies are required to represent a digit, this gives an error-detection capability. Combinations of two frequencies out of six give 15 possible digit values, so five extra signals are available in addition to the digits '1' to '0'. Since inter-register signalling precedes a conversation, signal imitation cannot occur. Thus, the VF receivers do not need speech immunity and this enables short signal pulses to be used.

The Bell R1 system [1,7] (which corresponds to the CCITT Regional Signalling System 1) uses six signal frequencies spaced at 200 Hz between 700 Hz and 1700 Hz. Its signals are unacknowledged, so no frequencies are

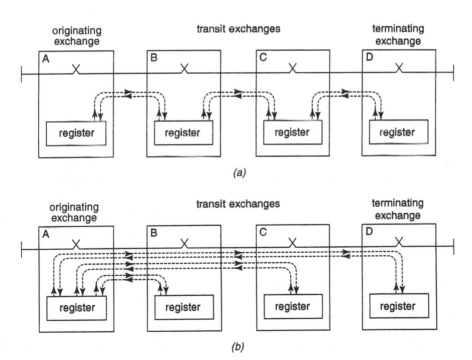

Figure 8.9   Link-by-link and end-to-end signalling between registers

    a Link-by-link signalling
    b End-to-end signalling

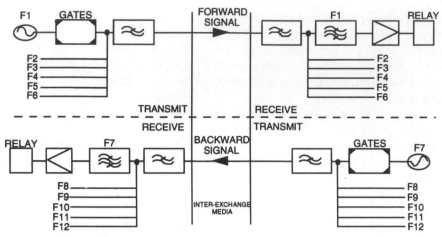

Figure 8.10   Multifrequency (MF) inter-register signalling system

(Directional filtersare only required for 2-wire circuits)

provided for backward signalling. The system transmits 12 forward signals as follows:

KP (start of pulsing)
digits 1 to 0
ST (end of pulsing)

The system employs link-by-link signalling.

The CCITT R2 system [1,8] (Regional Signalling System 2) provides both forward and backward signals. It normally uses continuous compelled signalling. However, a version with pulse signalling is used for satellite circuits because of their long delay. End-to-end working is employed, i.e. a connection is controlled by the originating register.

The R2 system uses signal frequencies spaced at 120 Hz, as follows:

forward: 1380, 1500, 1620, 1740, 1860, 1980 Hz
backward: 540, 660, 780, 900, 1020, 1140 Hz

In this system, more than 15 signals can be sent in each direction, because one signal frequency combination is used as a shift signal to enable each of the others to have two meanings. Thus, a very wide repertoire of signals is available. Some of the additional signals are specified for use on international calls. However, since the system was defined as a regional system, operating administrations were free to choose how to use the other signals. Consequently, there are different national versions of the R2 system. An example was the British MF2 system [1].

8.4 Common-channel signalling principles

8.4.1 General

In a network of SPC exchanges, a connection which is made through two exchanges requires call processing by a processor in each exchange. If channel-associated signalling is used for calls from exchange A to exchange B, *a* as shown in Figure 8.11*a*, it is necessary for the central processor of exchange A to send its outgoing forward signals to the individual speech circuit for transmission to exchange B. At this exchange, the signals must be

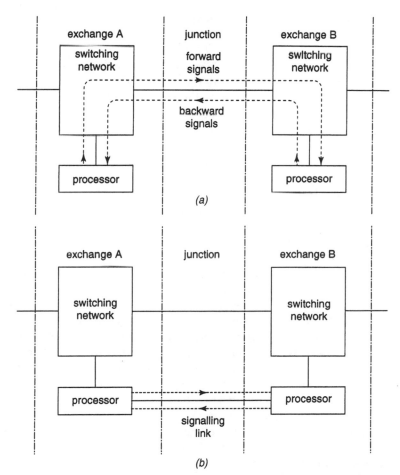

Figure 8.11   *Principle of common-channel signalling (CCS) between exchange central processors*

     (*a*) Channel-associated signalling
     (*b*) Common-channel signalling

detected on the speech circuit and passed to the central processor. Similarly, backward signals from processor B must be sent over the speech circuit, detected at exchange A and extended to its processor. This is an inefficient arrangement for signalling between the two processors!

If a high-speed data link is employed between the processors, as shown in Figure 8.11*b*, it can provide a channel for all signals between exchanges A and B. This is known as *common-channel signalling* (CCS). It gives the following advantages:

(*a*) Information can be exchanged between the processors much more rapidly than when channel-associated signalling is used.

(*b*) As a result, a much wider repertoire of signals can be used and this enables more services to be provided to customers.

(*c*) Signals can be added or changed by software modification to provide new services.

(*d*) There is no longer any need for line-signalling equipment on every junction, which results in a considerable cost saving.

(*e*) Since there is no line signalling, the junctions can be used for calls from B to A in addition to calls from A to B. (Both-way working requires fewer circuits to carry the traffic than if separate groups of junctions are provided from A to B and from B to A.)

(*f*) Signals relating to a call can be sent while the call is in progress. This enables customers to alter connections after they have been set up. For example, a customer can transfer a call elsewhere, or request a third party to be connected into an existing connection.

(*g*) Signals can be exchanged between processors for functions other than call processing, for example for maintenance or network management.

The error rate for common-channel signalling must be low and the reliability required is greater than for channel-associated signalling. Failure of the data link in Figure 8.11*b* would prevent any calls from being made between exchanges A and B, whereas failure of a line signalling equpiment, or even of an inter-register signalling system, would only result in the loss of a small fraction of the traffic. Consequently, duplicate CCS links are provided and either can carry the full signalling traffic. The error rate of each link is monitored and if it exceeds a preset rate (e.g. 10^{-4}) a fault report is generated and the link is automatically withdrawn from service.

When channel-associated signalling is used, the successful exchange of signals over a circuit proves that the circuit is working. CCS does not provide this checking facility, so a separate means (e.g. automatic routine testing) must be provided to ensure the integrity of the speech circuits.

CCS systems use *message-based signalling*. Successive messages exchanged between the processors usually relate to different calls. Each message must therefore contain a label, called the *circuit-identity code*, indicating to which speech circuit, and thus to which call, it belongs. Since messages pass directly between central processors, no connection is required to an incoming junction before an address signal is received. The address signal can therefore be the first message sent and there is no need for a seize signal. In a multilink connection, signalling takes place from one transit exchange to the next without involving the originating exchange. Thus, link-by-link signalling is inherent with CCS.

In a CCS system, messages from a processor queue for transmission over the signalling link. The number of speech circuits that can be handled by a CCS system is therefore determined by the acceptable delay. A signalling link operating at 64 kbit/s normally provides signalling for up to 1000 or 1500 speech circuits [1,9]. However, more may be handled (with increased delays) when the load of a link which has failed is added to the existing load on a back-up link.

The use of CCS for interexchange signalling has been followed by its application to customers' lines in integrated-services digital networks (ISDN). Common-channel signalling is also used between processor-controlled PBXs in digital private networks. The signalling methods used are described in Sections 8.6 and 8.7.

8.4.2 Signalling networks

Figure 8.12*a* shows a direct CCS link between two exchanges. This is known as *associated signalling*. In a multi-exchange network, there will be many CCS links between exchanges and these form a *signalling network*, as shown in Figure 8.12*b*. In principle, CCS signals can follow different routes from the connections which they control and they can pass through several intermediate nodes in the signalling network. This is called *nonassociated signalling*. Since, in general, signal messages entering the network may be destined for any other exchange, the messages must include labels containing their destinations. The network used for nonassociated signalling is thus a form of packet-switched network.

In practice, CCS messages are usually only routed through one intermediate node, as shown in Figure 8.12*b*. This is known as *quasi-associated signalling* and the intermediate node is called a *signal-transfer point* (STP). In Figure 8.12*b*, the CCS equipment in exchange C handles signalling for connections between exchanges A and B, in addition to signalling for connections between C and A and between C and B.

Since CCS signals may be routed via an STP, each message contains a *destination-point code* to enable it to be routed to the correct exchange. It also contains an *originating-point code* to enable the messages sent back to be routed correctly. If the CCS system in an exchange recognises the destination-point code of an incoming message as its own, the message is accepted and passed to the central processor. If the code is that of another exchange, the CCS system looks up a translation table to determine the route for onward transmission of the message.

Quasi-associated signalling is used when there are few circuits between A and B and thus little signalling traffic between them. It is then economic to share a single signalling link from A to C between the route from A to B and routes from A to other exchanges. When there are many circuits between A and B, and thus much signalling traffic, it is economic to use associated signalling. However, an alternative route via an STP is normally provided in case the associated-signalling link fails.

The transmission bearers used for a CCS network are channels in the main transmission-bearer network. The first generation of CCS systems [10,11] (CCITT no. 6) used modems to transmit at 2.4 kbit/s or 4.8 kbit/s over analogue telephone channels. A 4 kbit/s channel could also be provided over a 1.5 Mbit/s PCM system, as described in Section 8.3.4. Current CCS systems [9,10,12] (CCITT no. 7) use a 64 kbit/s channel provided by time slot 16 in a 2 Mbit/s PCM system or time slot 24 in a 1.5 Mbit/s PCM system.

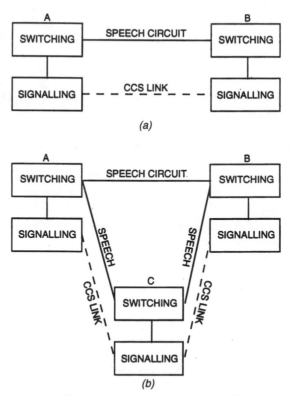

Figure 8.12   Association between common-channel signalling and traffic network

a Associated signalling
b Quasi-associated signalling

8.5 CCITT Signalling System 7

8.5.1 General

A block schematic diagram of the CCITT no. 7 signalling system [9,10,12] is shown in Figure 8.13. Signal messages are passed from the central processor of the sending exchange to the CCS system. This consists of three microprocessor-based subsystems: the signalling-control subsystem, the signalling-termination subsystem and the error-control subsystem. The signalling-control subsystem structures the messages in the appropriate format and queues them for transmission. When there are no messages to send, it generates filler messages to keep the link active. Messages then pass to the signalling-termination subsystem, where complete *signal units* (SU) are assembled using sequence numbers and check bits generated by the error-control subsystem. At the receiving terminal, the reverse sequence is carried out.

The system can be modelled as a stack of protocols like the ISO seven-layer model described in Section 1.7. However, the system was specified before the ISO model was published and the layers are referred to as *levels* in its literature. The levels are as follows:

level 1: the physical level
level 2: the data-link level
level 3: the signalling-network level
level 4: the user part.

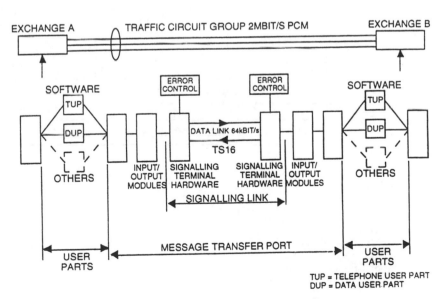

Figure 8.13   Block diagram of CCITT Signalling System 7

The relationship between these levels and the layers of the OSI model is shown in Figure 8.14. The user part encompasses layers 4–7 of the OSI model.

Level 1 is the means of sending bit streams over a physical path. It uses time slot 16 of a 2 Mbit/s PCM system or time slot 24 of a 1.5 Mbit/s system.

Level 2 performs the functions of error control, link initialisation, error-rate monitoring, flow control and delineation of messages.

Level 3 provides the functions required for a signalling network. Each node in the network has a *signal-point code*, which is a 14-bit address. Every message contains the point codes of the originating and terminating nodes for that message.

Levels 1–3 form the *message-transfer part* (MTP) of CCITT Signalling System 7. Level 4 is the *user part*. This consists of the processes for handling the service being supported by the signalling system. The message-transfer part is

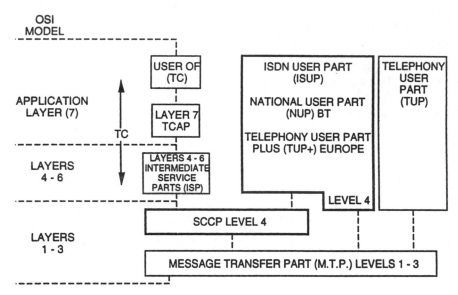

Figure 8.14   *Relationship between functional levels of CCITT Signalling System 7 and layers of OSI 7-layer model*

ISP   = intermediate service parts
ISUP = ISDN user part
MTP = message-transfer part
NUP = national user part
OSI   = open systems interconnection
SCCP = signal-connection-control part
TC   = transaction capabilities
TCAP = transaction-capabilities application part
TUP = telephony user part

capable of supporting many different user parts. These include: the telephone-user part (TUP), the data-user part (DUP), the mobile-user part (MUP)and the ISDN-user part (ISUP).

Because of the long time needed to develop the international standards for ISDN, some countries enhanced the TUP part to cater for some ISDN services. In the UK, this resulted in a national user part (NUP) [13]. Some countries use an enhanced TUP known as TUP+.

CCS systems are now being used for messages which are not directly associated with call establishment, for example to update a location register in a cellular-mobile-radio network (as described in Chapter 12), to interrogate a remote database in an intelligent network (as described in Chapter 16), for traffic management (as described in Chapter 18) and for operations, maintenance and administration. This led to the specification of the part known as the *transaction capabilities* (TC). Since this was designed after publication of the OSI model, its protocols were specified to conform with it [10]. As shown in Figure 8.14, a *signalling-connection control part* (SCCP) has been added to level 3 to make it fully compatible with layer 4 of the OSI model. The *intermediate service part* (ISP) performs the functions of layers 4–6 of the OSI model and the *transaction-capabilities application part* (TCAP) provides for layer 7.

8.5.2 The high-level data-link control protocol

The level-2 protocol used in the CCITT Signalling System 7 uses the international standard known as *high-level data-link control* (HDLC) [14]. Messages are sent by packets contained within frames having the format shown in Figure 8.15.

The beginning and end of each HDLC message is indicated by a unique combination of digits (01111110) known as a *flag*. Of course, this sequence of digits can occur in messages and must then be prevented from being interpreted as a flag. This is done by a technique known as *zero-bit insertion and deletion* or *bit stuffing and unstuffing*. When sending digits of a message between

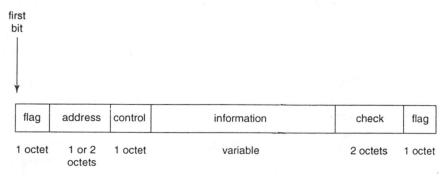

first
bit

flag	address	control	information	check	flag
1 octet	1 or 2 octets	1 octet	variable	2 octets	1 octet

Figure 8.15   Frame structure for high-level data-link-control (HDLC) protocol

the two flags, the sending terminal inserts a '0' after every sequence of five consecutive '1's. The receiving terminal deletes any '0' which occurs after five consecutive '1's and so restores the original message.

The opening flag is followed by bit fields for address and control information. These are followed by the data field containing the message information. Between this and the closing field, there is an error-check field, which enables the receiving system to detect that a frame is in error and request retransmission. The error-check field contains 16 bits generated as a cyclic redundancy-check (CRC) code [15].

8.5.3 Signal units

Information to be sent is structured by the signalling-control unit (level 2) into a *signal unit* (SU). The SU is based on the HDLC protocol described in Section 8.5.2.

Signal units, as shown in Figure 8.16, are of three types:

(*a*) *Message signal unit (MSU)*
This transfers information between two signalling nodes. The length indicator is set to three or more octets.

(*b*) *Link-status signal unit (LSSU)*
This is used to set up or close down the signalling link or indicate and adjust its status. For example, it can indicate that the stores at the distant end are congested or that the processor is faulty. A pair of these messages is used to reset the link when necessary. The length indicator is set to one or two octets.

(*c*) *Fill-in signal unit (FISU)*
When there are no information or status messages to send, fill-in messages are transmitted to establish that the link is activated and to provide a bit-error check on the link. The length indicator is set to zero.

The format of the MSU is shown in Figure 8.16*a*. Messages are of variable length and are sent in 8-bit bytes (octets) as follows:

(*i*) Opening and closing flags are used to delimit a signal unit. They have the code pattern '01111110'.
(*ii*) The forward sequence number (FSN), backward sequence number (BSN), forward indicator bit (FIB) and backward indicator bit (BIB) are used for error correction, as described below.
(*iii*) The length indicator (LI) gives the length of the signal unit. A value of LI greater than two indicates that the SU is a message-signal unit.
(*iv*) The service-information octet (SIO) indicates the user part appropriate to the message (e.g. TUP, ISUP, DUP etc.).
(*v*) The signalling-information field (SIF) may consist of up to 272 octets and contains the information to be transmitted. Its format is defined by the user part, as described in Section 8.5.4.

(*vi*) The error-check field is immediately before the closing flag. It contains 16 bits generated as a cyclic redundancy check code [15].

Each time an information message is sent, the forward sequence number (FSN) is incremented. However, for status units and fill-in units, the FSN is not incremented and the last FSN is carried again. Response messages carry the FSN number in the backward-sequence-number (BSN) position. When the signalling link is normal, the FIB and BIB in the transmitted SU are set to the same value (e.g. zero). Level 2 of the receiving CCS system analyses the check bits and discards the message if it detects an error. If the message is not discarded, the received FSN is compared with the expected value (i.e. one more than the previous FSN) and the message is passed to level 3. The FSN is sent back as a BSN and the BIB provides a positive acknowledgement.

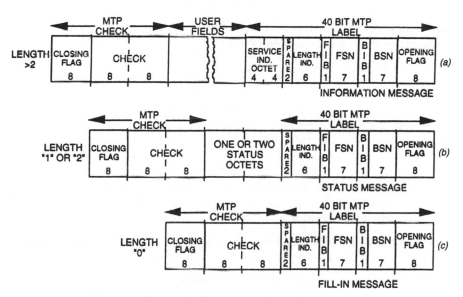

Figure 8.16   Formats of signal units in CCITT Signalling System 7

    a Message signalling unit
    b Link-status signalling unit
    c Fill-in unit

BIB	= backward-indicator bit
BSN	= backward-sequence number
FIB	= forward-indicator bit
FSN	= forward-sequence number
LI	= length indicator
SF	= status field
SIF	= signalling-information field
SIO	= service-information octet

If there has been an error and the message has been discarded, the comparison of FSNs at the receiving end indicates a discrepancy. The BIB is then inverted to provide a negative acknowledgement and the BSN has the value of the last correctly-received MSU. On receipt of the negative acknowledgement, the sending CCS system stops sending further SUs and retransmits previous MSUs in their original order.

8.5.4 The signalling information field

A message from the user part occupies the signal information field (SIF) of the MSU shown in Figure 8.16a. The relevant user part has been signalled by the preceding control field. The MSU can transmit a large number of different layer-3 messages. Those for the telephone-user part (TUP) are listed in Table 8.1.

An example is the initial address message (IAM). This is the first message to be sent, since there is no separate seize signal. The format is shown in Figure 8.17. Its fields, in order of transmission, are:

(*a*) A label of 40 bits (5 octets) containing:
   The destination-point code (14 bits)
   The originating-point code (14 bits)
   The circuit-identity code (12 bits).
(*b*) The H0/H1 octet. The H0 field (4 bits) identifies a general category of message (e.g. forward address message). The H1 field completes the definition of the message (e.g. variable-length IAM).
(*c*) The calling party category (e.g. ordinary customer or operator).
(*d*) The message indicator. This indicates any special requirements (e.g. whether a satellite link or echo suppressors will be used).
(*e*) The number of address signals (4 bits). This gives the number of address digits in the IAM.
(*f*) The address signal itself.

This signalling-information field is then followed by the check bits and the closing flag to complete the MSU.

8.6 Digital customer-line signalling

Digital transmission is used on customers' lines to provide access to an ISDN as described in Chapter 14. Each line may give access to several terminals on the customer's premises and common-channel signalling is used to serve them. Basic-rate access [16] provides, in each direction, two B channels at 64 kbit/s and a D channel at 16 kbit/s for signalling. Primary-rate access [17] provides, in each direction, 30 channels (in a 2 Mbit/s-based network) or 23 channels (in a 1.5 Mbit/s-based network), together with a 64 kbit/s signalling D channel.

The CCITT has defined the Digital Subscriber Signalling System No. 1

Table 8.1 Messages used in CCITT Signalling System 7

Abbreviation	Description
Used to establish a call or supplementary service:	
IAM	Initial Address Message
SAM	Subsequent Address Message
FAM	Final Address Message
IFAM	Initial and Final Address Message
ACI	Additional Call Information
SIM	Service Information Message
SNM	Send 'N' Digits/Send 'All' Digits
SASUI	Send Additional Set-Up Information
ASUI	Additional Set-Up Information
Indicates that the call has been successfully established:	
ACM	Address Complete Message
Indicates that the call has been answered:	
ANS	Answer
RE-ANS	Re-Answer
Indicates that the call has failed:	
CON	Congestion
TERM CON	Terminal Congestion
CNA	Connection Not Admitted
SUB-ENG	Subscriber Engaged
SUB-OOO	Subscriber Out Of Order
SUB-TFRD	Subscriber Transferred
RA	Repeat Attempt
Used to release a call:	
REL	Release
CLEAR	Clear
CCT-FREE	Circuit Free
Used to pass information during a call:	
UUD	User-User Data
NEED	Nodal End-to-End Data
Used to change from voice to data and vice-versa during a call:	
SWAP	Swap
Used to remove channels from service:	
BLKG	Blocking Messages
Miscellaneous:	
CFC	Coin-and-Fee Checking
HLR	Howler
TKO	Trunk Offer

first bit ⟶

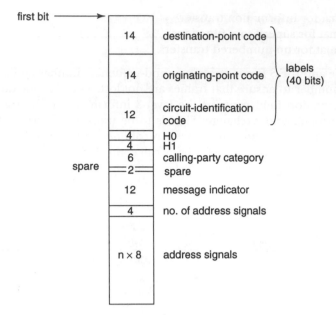

Figure 8.17   TUP initial-address-message format in CCITT Signalling System 7

The CCITT has defined the Digital Subscriber Signalling System No. 1 (DSS1) for signalling over the D channel [10,16]. The transfer of information in each direction between the customer's premises and the exchange is by messages, called *frames*, which are similar to the CCITT no. 7 signal units described in Section 8.5.3. The form of HDLC protocol used is known as LAP-D (link-access protocol for the D channel).

The format of a frame [10,18] is shown in Figure 8.15. Each frame begins and ends with an 8-bit flag and there is a 16-bit cyclic redundancy check field. The address field is not used for routing in the network; it selects ports at either end of the line. Its first octet contains the *service-access-point identifier,* which indicates the exchange terminal to be used (e.g. for a circuit-switched call or a packet-switched call). The second octet contains the *terminal-end-point identifier,* which identifies the equipment on the customer's premises to which the message refers. Each octet begins with an *extension bit.* If this is '0', it indicates that another octet is to follow; if it is '1', it indicates that the octet is the last. Normally, the extension bit of the first octet is '0' and that of the second octet is '1'. However, the latter can be changed to '0' to add a third octet, or more. The second bit in the address field is the *command/response bit* (C/R). If the frame is a command, it indicates the receiver (i.e. either the exchange or the customer). If the frame is a response, it indicates the sender.

The control field (one or two octets) indicates the type of frame being transmitted. There are three types of frame:

(*i*) I format for information transfer
(*ii*) S format for supervisory functions
(*iii*) U format for un-numbered transfers.

The I frame and the S frame contain a send-sequence number and a receive-sequence number to ensure that frames are not lost. The U frame does not.

The information field contains the layer-3 information to be transferred between customer and exchange. Signals to be transmitted for establishing and controlling connections have been recommended by the CCITT [19]. For example, I.451 messages are for basic call control [20] and I.452 messages to control supplementary services [21]. A list of messages used is given in Table 8.2.

Table 8.2 Messages used in I series

Abbreviation
Used to establish a call: SET-UP SET-UP ACKNOWLEDGE INFORMATION CALL PROCEEDING *Used to indicate that the called party is being alerted (e.g. rung):* ALERTING *Indicates that the call has been answered:* CONNECT CONNECT ACKNOWLEDGE *Used to pass information back to the originating exchange or PBX during call establishment:* PROGRESS *Used to release a call:* RELEASE RELEASE COMPLETE DISCONNECT

8.7 Private network common-channel signalling

Primary-rate access with common-channel signalling is used between digital PBXs and public exchanges. Common-channel signalling based on similar principles can also be used between PBXs in a digital private network. A system [22] known as DPNSS (digital-private-network signalling system) is widely used in the UK and other countries. When this system is applied to a private network, its PBXs appear to users as a single large PBX, with all its supplementary services available wherever the user is located. The messages used are listed in Table 8.3.

The DPNSS system was developed from a system known as DASS2, which was used for primary-rate access signalling in the UK until supplanted by I.421 in the early 1990s. The two forms of signalling can co-exist in channel 16 and inter-PBX private circuits can be mixed with circuits connected to the PSTN on the same 30-channel multiplex.

The lack of compatibility between DPNSS and I.421 is inconvenient and a successor is clearly needed. The European Computer Manufacturers Associa-

Table 8.3 Messages used in DPNSS1

Abbreviation	Description
Used to establish a call or supplementary service:	
ISRM	Initial Service Request
SSRM	Subsequent Service Request
Indicates that the call has been successfully established:	
NAM	Number Acknowledge
Indicates that the call has been answered:	
CCM	Call Connected
Used to release a call:	
CRM	Clear Request
CIM	Clear Indication
Used to pass information back to the originating PBX during call establishment:	
NIM	Network Indication
Used to signal between the originating and terminating PBXs without affecting intermediate transits:	
EEM	End to End
Used to change from voice to data and vice-versa during a call:	
SM	Swap
Used to signal between adjacent PBXs:	
LLM	Link by Link
LLRM	Link by Link Reject
Used to establish a second call on a channel, for example an enquiry call:	
RM	Recall
RRM	Recall Reject
ERM	End-to-End Recall
Used to release a second call on a channel:	
SCRM	Single-Channel Clear Request
SCIM	Single-Channel Clear Indication

tion (ECMA) has taken the lead in this, with a signalling system called QSig. This has been offered for international standardisation.

8.8 References

1. WELCH, S.: 'Signalling in telecommunications networks' (Peter Peregrinus, 1979)
2. ATKINSON, J.: 'Telephony. Vol. 2' (Pitman, 1950)
3. ATKINSON, J.: 'Telephony. Vol. 1' (Pitman, 1948)
4. BREEN, C., and DAHLBOM, C.A.: 'Signalling systems for control of telephone switching' *Bell Syst. Tech. J.,* 1960, **39,** pp. 1381–1444
5. NEWELL, N.A., and WEAVER, A.: 'Single-frequency signalling system for supervision and dialling over long-distance telephone trunks', *Trans. Am. Inst. Electr. Eng.,* 1951, **70,** pp. 489–495
6. GAZANION, H., and LEGARE, R.: 'Systemes de signalisation Socotel', *Commut. Electron.,* 1963, **4,** p. 32
7. DAHLBOM, C.A., HORTON, A. W., and MOODY, D.L.: 'Multi-frequency pulsing in switching', *Electr. Eng. (USA),* 1949, **68,** pp. 505–510
8. CCITT Recommendations Q.350–Q.368: 'Specifications of signalling system R2'
9. REDMILL, F.J., and VALDAR, A.R.: 'SPC digital telephone exchanges' (Peter Peregrinus, 1990)
10. MANTERFIELD, R.J.: 'Common-channel signalling' (Peter Peregrinus, 1991)
11. CCITT Recommendations Q.251–Q.300: 'Specifications of Signalling System No. 6'
12. CCITT Recommendations Q.700–Q.775: 'Specifications of CCITT Signalling System No. 7'
13. MITCHELL, D.C., and COLLAR, B.E.: 'CCITT Signalling System No. 7: National user part', *Br. Telecommun. Eng.,* 1988, **7,** pp. 19–31
14. ISO Standard 4335: 1993: 'Information technology — Telecommunications and information exchange between systems — high-level data-link control (HDLC) procedure — Elements of procedures'
15. BREWSTER, R.L.: 'Communication systems and computer networks' (Ellis Horwood, 1989)
16. CCITT Recommendation I.420: 'Basic user-network interface'
17. CCITT Recommendation I.421: 'Primary rate user-network interface'
18. CCITT Recommendation Q.921: 'ISDN user-network interface, data-link layer specification'
19. CCITT Recommendation I.450 (also Q.930): 'ISDN user-network interface layer 3 — general aspects'
20. CCITT Recommendation I.451 (also Q.931): 'ISDN user-network interface layer 3 — specification for basic call control'
21. CCITT Recommendation I.452 (also Q.932): 'Generic procedures for the control of supplementary services'
22. HIETT, A.E., and DANGERFIELD, W.: 'Private network signalling', *Comput. Commun.,* 1988, **11,** pp. 191–196

Chapter 9

Numbering, routing and charging

J. E. Flood

9.1 Introduction

When one telephone customer wishes to make a call to another, the number to be dialled is determined by the numbering plan of the operating administration. The route selected for setting up the connection depends on the call destination, and this must be determined by switching equipment from the number dialled. The charging rate for the call is also determined from its destination, as given by the digits dialled. Numbering, routing and call charging are therefore very closely related. It is essential that the plans developed by an administration for these three aspects of a public switched telecommunication network (PSTN) should be consistent with each other.

The plans must also be compatible with a national transmission plan and signalling plan (see Chapters 17 and 20). For example, either transmission or signalling standards may limit the size of a local area, or the number of links that may be used in tandem for a trunk connection.

The national plans for numbering, routing, charging, transmission and signalling thus form a set of inter-related standards which governs planning both for a national network and its constituent local networks. They must also cater for international calls passing between the network and the networks of other countries. In countries where there is competition between operating companies, these plans must also cater for interconnection of the rival networks [1].

9.2 Numbering plans

9.2.1 Basic principles

The object of a numbering plan is to allocate a unique number to each customer connected to the network. At first, each numbering scheme applied only to a single exchange, and exchanges were identified by the names of their towns. Following the introduction of automatic telephony, *linked numbering schemes* were applied in multi-exchange areas. In a linked numbering scheme, the 'local' numbering scheme covers a number of exchanges, so a call from any exchange in the area uses the same number to reach a particular customer. The first part of the directory number is an exchange code and the remainder is the customers' number on that exchange. For example, a 6-digit linked numbering scheme has a theoretical capacity for 100 4-digit exchanges. In practice, this is reduced by the need to allocate codes for access to various services.

The subsequent introduction of *direct distance dialling* [2] (DDD) or *subscriber trunk dialling* [3] (STD) for long-distance calls required the development of national numbering plans. Later, the introduction of *international subscriber dialling* (ISD) made it necessary for national numbering plans to conform to an international numbering plan [4].

A national numbering plan must make a generous allowance for growth in the number of customers for up to 50 years ahead. However, the number of digits in any *national number* must not exceed the maximum recommended by the ITU. The number plan must also be considered in relation to division of the country into areas for tariff purposes. The boundaries of numbering areas should thus coincide with call-charging boundaries. Consideration should also be given to the pattern of traffic flow to make the dialling procedure as simple as possible for the majority of calls.

Numbering plans may be either 'open' or 'closed'. An *open numbering plan* has no fixed number of digits. Such a plan results from a network of step-by-step exchanges without registers; the number of digits to be dialled for each call is determined by the number of switching stages it is routed through. A *closed numbering plan* has a fixed number of digits to be dialled for all calls, regardless of the geographical positions in the network of the calling and called customers. The need to use a fixed number of digits arose with the introduction of register-controlled systems. The registers had a limited digit-storage capacity, and relied on this being full to indicate that sufficient information had been received to enable a connection to be set up and the register released. More modern systems can cater for numbers of variable length. Many national numbering plans therefore contain numbers having several different lengths, up to a given maximum number of digits.

There must always be an upper limit to the maximum number of digits in a national number, to comply with international requirements. The ITU has recommended that the maximum number of digits dialled for an inter-national call* should be 12. The maximum number of digits in a national number is thus 12-N, where N is the number of digits of the country's code in the world numbering plan.

Since each customer has a directory number in his or her own exchange, a unique national number can be obtained by adding to this number a code identifying the exchange. These codes could be allocated in a purely arbitrary manner. This has the advantage of being able to accommodate the largest possible number of exchanges within the number of digits available, and

---

* This excludes any prefix digits used in the calling country to indicate an international call. Details of the international numbering plan are given in Section 9.2.4.

making it easy to cater for forecast changes and unforeseen developments in any area of the country. It has the disadvantage that each originating trunk exchange must be able to identify the code of every local exchange in the country to determine the routings and charges for calls. Alternatively, the country may be divided into geographical areas, each having a separate code and containing a number of local exchanges. The originating trunk exchange then only requires to identify the area code to determine the routing and charging rate for each call. The latter method is clearly preferable, unless the country is very small.

In general, a national number consists of three parts:

(*a*) *An area code*
This identifies the particular numbering area of the called customer, and thus determines the routing for a trunk call and the charge for it.
(*b*) *An exchange code*
This identifies the particular local exchange within the area, and so determines the routing from the incoming trunk exchange of the area to the terminating local exchange.
(*c*) *The customer's local-exchange number*
This selects the called customer's line at the terminating exchange.

In addition to dividing the country into numbering areas (each with an area code) and determining a local numbering plan for each area, it is necessary to determine appropriate dialling procedures for both trunk and local calls.

There are three possible approaches to dialling procedures:

(*a*) a single procedure, whereby local and trunk calls are obtained by dialling the national number;
(*b*) two procedures, whereby local calls are obtained by dialling the local number and all other calls by dialling the national number; and
(*c*) three procedures, in which special short dialling codes are used for adjacent numbering areas.

The use of a single dialling procedure (i.e. the full national number is dialled for both trunk and local calls) is only justified if trunk traffic represents a high proportion of the total traffic or if the country is sufficiently small for national numbers to contain few digits. The second method is therefore usually adopted. The third method also uses the two main procedures, but it supplements these with special short codes to avoid having to dial full national numbers for calls between exchanges in adjacent areas which have a large amount of traffic between them. Since these codes apply only at particular originating exchanges, customers at these exchanges must be given special dialling instructions. The use of short access codes also removes numbers from the numbering scheme of the area concerned, and thus reduces the total number of customers it can accommodate.

When different dialling procedures are used for local calls and trunk calls, it is obviously necessary for the originating local exchange to be able to differentiate between them. One method is for the originating local exchange to determine this from the area and exchange codes. This implies that no combination of digits can be used both in an area code and in an exchange code, thus reducing the number of areas that the number plan can accommodate and the number of exchanges in each. It is also necessary for the local exchange to be equipped with suitable registers to receive the code digits and make the necessary discrimination.

An alternative method, which has been adopted by most countries, is to make use of a *trunk prefix*. For a 'within-area' call, the calling customer dials only the exchange code followed by the number of the called customer. For a trunk call, the calling customer dials the prefix (e.g. the digit '0') before the full national number of the called customer. Recognition of the prefix causes the originating exchange to route the call to a trunk exchange. By this means, only codes beginning with the digit chosen for the prefix are excluded from the area numbering scheme. Also, registers to recognise area codes are required only in trunk exchanges. The method can thus be used with non-register local exchanges.

An additional prefix can also be used to differentiate between trunk calls and international calls. For example, if the *international prefix* is '00', the first digit (0) causes the call to be routed to a trunk exchange and the next digit (0) causes the call to be switched to the international gateway exchange. In dialling, the two-digit prefix is followed by the international code for the required country and the national number of the called customer in that country.

If numbering areas are large, wide variations will arise in the distances available for the same charge, which may lead to complaints from customers. Large areas necessitate long and expensive junctions from peripheral local exchanges to central trunk-switching centres. Also, an area must not be so large that the number of customers it will ultimately serve is beyond its numbering capacity. Numbering areas containing a large number of customers necessitate more digits being dialled for calls within the area, but fewer digits for trunk calls to other areas.

If numbering areas are small, junctions from local exchanges to trunk exchanges will be short and inexpensive. However, the number of trunk switching centres will be large. If areas are small, nearby exchanges having considerable traffic between them may be in different areas, thus complicating the dialling procedure for these calls. Moreover, if areas are small, there may be too many of them for the number of area codes available in the national numbering plan.

Historically, local-area numbering schemes have preceded national numbering plans. Numbering areas in national plans have therefore usually been based on existing local areas. An area may be served by a single local exchange, having an independent numbering scheme. Alternatively, the numbering schemes of a group of local exchanges may be integrated into a single linked numbering plan.

A linked numbering scheme allows a uniform dialling procedure to be used between all customers in the numbering area. However, it is necessary to apportion the numbering range to the individual exchanges so as to allow for the different growth rates likely to be encountered. The numbering plan must also allow additional exchanges to be added when the increase in telephone density in particular locations makes this desirable. The numbering range available for individual exchanges is reduced by the use of code digits for trunk and international prefixes and for special services (e.g. operator assistance, enquiries, etc.). For the convenience of customers, these codes should be standardised throughout the country (e.g. the use of '100' for operator assistance and '999' for emergency services).

In areas using register-controlled exchanges, either two-digit or three-digit exchange codes may be used, depending on the number of exchanges in the area. In the UK, director areas use seven-digit numbers, each comprising a three-digit exchange code and four digits for the customer's number. The theoretical maximum number of subscribers is thus 10 million. In practice, the need to allocate codes to special services etc. reduces the capacity to about four million.

In preparing numbering plans for areas, forecasts of the ultimate telephone population should make allowance for direct dialling-in (DDI) to PABX extensions. This facility is desirable for long-distance calls to PABX extensions, since the caller is not charged for the time waiting for a PABX operator to establish a connection to the wanted extension. There are two methods of providing for DDI:

(*a*) Sub-addressing may be used. One or two suffix digits may be added to the national number to give DDI access to 10 or 100 extensions, respectively. This effectively increases the length of the customer's number and reduces the number of digits available for exchange and area codes.

(*b*) A national number may be allotted to each PABX extension. This reduces the number of customers who can be accommodated within the numbering scheme of the local exchange.

In practice, growth is not uniform and numbering schemes in some areas become exhausted before others. For example, in the UK, the London area, which previously had a single code, has had to be split into two: '71' for inner London and '81' for outer London [5]. Similar changes have been made in a number of American cities. In France, the entire country was divided into two numbering zones: L'Isle de France covering the region around Paris and La Province covering the rest of France.

In countries where telecommunications has been 'liberalised' there are several competing telecommunication operators, each with its own network. It may then be necessary to use additional prefix digits to enable a customer to select a particular network for a call. For example, in the UK, a Mercury customer whose line is connected to a BT local exchange uses the code '132' to make a call via the Mercury network. In practice, these digits are generated automatically by a telephone set with a 'blue button', or a customer's PABX may be programmed to send them.

If a customer decides to obtain service from a different network operator having a separate access network, this results in the telephone being connected to a different local exchange; thus, the customer's number is changed. This is a disincentive to customers changing their network operators and this inhibits competition. Consequently, there is a demand for *number portability*, so that customers can change from one network to another without having to change their numbers. The customer's number is required to be ported unchanged from a donor network to a recipient network. The routing methods required to provide number portability are considered in Section 9.3.4.

9.2.2 UK numbering plan

The UK numbering plan [6] introduced with the advent of STD caters for about 700 numbering areas, each of which can contain up to 700 000 customers. The plan therefore has an ultimate capacity of about 500 million numbers. A typical area covers about 200 square miles. The numbering areas also serve as charging areas. However, in some areas having a large number of local exchanges, a single charging group is divided into two or more numbering groups.

The national numbering plan is now administered by the Office of Telecommunications (OFTEL) and is shared by competing network operators. The largest cities have a uniform seven-digit numbering scheme, consisting of a three-digit exchange code and a four-digit customer's number. These areas are identified by two-digit area codes. Other areas have three-digit area codes and numbering schemes with uniform six-digit numbers or mixed schemes not exceeding six digits. The total number of digits, excluding the trunk prefix, thus varies between eight and nine. The long-term objective is to have a national number of fixed length with six-digit or seven-digit local numbers [7].

In April 1995, the numbering scheme was expanded to allow for the continuing growth of competing services. The scheme for the fixed PSTN was retained (except that seven-digit numbers replaced six-digit numbers for five more cities). However, the trunk prefix changed from '0' to '01', to allow customers access to other networks by using prefixes of the form '0X', as shown in Table 9.1.

It will be noted that the scheme has capacity for the introduction of new services. Also, introduction of the prefix '02' allows for growth in the present PSTN. It has been proposed by OFTEL that the prefix used for numbers in London, Belfast, Cardiff, Portsmouth, Reading and Southampton shall change from '01' to '02' in the year 2000. Changes for other cities may follow later.

Table 9.1 UK prefix codes

01	Current geographic PSTN, some paging services
02	Future geographic scheme
03	Mobile services: cellular, PCN, paging
04	Unallocated
05	Freephone
06	Unallocated
07	Personal numbering
08	Specially-tariffed services: premium-rate, national calls charged at local rate, freephone, cellular
09	Unallocated
00	International access

9.2.3 North American numbering plan

A single numbering plan was designed to cover the whole of North America, including the USA, Canada, Mexico and the Caribbean [8]. The territory is divided into number-plan areas, each assigned a unique three-digit code. The size of an area depends on the population density and the forecast of ultimate telephone penetration. In the USA, each state contains from one to five. Within a number-plan area, each central office (local exchange) has a three-digit code and four-digit customer's-station numbers. Thus, each customer has a 10-digit national number. This plan accommodates 792 numbering areas, each of which can accommodate 792 local exchanges of 10 000 lines and it has capacity for 6300 million numbers.

Two prefixes are used for toll calls:

'1' for customer-dialled station-to-station calls
'0' for customer-dialled calls requiring operator assistance (e.g. person-to-person or reverse-charge calls).

In the USA, a customer may choose one of the competing long-distance carriers by dialling a three-digit *carrier-identification code* (CIC) after the toll prefix, e.g. '288' for AT&T or '222' for MCI. However, a customer on an SPC local office may choose a particular carrier in advance for all long-distance calls. This information is stored in the database of the office and causes toll calls to be routed to the chosen carrier. It may be necessary, at some future date, to increase the length of carrier-identification codes to four digits to accommodate further growth of competing services.

9.2.4 International numbering plan

The introduction of international subscriber dialling (ISD) made it necessary for every customer's station in the world to be identified by a unique number.

Table 9.2 World numbering zones

Code	Zone
1	North America (including Hawaii and the Caribbean Islands, except Cuba)
2	Africa
3 & 4	Europe
5	South America and Cuba
6	South Pacific (Australasia)
7	CIS (formerly USSR)
8	North Pacific (Eastern Asia)
9	Far East and Middle East
0	Spare code

The CCITT therefore prepared a *world numbering plan* [4]. Each world telephone number consists of a *country code* followed by the customer's national number. These country codes contain one, two or three digits.

For numbering purposes, the world is divided into zones, each given a single-digit code. Each country within a zone has the zone number as the first digit of its country code. However, the European numbering zone was allocated two codes because of the large number of country codes required within this zone. The codes for the world numbering zones are listed in Table 9.2.

Within each zone, every country has been allotted a single-, two- or three-digit code number. For example, within zone 3 (Europe), the Netherlands has the code '31' and Albania has '355'. The three-digit codes have been allocated to the smaller countries, having fewer digits in their national numbering plans, to minimise the total number of digits in customers' international numbers. An exception occurs where an integrated numbering plan already covers an entire zone; countries in these zones require only a single-digit code. Thus, '1' is the country code for all countries in the North American numbering plan.

The complete international number of a customer consists of the country code followed by the national number. Some examples of these international numbers are given in Table 9.3.

The existence of a world numbering plan places restrictions on the national numbering plan of each country to enable customers in other countries to make international calls into its national network. The restrictions necessary are as follows:

(*a*) The number of digits in a customer's world number is limited to a maximum of 12. The number of digits available for a national numbering plan is thus 12-N, where N is the number of digits in the country code.

(*b*) National numbers must contain all numbers and no letters. Countries

Table 9.3 Examples of typical international numbers

Zone	Country	Country code	Number of digits in national number	Total number of digits
1	USA	1	10	11
1	Canada	1	10	11
2	Egypt	20	8	10
2	Liberia	231	6	9
3	France	33	8	10
3	Portugal	351	7 or 8	10 or 11
4	UK	44	8 or 9	11 or 12
4	Switzerland	41	8	10
5	Brazil	55	9	11
5	Ecuador	593	7	10

which used letters on telephone dials chose different ways of allocating the 26 letters of the alphabet among the 10 decimal-number characters. Thus, area and exchange codes containing letters could be dialled as different numbers by subscribers making calls from outside the country. Administrations which previously used codes containing letters (e.g. the USA and the UK) therefore changed to all-number codes to permit ISD calls to and from their countries. (*c*) National numbering plans must include international prefix codes, to avoid ambiguity when national and international numbers have the same initial digits. For example, the UK uses the prefix '00' for ISD calls. North America uses '01' for overseas calls requiring operator assistance and '011' for overseas station-to-station calls.

9.2.5 Numbering plans for the ISDN era

In 1984 the CCITT made recommendations for an international numbering plan for integrated-services digital networks [9] (ISDN). This extends the existing numbering plans for PSTNs, since a customer's ISDN access will normally be indistinguishable from PSTN access and be provided by the same local exchange.

The maximum length of international numbers is increased to 15 digits. The area code becomes a *network destination code* (NDC). This recognises that some 'area' codes have been used for access to mobile networks and special services that are nongeographic. An international exchange will analyse six digits to choose a route. Thus, these can include digits within the NDC to route a call to the appropriate network in the destination country.

It is intended to provide for *subaddressing*. Additional digits following a customer's number will be sent to the destination for use on the customer's premises. These can select the appropriate ISDN terminal or provide DDI on a PBX without reducing the numbers available in the area numbering scheme. The subaddress field can range from four digits for normal applications to 40 digits in open-systems interconnection (OSI). For OSI, a global scheme for identifying network-service access points has been developed by the International Standards Organisation (ISO) [7].

The CCITT recommended implementation of this plan by the end of 1996. It was hoped that SPC digital exchanges will then have penetrated all countries and be able to handle 15-digit international numbers and the increased number of digits to be analysed.

9.2.6 Public data networks

For public switched data networks [10], an international number consists of two parts: the *data-network identification code* (DNIC) of four digits and the *network terminal number* (NTN) of 10 digits. This gives a total of 14 digits. The DNIC consists of a *data country code* (DCC) of three digits, followed by a 'network' digit, thus enabling a country to have up to 10 public data networks. Countries with more than 10 networks can have more than one DCC. For example, the UK has been allocated '234' to '237' and so can have up to 40 public data networks.

The format of the 10-digit NTN can be decided by the network operator, because only the DNIC needs to be analysed to route a call to that network from another country. The CCITT recommended that the last two digits should be a subaddress for use by the customer. In the British Telecom packet-switched service (PSS), the first three digits identify the packet-switching exchange and the next five identify the customer's network termination [7]. The final two digits are for a subaddress on the customer's premises.

9.3 Routing plans

9.3.1 Basic principles

National networks are hierarchical, as shown in Figure 1.3. The minimum configuration of star-connected exchanges is usually augmented by *direct routes* interconnecting those exchanges where a high community of interest results in sufficient traffic. There is thus a *backbone route* joining each switching centre at the lowest level to the highest-level centre via intermediate centres, together with *transverse branches* between centres at the same level. There may also be some other direct routes between centres at different levels which violate this pattern.

In some countries *automatic alternative routing* (AAR) is used. Direct routes are underprovided with circuits; when all circuits on a direct route are busy, traffic overflows to a fully-provided tandem route through a switching centre at a higher level in the hierarchy. An underprovided direct route is called a *high-usage* route and the fully-provided indirect route to which its traffic finally overflows is called a *final route*. Only a small proportion of connections use the complete backbone of final routes, since transverse routes are used whenever there are free circuits [11].

Large groups of circuits are more efficient than small groups because of their higher occupancy (i.e. traffic per circuit), as explained in Section 10.4.2. If there is a large amount of traffic between two exchanges, it is economical to provide a direct route between them. If there is less traffic, it becomes economical to use automatic alternative routing; the direct circuits are used efficiently and the overflow traffic shares tandem circuits with traffic to other destinations. If there is very little traffic between two exchanges, it is uneconomical to provide any direct circuits and all calls should be routed via a tandem exchange. The traffic levels for which these three different routing methods should be used depend on the relative costs of direct and tandem circuits (including associated switching and signalling equipment), as shown [12] in Figure 9.1. If automatic alternative routing is to be used, the numbers of high-usage and tandem circuits required can be determined by methods described in Section 10.5.2.

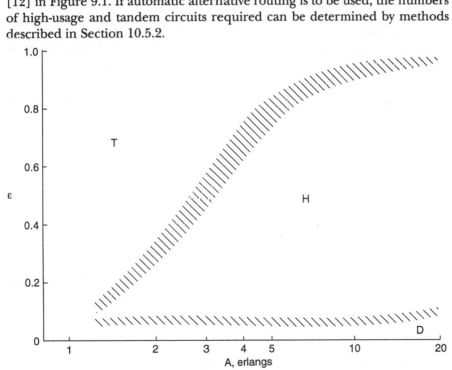

Figure 9.1   *Domain for employment of tandem (T), high-usage (H) and direct (D) circuits as a function of traffic (A) and cost ratio between high-usage and tandem routes (ε)*

B = marginal cost per circuit on direct route, B_1, B_2 = marginal costs of circuits forming a tandem route
Minimum number of circuits on high-usage routes = 4
$\varepsilon = B / (B_1 + B_2)$
From Reference 12

Some saving in line plant can be obtained by both-way working, since the total traffic requires fewer circuits when they are in a single both-way group than when they are divided into two unidirectional groups. However, when electromechanical exchanges are used, this is offset by the increased cost of signalling equipment and the possibility of calls failing owing to 'collisions' (when simultaneous attempts are made to set up connections from both ends of a circuit). Both-way working is therefore normally only used when small groups of circuits are required over routes that are long and expensive. When a network has SPC exchanges, their use of common-channel signalling obviates these difficulties and enables both-way routes to be provided extensively.

In addition to being influenced by cost and traffic considerations, the choice between direct and indirect routing may be affected by the transmission and signalling plans. These may limit the maximum number of tandem links that may be used to set up connections.

In networks using step-by-step exchanges, routing is also affected by the numbering plan, since each switching stage in a connection corresponds to a digit of the called customer's number. The use of registers divorces routing from numbering, provided that the storage and translation capacities of registers are not exceeded. For this reason, most administrations have adopted register control for their trunk network, even when local areas use nonregister systems.

If it is assumed that a country has already been divided into local-exchange areas and the locations of their exchanges have been decided, a routing plan should be developed to determine:

(a) which local exchanges should be interconnected by direct junctions, and which connections made indirectly via tandem switching centres;
(b) the number and location of tandem switching centres;
(c) the number of levels of tandem switching to be used in the network; and
(d) whether automatic alternative routing is to be used and, if so, under what conditions.

In designing a routing plan, consideration must be given to the following factors:

(i) the transmission plan
(ii) the signalling plan
(iii) the numbering plan
(iv) the grade of service required end to end
(v) the capabilities of the type of switching equipment employed
(vi) the economics of direct and indirect routing.

In a national plan, the routing of traffic can usually be considered separately for the trunk and local networks. However, the routing plan for a local area must be compatible with the national transmission and numbering plans, because long-distance connections are set up through both the local network and the trunk network, possibly involving the maximum number of links in tandem.

The interface between a local network and the trunk network is the highest-ranking exchange in the local-network hierarchy, which is also the lowest-ranking exchange in the trunk network. In CCITT terminology, this exchange is called a *primary centre*. Different nomenclature is used in various national networks. For example, in the BT network it is called a digital main switching unit (DMSU) and in North America it is called a toll centre or class-four office.

9.3.2 Local-area routing

Within a number-plan area, the routing plan is influenced by the size of the area, the volume of forecast traffic to be handled, the locations of exchanges and the community of interest between them. The routing plan is also influenced by the choice of switching system. When a nonregister step-by-step switching system is used, the routings that can be used are limited by the number of digits in the numbering plan for the area.

In a small isolated town there may be a single telephone exchange. If this exchange is small, its trunk traffic will be routed to a primary centre in a larger town, which may also provide operator service. The primary centre will also provide tandem switching for the majority of traffic from the local exchange, but there may also be a small number of direct routes to other local exchanges. If a step-by-step switching system is used, all these routings need to be accommodated in the exchange numbering scheme. Since a single-digit trunk prefix is used, there must be a direct route to the primary centre.

In a larger area, there may be several local exchanges in a linked numbering scheme. These may comprise a main exchange with several satellites. In addition to catering for its own customers, the main exchange acts as an *area tandem exchange* and as a primary centre providing access to the trunk network. In a linked-numbering area with nonregister step-by-step exchanges, a call from one satellite exchange to another was routed to the main exchange, where the first one or two dialled digits selected the required outgoing route. In this simple form, known as *trombone working*, an own-exchange call thus used two junctions between the satellite and the main exchange. This could be avoided by using the *drop-back* principle. A *discriminating selector* in the satellite exchange recognised an own-exchange call by the first one or two dialled digits and released the junction to the main exchange. The subsequent digits then completed the connection to the called line.

In a city, there may be a number of large exchanges, each having a 4-digit number scheme. Each exchange will have direct junctions to some others and an area tandem exchange will cater for connections between those exchanges which do not have direct junctions between them. Since each local exchange will have a cable route to the tandem exchange, this is an obvious economic choice for the location of the primary centre.

In the largest cities, the number of local exchanges makes it economic to use several tandem exchanges and the level of trunk traffic may justify each of these acting as a primary centre for its sector of the city [13,14]. Outgoing trunk calls are dealt with in the normal way. Incoming trunk calls must be routed to the correct sector, but they all share the same area code. However, it is possible to arrange for all local-exchange codes in a sector to have the same initial digit. If the city has a two-digit area code, a distant trunk exchange can analyse the first three digits of a customer's national number to determine both the city and the appropriate sector.

The equipment for a digital switching system is smaller than for an electromechanical system, so an exchange building can serve more customers' lines. If customer's line units can be located remote from the exchange, still more customers can be served (e.g. 50 000 lines rather than 10 000). This favours the use of remote switching units, concentrators and multiplexers in access networks. Consequently, as shown in Figure 2.9, this results in fewer and larger exchanges, serving much larger distribution areas. Since these larger exchanges generate more junction and trunk traffic, this justifies each having junction routes to two or more primary centres, as shown in Figure 2.10. This improves the resilience of the network, i.e. its ability to cope with failures and overload.

9.3.3 Trunk-network routing

The number of levels in a trunk network depends on the size of the network and on the relative costs of switching and transmission. In a small densely-populated country, transmission costs are low and traffic levels between centres are high. Consequently, there are many direct routes between centres and few levels in the hierarchy. In a large sparsely-populated country, transmission costs are higher, traffic levels between centres will be lower and fewer direct circuits are justified. The network will therefore conform more closely to a hierarchical structure.

During recent years, there have been enormous reductions in transmission costs. For example, the introduction of high-capacity optical-fibre systems has greatly reduced the per-channel-kilometre cost of transmission and the introduction of digital switching has eliminated the cost of channelling equipment to terminate transmission links at switching centres. This justifies many more direct routes and results in a hierarchy with fewer levels. It also makes it economic for an exchange to have routes to two (or more) switching centres at the next higher level, instead of the single backbone route of a traditional analogue network. If one of these routes breaks down, the other can still carry traffic, thus increasing the resilience of the network.

The integrated digital trunk network of British Telecom is shown in Figure 2.10. It is nonhierarchical; there is a single level of trunk exchanges, called digital main switching units (DMSU), and these form a fully-interconnected mesh. Main local exchanges, called digital cell-centre exchanges (DCCE), each have direct routes to two or more DMSUs. The integrated digital long-distance network of AT&T is also nonhierarchical [15]. As shown in Figure 9.2, one class of tandem switching centre replaces both the sectional and regional centres of the previous analogue network. However, traffic entering this network continues to be concentrated by a hierarchy of toll centres and primary centres.

In a traditional hierarchical network, each exchange normally has only one backbone route to others at higher levels. However, an integrated digital trunk network is usually nonhierarchical. If the network is fully interconnected, a direct route between two exchanges can have an alternative via any other exchange. Thus, each direct route between two exchanges can also form part of a tandem route between other pairs of exchanges. This gives a very large number of options. Each exchange has a routing table containing all its permitted options and will select from these in some preferred order.

Common-channel signalling is employed between the processors of the exchanges. This enables drop back, which is also known as *crank back*, to be used. If a connection is required between exchanges A and B and all circuits

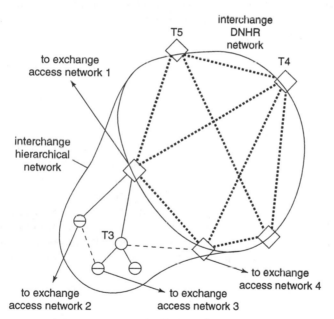

Figure 9.2   AT&T digital integrated toll network

◇ DNHR tandem
◯ primary centre
⊖ toll centre
● end office
· · · DHHR routes
---- high usage (first choice)
—— final route (last choice)
From Reference 15

on the direct route A–B are busy, then exchange A attempts to route the call via exchange C if there are free circuits on route A–C. However, the attempt will be unsuccessful if there is congestion on route C–B. Exchange C signals this information back to A. On receipt of the message, exchange A releases the connection A–C and makes a new attempt, say over route A–D–B.

Since modern networks have SPC exchanges, their routing translations can readily be changed. Moreover, the use of common-channel signalling enables the routing tables of exchanges to be changed remotely. The common-channel signalling network can link the processors of the exchanges to a central *network-management centre* (NMC). As described in Chapter 18, this can monitor the traffic on all the routes at frequent intervals to enable its staff to change routings in order to bypass failures and congestion.

The ability to change routing tables in exchanges permits *dynamic routing*, whereby the preferred choices of routes are changed from time to time. Noncoincident busy hours make it advantageous to use indirect routes over links that are highly loaded by direct traffic. For example, the different time zones of the east and west of the USA result in little traffic between New York and Los Angeles and between Washington and Los Angeles in the morning when the route between New York and Washington gets busy. Thus, circuits on the indirect route New York–Los Angeles–Washington can supplement those on the direct route New York–Washington. It has been estimated [16] that the use of dynamic routing in the AT&T long-distance network provides a cost saving of about 15%.

In the form of dynamic routing described above, the routes are predetermined on the basis of traffic forecasts for different times of the day. More sophisticated methods are also possible, in which the choice of routes is automatically adapted according to the traffic conditions actually encountered. Bell Canada [17] has used a method in which a central routing processor periodically receives information from exchanges as to the number of idle trunks on each of their outgoing routes. Routes are then selected centrally and the choices signalled back to the originating exchanges.

The AT&T toll network uses a distributed dynamic alternative-routing scheme [18] known as *real-time network routing* (RTNR). Every switching centre has a table of data containing the load condition of each of its direct routes and the two-link alternatives. These tables are updated by messages sent between the centres over the common-channel signalling network. When a direct route is busy, the switching centre checks the available capacity of each of the alternatives and selects that which is least-heavily loaded. In making the selection, account is taken of the class of service of the call, which determines the bandwidth required. An ISDN call may require 64 kbit/s for voice or switched data, 384 kbit/s for an H_0 channel or 1536 kbit/s for an H_{11} channel connection.

British Telecom has a distributed dynamic alternative-routing scheme which uses a learning approach [19] and requires no central control. When all trunks are busy on a direct route, the alternative two-link route which was last used successfully is selected again. If this is busy, a new alternative route is selected at random.

Use of automatic alternative routing increases the traffic capacity of a nonhierarchical network under normal conditions. However, once the total traffic exceeds a certain level, the traffic actually carried begins to decrease with further increase in offered traffic. An analysis of this behaviour is complex [20,21,22]. However, a simple explanation is that a large number of calls are then using alternative routes. Each such connection uses two trunks in tandem; thus one call being carried can prevent two new direct connections from being made. Clearly, this situation would be exacerbated if connections were made over more than two links in tandem. In practice, therefore, AAR in nonhierarchical networks is restricted to routings with only two tandem links.

Measures must be taken to prevent overload reducing network capacity. One method is to use trunk reservation. Calls overflowing from a direct route are blocked whenever there are fewer than a certain number of circuits free on the alternative route. However, first-choice traffic always has access to all circuits on that route. It has been seen that use of common-channel signalling allows both-way working on routes between exchanges. Thus, an overload of traffic from A to B may prevent calls from B to A. This can be avoided by reserving some of the trunks for use in one direction only.

Another strategy for preventing overload is *call gapping*. This is useful in the case of a *focused overload*, due to failure of an exchange or transmission link. It may also be due to an exceptionally large amount of traffic to one destination. This may be caused, for example, by a local disaster or by a 'phone in' programme on radio or television inviting people to make calls to a particular telephone number. If there is severe congestion in one part of the network, most attempts to route calls into it will be unsuccessful and will only cause congestion to spread elsewhere. It is therefore better to stop these calls where they originate. The NMC can instruct the processors of originating exchanges to throttle back this traffic by allowing through only one call every t seconds, where t is varied according to the capacity of the destination to receive calls.

In a country where there are competing long-distance networks, customers obtain access to each of them via a trunk exchange. In the UK, this is the DMSU. In the USA, customers in a *local-access and transport area* (LATA) obtain access to the long-distance operators, known as *inter-exchange carriers* (IXC), at a single switching centre known as a *point of presence* (POP).

9.3.4 Call routing for number portability

In countries, such as the UK, where there is competition between network operators, the rival networks are usually interconnected at trunk exchanges. If number portability is provided for customers who change from one network to another, a local exchange must recognise if a number that it receives for a call has been ported, i.e. it is a number of a customer who has transferred to another network. The call must then be routed to the other network via a trunk exchange. Thus, even calls which would previously have been junction calls or own-exchange calls require trunk connections.

When a number has been ported to another network, the identity of the exchange to which the number has been transferred is recorded in the database of the donor local exchange. On receiving an incoming call, the local exchange sends the customer's number to the trunk exchange preceded by a code (e.g. 5xxxxx) in which the first digit ('5') indicates that the number has been ported and the subsequent digits ('xxxxx') identify the exchange to which it has been ported. The trunk exchange routes the call to the recipient network, which then routes it to the correct local exchange. Here, the original customer's number is translated to connect the call to the customer's line.

If the call to the donor local exchange is an incoming trunk call, there are two methods of routing it to the recipient exchange. The first method is tromboning. The connection from the trunk exchange to the local exchange is retained and a further connection is set up from the local exchange to the trunk exchange to route the call into the recipient network. Thus, two trunk connections are held for the duration of the call. The second method uses drop back. On receiving the code '5xxxxx', the trunk exchange routes the call directly to the recipient network and releases the connection to the donor local exchange. Thus, only one trunk connection is retained. This results in cost savings, but additional software is required to implement the method. There are, of course, no cost savings for own-exchange and junction calls, since these require only one trunk connection.

As described above, the databases containing the routing translations for ported calls are located at local exchanges. In an intelligent network, as described in Chapter 16, exchanges can obtain access to a remote centralised database via a signalling network. This can be used to provide number portability. It would then be unnecessary to route a trunk or junction call as far as the donor local exchange before obtaining the '5xxxxx' translation. However, compared with tromboning or drop back, it would introduce many additional signalling messages and database accesses, since these would also be required for calls to customers whose numbers have not been ported.

9.3.5 International routing

A world network was developed for international subscriber dialling [4]. It had a hierarchy with three levels of international exchanges, known as *centres du transit* (CT) and designated CT1, CT2 and CT3. National networks were connected to the international network at centres CT3 (international gateways). Centres CT1 and CT2 connected only international circuits and the centres CT1 were fully interconnected. The centres CT1 were few and served a very large geographical area, such as a whole continent. Centres CT2 were located in the principal countries of each CT1 zone and every country had one or more centres CT3.

The introduction of integrated digital transmission and switching, together with common-channel signalling, has now produced similar changes in international routing to those described in Section 9.3.3 for routing in national networks. The great reduction in transmission costs (for example by the introduction of satellite links with small earth stations) has led to many more direct routes between countries. Thus, the international network has become mainly nonhierarchical and fewer calls require tandem routing. Automatic alternative routing is now used extensively and common-channel signalling enables crankback to be employed.

In countries with more than one gateway exchange, the *proportional-traffic-distribution facility* (PTDF) may be used. For example, if there are 100 circuits to another country and 25 of these are on gateway switch A, 40 on switch B and 35 on switch C, then traffic from the national network is distributed so that 25% is sent to A, 40% to B and 35% to C. This ensures that the international circuits are used to maximum effectiveness.

In the international network, it is particularly advantageous to make use of time-zone differences by employing dynamic routing. For example, it is economic to route traffic between the UK and other European countries via the USA in the morning when the time difference ensures that there is little direct traffic between Europe and North America. Similarly, the USA may use the UK as a tandem to route calls to Pacific-rim countries while it is night time in Europe.

9.4 Charging plans

9.4.1 Basic principles

The expenses of providing a telecommunications service consist of capital costs and current operating costs. The capital costs include those of line plant, switching equipment, buildings and land. The operating costs include staff salaries, the cost of maintaining equipment and the cost of energy consumed. All these costs must be met by the income obtained by the operating administration from its customers.

It is equitable that the charges paid by each customer should be related, if possible, to that proportion of each of the above costs incurred in providing the service. For this reason, the charges that are made to the customer are levied in the following ways:

(*a*) an initial charge for installing the customer's termination and connecting it to the network;
(*b*) an annual rental or leasing charge; and
(*c*) charges for individual calls made.

The customer's share in capital costs may either be covered by the connection charge or part of the rental may cover the necessary interest and repayment charges. Part of the operating costs are incurred even if the network carries no traffic; these should therefore be covered by the rental. That part of both capital and operating costs which is proportional to the amount of traffic carried should be recovered from the charges made for customers' calls.

In local areas, all the costs of customers' access networks and a large proportion of the cost of local exchanges is independent of traffic. These costs should therefore be recovered by means of customers' connection and rental charges. Part of the cost of local exchanges and all of the cost of tandem switching centres and the junction and trunk networks are dependent on traffic. It is therefore appropriate for these costs to be recovered from charges for the calls made by customers.

In practice, actual tariffs rarely comply exactly with these principles. In a large complicated network, it is impossible to apportion costs exactly to different services. It may also be considered desirable to depart from these principles for reasons of marketing policy. For example, it is common for revenue from a trunk network to subsidise the cost of local areas, because technical progress in the former produced cost reductions that were not available to the latter. Incidentally, this is why, when competition was allowed in the USA and the UK, it was profitable for new carriers to enter the long-distance market with lower call charges than AT&T and BT. Tariffs are always subject to government regulation, even when operating organisations are private companies.

The traffic carried by a network varies throughout the day and from day to day during the week. The quantity of switching equipment and transmission plant required depends on the traffic level in the busy hours. Calls made at off-peak times incur virtually no capital cost, since no plant would be saved if these calls were not made. Because of this, it is common to make call charges lower at off-peak times than during peak periods of traffic. This encourages off-peak demand and discourages peak demand.

Normally, the calling customer pays for a connection, but there are exceptions, for example free calls to emergency services. For calls to a 'freephone' (800 in the UK) number, the charge is billed to the called customer instead of to the caller. Calls to various service providers (and to mobile stations) may be charged at a *premium rate* and the higher charge is shared between the network operator and the service provider. Special charging arrangements are therefore required for freephone and premium-rate calls. These can be provided by an *intelligent network*, such as the BT Digital Derived Services Network, as described in Chapter 16.

9.4.2 Local-call charging

If a separate fee is charged for each call made by a customer, this is known as *message-rate charging*. The simplest method is to equip each customer's line circuit with a meter (message register) and to operate this (to make a unit charge) when the called customer answers. Alternatively, a common electronic data store may be used instead of individual meters. In SPC exchanges, the processors which set up the connections can generate records of the calls for charging purposes.

Junction calls use more plant than own-exchange calls. Some administrations in the past therefore adopted a policy of making more than one unit of charge for these calls. This is known as *multimetering.* The supervisory apparatus in the originating exchange operates the calling customer's meter a number of times determined by the routing digits used for setting up the connection (e.g. corresponding to group-selector levels in a step-by-step exchange). It is now more usual, however, for a uniform charge to apply throughout a numbering area, or even across adjacent areas.

The traffic generated (and thus the amount of plant required) depends on the duration of calls as well as their number. Message-rate charging schemes therefore often make the charge for each call proportional to its duration. In electromechanical exchanges, a local-call timing circuit is fitted in the supervisory unit. This is operated by a periodic train of pulses from a common pulse generator and operates the calling customer's meter at the appropriate intervals [23] (e.g. once every three minutes). The method is called *periodic pulse metering* (PPM). As explained in Section 9.4.1, it is common to make the charging rate vary with the time of day. This can be done simply by changing the pulse-repetition frequency of the pulse generator under the control of a clock.

Because the major proportion of the cost of a local network is independent of traffic, some administrations do not make a separate charge for local calls but include them in the customers' rental. This is known as a *flat-rate tariff.* It avoids the capital cost of providing customers' meters and the operating costs of reading them at regular intervals and preparing the accounting information from them. The method is unfair to customers who make few calls, since their rental covers the average number of calls per customer. However, some account is taken of different calling rates by charging higher rentals to business customers than to residential customers. When flat-rate charging is used, customers naturally make more calls than when message-rate charging is used. Local exchanges therefore need to be designed for a higher traffic level. The increased number of calls is, to some extent, offset by shorter duration. However, common-control switching systems in flat-rate areas require more common-control equipment than in message-rate areas to handle the increased number of calls. Recently, problems of traffic congestion have arisen in flat-rate areas because of calls of very long duration being made for data transmission. This has necessitated the installation of additional plant without bringing in any additional revenue.

Some administrations combine the features of flat-rate and message-rate charging. The customer's rental payment covers a particular number of calls per annum and only calls above this number are charged for. This method retains the disadvantages of message-rate charging and loses the advantages of the flat rate.

When message-rate charging is used, customers are billed at the end of each accounting period. For flat-rate customers, there is no billing for local calls. In either type of area, however, it is necessary to collect money at payphones, at the time each call is made. In order to control the charging operation, it was necessary to send and receive signals between the exchange and the payphone. This required a different type of supervisory unit from that used for calls originated by ordinary customers. Thus, in an electro-mechanical exchange, payphone lines had to be segregated into a separate group for connection to coin-and-fee-checking relay sets. Now, microprocessors have been incorporated in payphones, so the charging function can be performed at the payphone instead of at the exchange [24].

9.4.3 Trunk-call charging

A charging plan for trunk calls should satisfy the following criteria:

(*a*) The revenue from trunk traffic should recover the capital charges and operating costs of the trunk network (since these costs are almost entirely traffic dependent).

(*b*) Charges for calls to and from customers who are geographically close should be similar, to give equitable treatment and avoid complaints (i.e. anomalous treatment of customers near the boundaries of charging areas should be minimised).

(*c*) The charging plan should be easily understood by customers.

(*d*) The charging plan should be suitable for implementation using automatic equipment (it should therefore be compatible with the numbering and routing plans).

To recover the cost of providing and operating a trunk network from the revenue obtained from traffic, it is general practice to make a separate charge for each trunk call. The cost of a trunk call depends both on distance and duration, so charges are often roughly based on the product distance × time.

When a trunk call is established manually, the operator writes the details of the call on a form known as a *ticket*, from which the charge is subsequently calculated. Since the ticket contains the name of the terminating local exchange, it is possible for the charging rate to be found from a table based on straight-line distances from the originating exchange to every other exchange in the country. The labour cost of a call set up by an operator is high, so it is necessary to make some minimum charge for any call, however short its duration. A typical tariff is based on a minimum charge for up to three minutes.

When trunk calls are directly dialled by customers, the charges must be determined by automatic equipment. This equipment would be complex if required to identify the charging rates for calls to every local exchange in the country. Charges are therefore based on distances between the centres of the numbering areas containing the calling and called customers (i.e. distances between a relatively small number of trunk exchanges instead of a large number of local exchanges). The exchanges in a charging area are called a *charging group*. Since subscriber trunk dialling eliminates the labour costs of manual operation, there is no longer any need to make a comparatively high minimum charge. The minimum can be the unit charge used for local calls, but for a shorter call duration. For longer calls, the charge is proportional to duration in multiples of the unit charge.

If charges are determined by distances between numbering areas, the information needed to determine the charging rate for a call is the same as that needed to determine the routing, i.e. the area code of the called customer. At the outgoing trunk exchange, the register can obtain from the dialled code a translation which includes a digit giving the charging rate for the call in addition to the required routing digits. The register then transfers the charging-rate digit to charging equipment associated with subsequent supervision of the call.

When periodic-pulse metering is used, this supervisory equipment generates metering pulses at intervals corresponding to the charging rate. However, the charging equipment is at the outgoing trunk exchange, but the customer's meter is at the originating local exchange*. It is therefore necessary for the trunk exchange to send the pulses back over the junction to the local exchange [25].

A form of ticketing, known as *automatic toll ticketing* (ATT) has also been used [26]. The supervisory equipment records the charging rate, the time of answer and the time of termination of the call. Since the supervisory equipment is at the outgoing trunk exchange, the calling customer's number must be sent forward to the trunk exchange from the originating local exchange. To do this automatically, the local exchange requires apparatus known as *calling-line-identification* (CLI) or *automatic-number-identification* (ANI) equipment. When a local exchange was not equipped with CLI, it was necessary for trunk calls to be intercepted by an operator to obtain callers' numbers [27]. The ticketing information is read out to a data store at the end of the call and used subsequently by the accounting system for billing the customer. In a network with SPC exchanges, the information required can be generated by the central processor at the local exchange [28]. This consists of: calling number, called number, time of answer and time of call termination. The charging rate is not required, since this can be determined subsequently from the called number by the accounting system. It is no longer necessary for call charges to vary in relatively coarse steps corresponding to a unit fee. The charge can correspond more closely to the exact

---

* Some customers, such as hotel proprietors, find it essential to be notified immediately of charges for individual calls, so that they can bill these charges to their own customers. This minority requirement was met by providing (at additional rental) a meter on the customer's premises. This necessitates signals being sent back over the customer's line during a call, to operate the private meter in step with the meter at the exchange.

duration of a call, e.g. to the nearest second.

The PPM method was usually employed by administrations which used message-rate charging for local calls, since customers' lines were already equipped with meters. The ATT method was usually employed by administrations having a flat-rate local tariff, since their customers did not have meters. When PPM is used, a customer receives a *bulk bill*, covering both trunk and local calls. The use of automatic ticketing enables an *itemised bill* to be prepared, giving separate details for each call. The latter is preferred by customers and the introduction of SPC exchanges enables it to be done for both trunk and local calls. Consequently, it is now being introduced more widely.

In countries where there are competing network operators, trunk calls may traverse parts of two or more networks. Call-charging equipment must therefore be provided at the interface nodes between the networks in order that charges may be apportioned between the operators involved.

For charges to be based on distances between trunk exchanges, boundaries between charging areas must coincide with boundaries between number-plan areas. Charges for local calls between exchanges within an area are then equal, and charges for trunk calls from all these exchanges to all local exchanges in another area are also equal. These areas should be of approximately equal size, so that calls over similar distances result in similar charges, although originating and terminating in different parts of the country. This can be done more accurately if the areas are small. However, this increases the complexity of both routing and charging equipment by increasing the number of area codes to be identified. It also reduces the capacity of the national numbering plan.

Any system of charging based on areas is bound to introduce anomalies between adjacent exchanges on opposite sides of an area boundary, as shown in Figure 9.3a. The effect of these anomalies can be reduced by charging the same fee for calls within the same area and to adjacent areas [29], as shown in Figure 9.3b. The boundary of the customer's own charging area is then less important than the outer boundaries of adjacent areas. Since these are more distant, anomalies are less significant.

The requirements that charging areas shall be approximately equal in size and that their boundaries shall coincide with those of numbering areas may be in conflict. Numbering areas are usually small in urban regions, where the telephone density is high and local-exchange areas are compact. This situation can be accommodated by including two or more numbering areas within a single charging area [29] as shown in Figure 9.4b. Numbering areas tend to be large in a rural region, where subscribers are sparse and each local exchange serves a large area. This necessitates dividing the region into different numbering areas, to provide different charging areas. However, these areas can be served by a common trunk exchange, as shown in Figure 9.4c. This retains economies in routing costs. However, it makes wasteful use of the numbering plan, since two or more area codes are used for routing

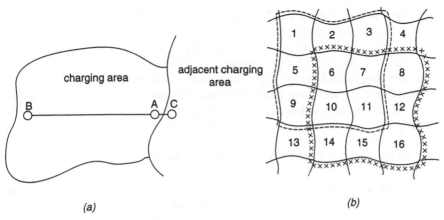

(a)   (b)

Figure 9.3   *Unit-fee boundaries in relation to charging areas*

 a Typical anomaly when unit-fee range covers only one charging area
  Call A to B: fee 1 unit
  Call A to C: fee 2 units
 b Unit-fee range covering adjacent charging areas
  ----- unit-fee boundary for customers in area 6
  xxx unit-fee boundary for customers in area 11

incoming calls to the same trunk exchange. The exchanges in an area which does not have its own trunk exchange are called a *dependent charging group*.

 The facility of being able to make the same charge for calls to several different numbering areas also enables the size of area covered by the same

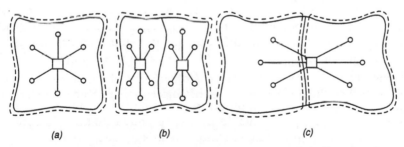

 (a)   (b)   (c)

Figure 9.4   *Arrangements for obtaining charging areas of similar sizes in regions having different telephone densities*

 a Charging area and number area with same boundaries
 b Charging area divided into two number areas in region of high telephone density
 c Two charging areas served by single trunk exchange in area of low telephone density

  ☐ trunk exchange
  ○ local exchange
  — number-area boundary
  --- charging-area boundary

fee to increase with distance. This reduces the fee per unit distance as the distance increases, which is consistent with the economics of long-distance transmission.* In the UK, the same charging rate is used for all telephone calls over distances greater than 56.45 km (35 miles). Call charges for the BT packet-switched data network are independent of distance.

Even when the majority of trunk calls are made by direct dialling, operators are still required on a minority of calls, to provide special services. Examples include:

(*a*) calls to be made to a particular person, rather than a particular station ('personal' or 'person-to-person' calls)

(*b*) calls to be charged to the called subscriber instead of the calling subscriber ('reverse-charge' or 'collect' calls).

These calls are subject to a minimum-duration charge (e.g. for three minutes), together with an additional charge for the extra time used by the operator in providing the service.

9.4.4 International charging

When international subscriber dialling (ISD) is used, call charging is carried out by the same methods as for subscriber trunk dialling (STD). Thus, either metering or automatic ticketing may be used. However, difficulties arose with periodic-pulse metering. On the longest (and most expensive) international calls, the pulses were so frequent (two or three per second) that a sluggish meter could fail to operate. When normal four-digit meters were used, a business customer who made many international calls could incur more than 9999 units of charge in the interval between meter readings, and thus fail to be charged correctly. It was found necessary to equip such customers' lines with five-digit meters. These difficulties are, of course, eliminated by SPC exchanges providing an electronic record of call charges.

For international connections, it is necessary to associate call-charging equipment with incoming and outgoing international circuits in addition to customers' lines. This is to enable operating administrations to apportion charges for calls passing through more than one country and to collect charges from other administrations as required.

---

* When circuits are provided by multiplex transmission systems, the total cost of each circuit depends on the cost of the line or radio-relay plant (which is proportional to distance, but is shared between many channels) and the cost of the multiplexing equipment (which is independent of distance). The cost of switching and signalling equipment is also independent of circuit length. The cost per kilometre of each circuit thus decreases with the length of the route.

9.5 References

1. ORBELL, A.G.: 'Network aspects of interconnection'. First UK Conference on *Telecommunication Networks*, London, 1987, *IEE Conf. Pub.* 279
2. CLARK, A.B., and OSBORNE, H.S.: 'Automatic switching for nation-wide telephone service', *Bell Syst. Tech. J.*, 1952, **31**, pp. 823–831
3. BARRON, D.A.: 'Subscriber trunk dialling', *Proc. IEE.*, 1959, **106B**, pp. 341–354
4. MUNDAY, S.: 'New international switching and transmission plan recommended by the CCITT for public telephony', *Proc. IEE*, 1967, **114**, pp. 619–627
5. BANERJEE, U., RABINDRAKUMAR, K., and SZCZECH, B.J.: 'London code change', *Br. Telecom Eng.*, 1989, **8**, pp. 134–143
6. FRANCIS, H.E.: 'The general plan for subscriber trunk dialling' *Post Off. Electr. Eng. J.* 1959, **51**, pp. 258–267
7. MCLEOD, N.A.C.: 'Numbering in telecommunications' *Br. Telecom Eng.* 1990, **8**, pp. 225–231
8. NUNN, W.H.: 'Nationwide numbering plan', *Bell Syst. Tech. J.*, 1952, **31**, pp. 851–859
9. CCITT Recommendation E.164: 'Numbering for the ISDN era'
10. CCITT Recommendation X.121: 'International numbering plan for public data networks'
11. CLOS, C.: 'Automatic alternate routing of telephone traffic', *Bell Lab. Rec.* 1954, **32**, pp. 51–57
12. RAPP, Y.: 'Planning a junction network in a multiexchange area' *Ericsson Tech.*, 1964, **20**, pp. 77–130
13. WHERRY, A.B. *et al.*: 'The London sector plan' *Post Off. Electr. Eng. J.* 1974, **67**, pp. 1–25
14. HAIGH, B., and MEDCRAFT, W.: 'Modernisation of the London network', *Br. Telecom Eng.*, 1990, **9**, pp. 20–25
15. ASH, G.R., and MUMMERT, V.S.: 'AT&T carves new routes in its nationwide network', *AT&T Bell Labs Rec.*, 1984, **62**, pp. 18–22
16. ASH, G.R., CARDWELL, R.H., and MURRAY, R.P.: 'Design and optimisation of networks with dynamic routing', *Bell Syst. Tech. J.*, 1981, **60**, pp. 1787–1820
17. CAMERON, W.H. *et al.*: 'Dynamic routing for intercity telephone networks', 10th *International Teletraffic Congress*, Montreal, 1983
18. ASH, G.R., *et al.* 'Real-time network routing in the AT&T network'. Proceedings of IEEE *Global Telecommunications Conference*, 1992, pp. 802–809
19. STACEY, R.R., and SONGHURST, D.J.: 'Dynamic routing in British Telecom network', 12th *International Switching Symposium*, 1989
20. AKINPELU, J.M.: 'The overload performance of engineered networks with hierarchical and nonhierarchical routing', 10th *International Teletraffic Congress*, Montreal, 1983
21. SCHWARTZ, M. 'Telecommunication networks: protocols, modelling and analysis' (Addison-Wesley, 1987)
22. GIRARD, A.: 'Routing and dimensioning of circuit-switched networks' (Addison-Wesley, 1990)
23. WALKER, N.: 'Periodic pulse metering' *Post Off. Electr. Eng. J.*, 1959, **51**, pp. 320–324
24. EMPRINGHAM, J.B.: 'Developments in payphones', *Br. Telecom Eng.*, 1992, **11**, pp. 77–84.
25. HEPINSTALL, D.L., and RYAN, W.A.; 'Metering over junctions', *Post Off. Electr. Eng. J.*, 1959, **51**, pp. 335–337
26. SEIBEL, C.F.: 'Automatic message accounting', *Bell Lab. Rec.*, 1951, **29**, pp. 401–404
27. KING, G.V.: 'Centralised automatic message accounting', *Bell Syst. Tech. J.*, 1954, **33**, pp. 1331–1342
28. REDMILL, F.J. and VALDAR, A.R.: 'SPC digital telephone exchanges' (Peter Peregrinus, 1994), 2nd Ed.
29. BREARY, D. and CHILVER, L.W.J.: 'Laying out of charging groups and rearrangement of line plant and switching equipment for subscriber trunk dialling', *Post Off. Electr. Eng. J.*, 1959, **51**, pp. 353–359

Chapter 10
Teletraffic engineering
D.J. Songhurst

10.1 Introduction

Traffic problems arise in systems which are subject to unpredictable fluctuating demands and which do not have the capacity to handle the maximum possible demand adequately. Teletraffic theory is therefore based on probability theory, and its application is essentially by mathematical modelling and computer simulation.

Although the mathematical theory underlying teletraffic engineering may be regarded as specialist expertise, the range of application is enormous. It has developed from the traditional areas of network dimensioning and optimisation and the design of multistage switching systems, through the modelling of processor-controlled exchange systems and packet-switched data networks, and into local-area networks (LANs), multiservice networks, mobile radio, cable television, network management and intelligent networks.

In modern networks with relatively cheap transmission capacity, the emphasis has moved away from speech-network dimensioning, although capacity requirements may still be important where there are inherently limited resources (as in mobile radio) or where there are large bandwidth demands (e.g. video, high-rate data). More typical of today's traffic-performance issues are call-set-up delays and control of overloads in intelligent networks, and the design of traffic controls for asynchronous-transfer-mode (ATM) multiservice networks.

The classic textbook on teletraffic is by Syski [1]. A good general introduction is given by Bear [2] and the economics aspects of teletraffic engineering are covered by Farr [3]. The objective of this Chapter is first to introduce the basic concepts and principles of teletraffic engineering (in the context of the simplest circuit-switched application), and then to illustrate these principles in a range of other application areas.

10.2 The unit of traffic

As shown in Figure 10.1, the number of calls in progress on a group of circuits fluctuates randomly as individual calls begin and end. The traffic carried is defined as the mean number of calls in progress. Although this is a

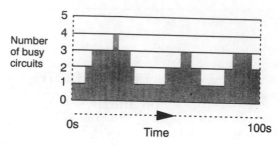

Figure 10.1   Variation of number of calls in progress

dimensionless quantity, the unit of traffic is called the erlang, after A.K. Erlang of the Copenhagen Telephone Company who put traffic engineering on a sound mathematical footing 75 years ago.

Figure 10.2 shows, as an example, how calls on a group of five circuits could generate two erlangs of traffic. In general:

$$A = \lambda H = \frac{CH}{T} \tag{10.1}$$

where

A = traffic in erlangs
λ = mean call-arrival rate
H = mean call duration (holding time)
C = number of calls arriving in time T.

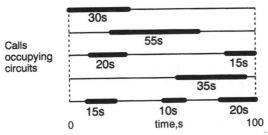

Figure 10.2   Example of 2 erlangs of traffic carried by five circuits

traffic = average occupancy
        = call-arrival rate × call-holding time
total occupancy = 200 s
traffic carried = 200/100 = 2 erlangs

10.3 Basic traffic model

To obtain analytical solutions to traffic problems, it is necessary to have a mathematical model of the traffic offered to a system. In the basic model, traffic is offered to a single group of circuits, for example representing a route between two exchanges. The following assumptions are made:

(a) Calls arrive in a Poisson process with a constant rate. (This is usually referred to as 'random traffic'.)
(b) Call-holding times have a negative exponential distribution. (Essentially, calls terminate at random regardless of how long they have been in progress. In fact, this is quite a good model.)
(c) The system is in statistical equilibrium. (That is it has been operating with the same call-arrival rate for sufficiently long that the effects of initial conditions can be disregarded.)

10.4 Lost-call systems

10.4.1 Grade of service

In a circuit-switching system, such as a telephone exchange, it is uneconomic to provide sufficient trunks to cater for the greatest possible demand that could arise. When there is *congestion* on a route, i.e. all circuits on the route are busy, some calls are unsuccessful. The traffic offered A (measured in erlangs) is defined as the product of the call-arrival rate and the mean call-holding time (both expressed in the same time units). Clearly, if the circuit group is sufficiently large that no calls are lost, then A equals the traffic carried (i.e. the mean number of calls in progress). If not, the traffic lost is the difference between the traffic offered and the traffic carried.

The route is dimensioned to provide a specified grade of service B, which may be defined as:

B=probability that a call will be lost
  =proportion of lost calls (known as *call congestion*)

If the traffic is random, we also have

B=proportion of time that congestion exists (known as *time congestion*)

Consequently, the traffic carried is $A(1 - B)$.

The grade of service is normally specified for the time of day when the traffic is at its peak. This is known as the *busy hour.*

Erlang derived the probability distribution of the number of busy circuits in a group on the basis of the traffic assumptions given in Section 10.3, together with the following assumptions about the system:

(a) The circuit group has *full availability*, i.e. an arriving call can seize any free circuit.

(*b*) Calls which arrive when all circuits are busy are lost and have no effect on the system. In particular, they do not give rise to repeat attempts.

Erlang's loss formula gives the probability B_N that all the circuits in the group are busy:

$$B_N = \frac{A^N}{N!} \bigg/ \sum_{i=0}^{N} \frac{A^i}{i!}$$

(10.2)

where N is the number of circuits in the group.

Thus, B_N is the grade of service of the full-availability group of N circuits when offered traffic A erlangs.

Assumption (*b*) of the traffic model (for holding-time distribution) is not necessary for Erlang's loss formula, but it simplifies the derivation. In more complicated loss systems, and particularly in queueing systems, the call-holding time does affect the congestion probability.

Erlang's loss formula is usually computed by the recursion

$$B_N = \frac{AB_{N-1}}{N + AB_{N-1}} \text{ (where } B_0 = 1)$$

(10.3)

For large groups of circuits (e.g. $N > 500$) there are faster methods, such as

$$\frac{1}{B_N} = 1 + \frac{N}{A} + \frac{N(N-1)}{A^2} + \cdots$$

(10.4)

This sum can be terminated when the terms become sufficiently small.

10.4.2 Dimensioning

Telecommunication systems and networks are generally planned by the separate dimensioning of individual component parts. Dimensioning involves determining the maximum traffic capacity for a given grade-of-service (GOS) objective. In telephone networks, GOS is normally defined in terms of the proportion of call attempts rejected because there is insufficient equipment (i.e. congestion), although delay criteria such as delay to receive dial tone may also be used.

The allocation of GOS criteria to component parts of a network is something of an art. In principle, this should involve an economic optimisation subject to constraints on the overall GOS performance of the network. In practice, the relationship between overall GOS and the GOS objectives for individual components is difficult to quantify realistically because of the multidimensional variability of traffic in the network.

An important feature of traffic-carrying systems is that large traffic streams can generally be carried more efficiently than small ones. This is illustrated in

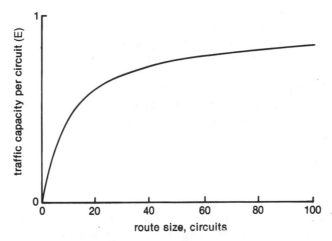

Figure 10.3   Variation of traffic efficiency with route size (for 1% loss probability)

Figure 10.3, which shows how traffic efficiency (in terms of traffic offered per circuit for a loss probability of 1%) increases with route size. This has major consequences for network design. It is cost effective to concentrate traffic into large streams which can be carried efficiently even though they may be routed over larger distances.

However, it also follows that large systems are more sensitive to traffic overload, as illustrated in Figure 10.4. Here two circuit groups of different sizes have both been dimensioned for a loss of 1% at normal load (=0% overload). The smaller circuit group has a traffic efficiency of only 68% at

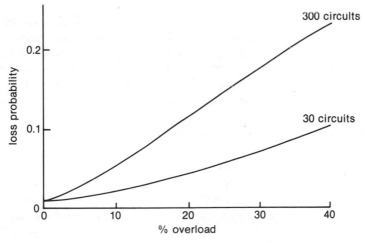

Figure 10.4   Overload sensitivity (for 1% loss at normal load)

normal load. With a 20% overload the loss probability degrades to 4.5%. The larger circuit group has a traffic efficiency of 92% at normal load. However, with a 20% overload the loss probability degrades to 12%.

In practice, traffic levels are variable and not predictable with accuracy, so it is desirable that moderate overloads should not cause excessive degradation in service. To achieve this, it is normal to dimension for overload conditions. To protect against overload, 'high-load' criteria are generally used. A route in a telephone network would typically be dimensioned to meet two criteria, such as the following:

normal load: 1% loss
20% overload: 5% loss

In this example, the normal-load traffic capacity of the larger group would be constrained to 84% efficiency at normal load.

Lead times for provision of equipment may be measured in years rather than months, so dimensioning is done on the basis of traffic forecasts. The accuracy of traffic measurement and forecasting procedures is therefore important. Traffic measurements are generally taken within busy hours on as many days as practicable. It is important to take a large measurement sample over the year so as to obtain good estimates of the mean, variability and seasonality of traffic streams. Forecasting over periods of more than one year involves a variety of economic and other factors. Measurement and forecasting procedures are presented in some length in the ITU-T E.500 series of Recommendations [4].

10.5 Routing in telephone networks

10.5.1 General

Large telephone networks are invariably based on hierarchical structures, but with varying degrees of nonhierarchical interconnection. The top tiers of such networks are often fully interconnected. Many networks use fixed routing, either throughout or with automatic alternative routing (AAR) restricted to part of the network. The use of fixed routing simplifies dimensioning and planning, but end-to-end GOS performance is sensitive to overloads and equipment failures on individual routes, thus requiring all routes to be dimensioned to overload criteria.

10.5.2 Automatic alternative routing

Figure 10.5 illustrates a hierarchical routing scheme. The top tier (trunk switches) is fully interconnected. The lower tier (local switches) has some direct interconnections but otherwise relies on routing via the top tier. With AAR in operation, calls are offered first to a direct route where appropriate, e.g. AB or AC. When that is busy they can overflow to the hierarchical route,

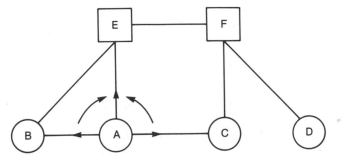

Figure 10.5   Automatic alternative routing in a hierarchical network

e.g. AEB or AEFC. This type of routing allows direct routes such as AB and AC to be provided efficiently on a high-usage basis.

Any routing scheme which uses overflow makes the mathematical-dimensioning problem more difficult. Overflow traffic is nonrandom, so Erlang's loss formula does not apply to routes offered overflow traffic (such as the circuit group AE in Figure 10.5). The dimensioning of hierarchical AAR networks can generally be tackled using decomposition and moment-matching techniques.

We use the fact that, when random traffic is offered to a full-availability route, the statistical moments of the overflow traffic (in particular the mean M and variance V) are readily calculated.

If A erlangs of traffic is offered to a direct route of N circuits, the mean of the overflow traffic is

$$M = AB_N(A) \tag{10.5}$$

and its variance is

$$V = M\left(1 - M + \frac{A}{N+1+M-A}\right) \tag{10.6}$$

(This variance is a theoretical parameter referring to independent observations of the traffic process. It is used for calculation purposes and is not readily observable in practice.)

Direct high-usage routes (AB and AC in Figure 10.5) are provided according to economic criteria. The moments of traffic overflowing from these routes are calculated from eqns. 10.5 and 10.6. These can be accumulated (on the reasonable assumption of statistical independence) to give the total mean and variance of traffic offered to hierarchical routes (such as AE). Traffic which is offered directly to hierarchical routes (e.g. traffic from A to D) is random and so has variance = mean.

Given the statistical characteristics of traffic offered to a final-choice route, a number of approximate methods are available to calculate blocking and hence to dimension the route. The best-known is Wilkinson's equivalent-random-traffic (ERT) method [5]. A single 'equivalent' high-usage route is modelled. This is a hypothetical group of $N*$ circuits offered $A*$ erlangs of random traffic, whose overflow moments (M, V) match those of the total traffic offered to the final-choice route.

$A*$ and $N*$ can be calculated from an iterative search, but Rapp's approximation [6] provides good estimates. The mean traffic $M*$ overflowing from the final-choice route is obtained from Erlang's loss formula for $N+N*$ circuits offered traffic $A*$. Thus, $M*=A*B_{N+N*}(A*)$ and the GOS on that route is then given by $M*/M$. The route is dimensioned by choosing N to give the required blocking.

A simpler method has been developed from a suggestion by Hayward [7,8]. If the total traffic offered to the final route has mean M and variance V, then the *peakedness* is defined as $z = V/M$. Random traffic has peakedness$=1$ (assuming negative exponentially distributed call-holding times) and overflow traffic has peakedness >1. Final-route blocking is estimated from Erlang's loss formula for N' circuits offered A' erlangs, where $N'=N/z$ and $A'=M/z$. This approach requires extension of Erlang's loss formula to noninteger circuits, but it works surprisingly well and is adaptable to more complex situations.

10.5.3 Service protection

In networks using AAR, unrestricted overflow is rarely permitted, since this can give very poor performance in overload conditions. In Figure 10.5, if traffic from A to B or from A to C overloads, the extra traffic almost all overflows to the hierarchical path (AE). This has two consequences. First, traffic which only has access to a hierarchical route (e.g. traffic from A to D) receives a very poor GOS. Second, calls overflowing from a direct route are using more circuits on the hierarchical path, hence spreading the overload and potentially causing many other calls to fail. The solution to this problem is to restrict overflow traffic, for which purpose two methods, illustrated in Figure 10.6, are well known.

The traditional method is to split the final route into two components, as shown in Figure 10.6a. One component forms a separate high-usage route reserved for first-choice-final traffic and the other forms a common component to which all traffic overflows. This arrangement can be analysed using the techniques described above.

A more modern method, implementable in stored-program-control (SPC) led systems, known as *trunk reservation*, is shown in Figure 10.6b. Calls overflowing from a high-usage route to a final route are blocked whenever the final route has few free circuits (i.e. less than a specified bound). First-choice final traffic always has full access to the final route. This approach is more robust than the split-route method, but the analysis techniques described

split final route

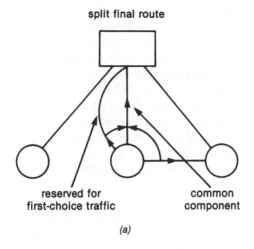

reserved for
first-choice traffic

common
component

(a)

trunk reservation

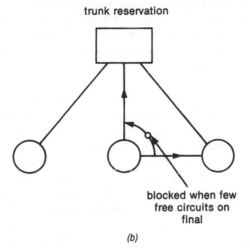

blocked when few
free circuits on
final

(b)

Figure 10.6   *Service-protection methods with automatic alternative routing*

     a Split final route
     b Trunk reservation

previously are not readily adaptable. More sophisticated techniques, such as the use of interrupted Poisson processes [9], are necessary for accurate analysis.

10.5.4 Nonhierarchical routing

There is growing interest in the use of nonhierarchical routing methods, which include *crankback* (i.e. returning control to the previous exchange on meeting congestion), load sharing, dual parenting, and dynamic routing.

These can offer more flexibility in dealing with traffic overloads and equipment failures, but they can be more difficult to plan and dimension. The major problem is that the network can no longer be decomposed and analysed sequentially, as with hierarchical AAR. Iterative methods using one-moment models are generally used [10,11,12].

Dynamic routing methods range from simple time-switched routing schedules, which formed the basis of AT&T's Dynamic Non-Hierarchical Routing system (DNHR) [13,14], to adaptive methods based on central control [15]. Dynamic alternative routing (DAR) is a more recently developed adaptive method, which requires no centralised network control and a minimum of inter-exchange signalling [16,17].

10.6 Queueing systems

10.6.1 General

In a queueing system, several waiting places are provided where demands are queued until a server becomes free. The traditional example of a tele-communication queueing system is the operator switchboard; however, the most important applications of queueing theory are now to packet-switched data networks and to the modelling of software tasks in processor-controlled switching systems.

Real systems can never provide unlimited waiting places and must therefore include some form of loss mechanism (even if that is simply the failure of the system under overload conditions!). Nevertheless, pure delay models are very useful; they can be much simpler than delay/loss models and can often represent the performance of real systems adequately in the ranges of interest. Kleinrock's two-volume work [18] provides a thorough survey of queueing theory and its application to data networks, although it lacks the more recent developments in queueing network theory.

10.6.2 Simple queues

In the Kendall notation the simplest queue is denoted $M/M/1$, where M stands for Markovian (indicating the lack-of-memory property of the call-arrival and termination processes which is sometimes referred to as randomness). The first M denotes a Poisson arrival process, the second M denotes a negative exponential service-time distribution, and the 1 refers to a single server. We also assume an unlimited number of waiting places, and service in order of arrival (first-come first-served). We then find that, in statistical equilibrium, the number of demands in the system has a geometric distribution, and the delay time has a negative exponential distribution with a discrete peak at zero (corresponding to demands which are served immediately). The following results can be obtained:

$$\text{probability of delay} = \text{server occupancy} = A \qquad (10.7a)$$

$$\text{mean number in system} = \frac{A}{1-A} \tag{10.7b}$$

$$\text{mean delay time} = \frac{AH}{1-A} \tag{10.7c}$$

where

H = mean service time
A = mean arrival rate of demands per mean service time
= offered traffic.

Now, a pure delay system cannot attain statistical equilibrium (i.e. it is unstable) unless the offered traffic is less than the number of servers. For the single-server queue this requires $A < 1$.

Relatively simple explicit results for delay distribution are also available for the first-come-first-served multiserver queue $M/M/N$ with limited or unlimited waiting places. One of the few other simple results in queueing theory is the Pollaczek–Khinchine equation, which applies to the $M/G/1$ queue, i.e. a single-server queue with a general service-time distribution:

$$\text{mean delay time} = \frac{A(1+C)H}{2(1-A)} \tag{10.8}$$

where C is the coefficient of variation of the service-time distribution, i.e. the variance divided by the square of the mean.

For the $M/M/1$ queue, $C=1$. For the $M/D/1$ queue with deterministic (i.e. constant) service times, $C=0$, and therefore the mean delay is precisely half that of the $M/M/1$ queue.

The performance of some simple queues is illustrated in Figure 10.7. This shows that the 10-server queue $M/M/10$ is more efficient than the single-server queue $M/M/1$. It is correspondingly more sensitive to overload.

Also shown is the 10-server queue with only 10 queue places, $M/M/10/10$. This is a combined loss/delay system, where calls which arrive to find all queue places occupied are rejected. This system has no stability problems. As the offered load per server increases, the proportion of calls rejected increases, but the mean delay experienced by accepted calls is held to a finite limit. This type of behaviour is often desirable in practice, for example in the processor unit of a switching system where call-setup tasks must be handled within given time constraints. For this reason, limited queueing facilities are often deliberately imposed as a load-control facility. However, in practice, even rejected tasks are likely to generate a small processor load, for example

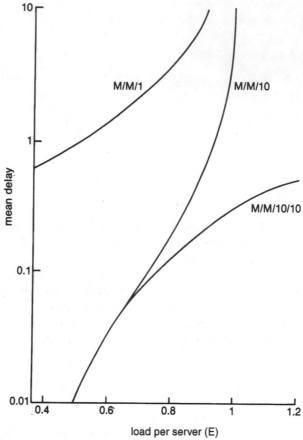

Figure 10.7   Performance of simple queueing systems

to check dialled digits for emergency calls, to return tone and to collect statistics.

10.6.3 Networks of queues

Many systems of practical interest are effectively networks of queues, where demands may move from one queue to another, receiving service at each. Although such systems can be exceedingly complex to analyse, a number of remarkably simple results have appeared in recent years. These show that, under a surprisingly wide range of conditions, the total mean throughput time across a queueing network can be calculated by assuming that each individual queue is an $M/M/N$ queue.

Reiser [19] has developed a method of mean-value analysis which applies to closed networks (closed in the sense that no fresh demands arise; there is simply a fixed number of demands circulating in the network). This is useful as an approximation for more-general networks.

10.6.4 Packet-switched data networks

Many countries now have public data networks based on packet switching. Such networks operate on the store-and-forward principle. Each packet-switching exchange has buffer space in which packets are stored while awaiting or receiving processing, while queueing for onward transmission and while awaiting acknowledgment from the next exchange. A packet-switched network is therefore a complex queueing network, whose primary resources are buffer space, processor time and link bandwidth. Queueing theory is widely used for the analysis of data networks, to tackle problems such as the following:

(*a*) optimisation of buffer space allocation;
(*b*) analysis of processing delays;
(*c*) analysis of protocols (such as X.25) which determine the effective link-transmission times dependent on error rates and acknowledgment procedures;
(*d*) analysis of end-to-end packet-transmission times;
(*e*) design of network structures and routing procedures; and
(*f*) analysis of flow-control procedures which protect the network from overload by limiting the rate at which packets or new calls are accepted.

10.6.5 Processor-system modelling

When setting up a call, an SPC system uses a number of sofware processes. These run on processors (possibly only one) and interact with storage devices and various input/output media such as inter-exchange signalling systems. Queueing models can often be used to analyse problems in such systems, such as store contention. A simple queueing-network model for a processor system is shown in Figure 10.8. Queueing-network models can generally be applied to provide estimates of processor occupancy and mean throughput time. However, many such systems are complex multiprocessor systems with multipriority processes. Moreover, proper dimensioning requires analysis of delay distributions, for which computer-simulation methods are often necessary.

10.7 Switching-system dimensioning

Switching systems are made up of various subsystems, and arriving calls make differing demands on these subsystems. For example, the processor control will typically be used only for call setup and clear down, whereas a path

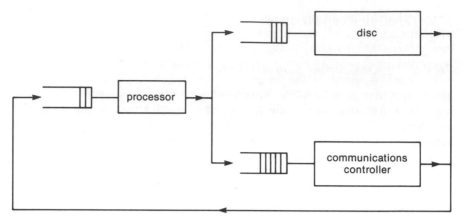

Figure 10.8   Queueing model for simple processor system

through the switch block will be occupied for the duration of a call. Different types of call make differing demands; for example, an intelligent-network call may need more processor effort than an ordinary call and use different peripherals.

Performance requirements for a processor-controlled exchange are specified in terms of loss probabilities and delay times. Losses may be due to switch blocking or to the processor control. Switch blocking is a significant factor in the concentration stages of local exchanges, whereas large digital tandem switches tend to be designed as virtually nonblocking.

A variety of delay criteria are specified, including dial-tone delay (in local exchanges), incoming-response delay and switching delay. These delays are a function of processor performance, and, since processors act as queueing systems, they are sensitive to overloads. Well-designed processor-controlled systems have a load-control mechanism which rejects call attempts before delays become unmanageable. The processor load is determined by the call-attempt rate and the switch load is determined by the offered traffic; therefore the mean call holding time is an important factor in exchange dimensioning. In overload conditions more calls are ineffective and generate repeat attempts, so it is usual to specify higher overload levels for call attempts than for offered traffics (typically 40% and 20%, respectively). There are therefore a number of requirements for switching-system dimensioning:

(a) Performance requirements must be specified for normal load, overload and partial-failure conditions.
(b) A good model for exchange behaviour must be available.
(c) Information is required on mean call-holding time, call mix (i.e. the relative proportions of different types of calls) and the characteristics of traffic overloads.

10.8 Customer repeat attempts

In network-overload conditions, where serious congestion exists, customer-generated repeat attempts are a critical factor. Figure 10.9 shows a flow model which enables some of the important results to be calculated. The key parameters are *B*, the blocking probability of the appropriate part of the network, and *R*, the re-attempt rate for customer-call attempts (also known as the *persistence*).

Equating flows around the diagram leads to the following expression for the total call-attempt rate in the network (i.e. including repeat attempts), and for the probability that the customer abandons a call:

$$\text{total call-attempt rate} = \frac{C}{1-BR} \tag{10.9}$$

$$\text{proportion of calls abandoned} = \frac{B(1-R)}{1-BR} \tag{10.10}$$

where *C*=arrival rate for new calls (intents).

The customer persistence is difficult to measure in practice and is not a widely known parameter. However, special measurement studies have shown that it may be higher than 90% in some circumstances. Clearly, a combination of high blocking in the network and high customer persistence leads to very high call-attempt rates being offered to the network.

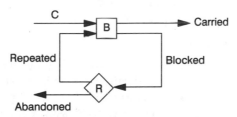

Figure 10.9   Flow model for customer-repeat attempts

   B=blocking probability
   C=call-arrival rate for new calls (intents)
   R=probability of re-attempt

10.9 Other applications

10.9.1 Common-channel signalling networks

The CCITT No.7 common-channel signalling system is now widely used in digital telephone networks, and is essentially a packet-switched data network which is closely linked to the speech network. Common-channel signalling networks have been used primarily for circuit-related signalling, i.e. to establish and release end-to-end connections for simple calls. In this application the traffic issues are relatively straightforward; the traffic offered to the signalling links is closely related to speech-link traffic and the signalling links operate at low occupancy.

An increasing number of ISDN and intelligent-network services require the use of noncircuit-related signalling, as do the introduction of GSM and more-advanced mobility services. These services generate signalling traffic with very different characteristics. It is more variable, it is not constrained by the size of an associated speech route, the message-arrival process may be highly correlated and the signalling messages are longer. Hence, traffic-performance issues within these signalling networks are becoming more and more important.

Relatively simple $M/G/1$ queueing models have been used to analyse message-delay performance on individual signalling links. However, it is difficult to extend these models to end-to-end performance across a network. Network delays, such as connection-establishment and connection-release delays, are dependent on the complexities of the signalling protocol, the characteristics of the signalling traffic and the message-processing delays incurred in switching systems.

Performance issues (delay, reliability and resilience) are now significant factors in signalling-network design. Advanced services may require extensive signalling to various peripheral devices, and the network topology needs careful design to ensure that signalling delays stay within acceptable limits and that the network is resilient to overloads and failures.

10.9.2 Multiservice networks

The developing ISDN is bringing a whole range of new traffic problems, many of which cannot yet be foreseen in detail. Areas which have already received substantial study include the performance of access protocols and the dimensioning of circuit groups carrying traffics with multiple bandwidths [20]. As described in Chapter 15, asynchronous-transfer-mode (ATM) switching is being developed to provide the future broadband ISDN. This is a fast packet-switching technology designed to carry a wide variety of services (voice, data and video) all packetised into 53-byte cells. Many traffic-performance issues are under study [21,22]. These include:

(i) *Source-traffic characterisation*
A lot of work has been done on modelling speech bursts in voice calls. For video services, increasingly efficient codecs are being developed which can transmit at variable bit rates dependent on the amount of information in the source image and its rate of change. The statistics of cell arrivals from such sources will be complex with strong autocorrelations.

(ii) *Connection-acceptance control*

A new call may be accepted or rejected by the network dependent on a declaration of its expected traffic characteristics (e.g. peak and mean bit rates). This must be followed up by some form of source policing to check that the characteristics are not exceeded. These issues, together with charging policy and customer-perceived quality of service, are all closely inter-related and fall within the scope of the contract with the customer.

(iii) *Switch design*

Various fast packet-switch designs are under development for ATM networks. These switches will need to switch cells at a very high rate and with low delay and loss probabilities. Most designs are based on multistage self-routing interconnection networks with buffering, and various combinatorial-analysis techniques have been developed to evaluate cell delay times and loss probabilities.

(iv) *Bandwidth management*

In addition to acceptance control, this includes routing and network-traffic controls. Bandwidth-reservation schemes may be required, analogous to trunk reservation in circuit-switched networks.

(v) *End-to-end performance*

This applies both to loss performance at the call level and to loss and delay performance at the cell level. Both areas are complex. At the cell level, delay performance is the critical factor for voice and video services, whereas cell losses are more important for many data services.

(vi) *Cell-delay variation*

Cells experience differing delays as they pass through various buffers within multiplexers and switches, and this process imparts a certain amount of jitter to the cell stream. This makes it difficult to police peak cell rates, and it may also increase buffering requirements.

10.9.3 *Local-area networks*

Local-area networks (LANs) are based on either bus designs (e.g. Ethernet) or ring designs (e.g. Cambridge and Orwell Rings) [23]. Random-access protocols, such as that used in Ethernet, have interesting traffic properties and are used in a variety of other applications such as cellular-mobile-radio signalling systems [24]. Ring designs are basically cyclic queueing systems, which tend to be very complex to analyse. LANs are generally designed to operate at relatively low occupancies.

10.9.4 *Cellular mobile radio*

In addition to the problems of random-access signalling systems referred to above, the optimal allocation of radio channels to cells can be a difficult problem [24]. Call handover from one cell to another is a form of overflow traffic whose characteristics depend on the mechanism for initiating it, and where blocking is not allowable. The analysis of advanced systems using dynamic channel allocation involves particularly complex topological considerations.

10.10 Tools for teletraffic analysis

We have seen that mathematical methods are available for a wide range of teletraffic problems. Although many of these methods may sometimes be regarded as 'exact', there are a number of reasons why they will never predict precisely what happens in practice:

(*a*) The models are statistical and give mean values which correspond to indefinitely long periods of observation. Congestion performance tends to be highly variable within limited periods (e.g. 1 h), although this variability can itself be predicted from mathematical models.

(*b*) Real traffic behaviour is unpredictable and may not match the assumptions of the model. For example, the effects of day-to-day traffic variability are often not included in dimensioning models.

(*c*) It is often necessary to use simplified models of system behaviour to enable mathematical analysis to be undertaken.

This last problem can be attacked by using computer simulation. The accuracy of the system model is then limited only by the information available from system designers, the effort available for program development, and the number of bugs in the program. It is important to remember that, even with an accurate system model, a simulation is still an attempt to solve a mathematical problem: i.e. how does the system perform under a certain mathematical model of traffic behaviour? Furthermore, simulation is an inherently statistical method, and long runs are often required to give results with sufficient accuracy. Nevertheless, simulation is virtually essential for detailed analysis and dimensioning of all but very simple systems.

Ideally, simulation should always be supplemented by simplified analytical models. These can usually be programmed and run far more quickly and cheaply, to give a better understanding of how the system performance depends on different factors and to help uncover bugs in the simulation. Analytical models can be calibrated by simulation, and then used for system dimensioning after the design and evaluation phases are complete. A number of packages are available commercially, such as BONES, OpNET and SES Workbench.

10.11 References

1. SYSKI, R.: 'Introduction to congestion theory in telephone systems' (Oliver and Boyd, 1990), revised ed.
2. BEAR, D.: 'Principles of telecommunication traffic engineering' (Peter Peregrinus, 1988), 2nd. ed.
3. FARR, D.: 'Telecommunications traffic, tariffs and costs' (Peter Peregrinus, 1988)

4. ITU-T E.500 Series of Recommendations: 'Traffic engineering'
5. WILKINSON, R.I.: 'Theories for toll traffic engineering in the USA', *Bell Syst. Tech. J.*, 1956, **35**, pp. 421–514
6. RAPP, Y.: 'Planning a junction network in a multi-exchange area', *Ericsson Tech.* 1964, **20**, pp. 77–130
7. FREDERICKS, A.A.: 'Congestion in blocking systems: a simple approximation technique', *Bell Syst. Tech. J.*, 1980, **59**, pp, 805–827
8. LINDBERGER, K.: 'Simple approximations of overflow system quantities for additional demands in the optimisation', 10th *International Teletraffic Congress*, 1983, paper 5.3.3
9. SONGHURST, D.J.: 'Protection against traffic overload in hierarchical networks employing alternative routing'. *Telecommunications Networks Planning Symposium*, Paris, 1980, pp. 214–220
10. GIRARD, A., and COTE, Y.: 'Sequential routing for circuit-switched networks', *IEEE Trans.*, 1984, **COM-32** (12), pp. 1234–1239
11. LEGALL, F., and BERNUSSOU, J.: 'An analytical formulation for grade of service determination in telephone networks', *IEEE Trans.* 1983, **COM-31**, (3), pp. 420–424
12. LEBOURGES, M., BECQUE, C.R., and SONGHURST, D.J.: 'Analysis and dimensioning of non-hierarchical telephone networks'. 11th *International Teletraffic Congress*, 1985, paper 2.2B–4
13. ASH, G.R., CARDWELL, R.H., and MURRAY, R.P.: 'Design and optimisation of networks with dynamic routing', *Bell Syst. Tech. J.*, 1981, **60**, pp. 1787–1820
14. ASH, G.R., KAFKER, A.H., and KRISHNAN, K.R.: 'Intercity dynamic routing architecture and feasibility', 10th *International Teletraffic Congress*, 1983, paper 3.2.2
15. CAMERON, W.H., *et al.*: 'Dynamic routing for intercity telephone networks'. 10th *International Teletraffic Congress*, 1983, paper 3.2.3
16. STACEY, R.R., and SONGHURST, D.J.: 'Dynamic alternative routing in the British Telecom trunk network', *International Switching Symposium*, 1987
17. GIBBENS, R.J., KELLY, F.P., and KEY, P.B.: 'Dynamic alternative routing, modelling and behaviour'. 12th *International Teletraffic Congress*, 1988, paper 3.4A.3
18. KLEINROCK, L.: 'Queueing Systems'. Vol. 1: Theory; Vol. 2: Computer applications' (Wiley, 1975)
19. REISER, M., and LAVENBERG, S.S.: 'Mean value analysis of closed multichain queueing networks' *J. Assoc. Comput. Mach.*, 1980, **27**, pp. 312–322
20. JOHNSON, S.A.: 'A performance analysis of integrated communication systems', *BT Technol. J.*, 1985, **3**, pp. 36–45
21. APPLETON, J.: 'Performance related issues concerning the contract between network and customer in ATM networks', *BT Technol. J.*, 1991, **9**, pp. 57–60
22. KAWASHIMA, K., and SAITO, H.: 'Teletraffic issues in ATM networks', *Comput. Net. ISDN Syst.*, 1990, **20**, pp. 369–375
23. FALCONER, R.M., and ADAMS, J.L.: 'Orwell: a protocol for an integrated services local network', *BT Technol. J.*, 1985, **3**, pp. 27–35
24. MACFADYEN, N.W., and EVERITT, D.E.: 'Teletraffic problems in cellular mobile radio systems'. 11th *International Teletraffic Congress*, 1985, paper 2.4B-1

Chapter 11

Private telecommunication networks

R.K. Bell

11.1 Introduction

Modern business requires effective communication facilities [1,2]. The communication requirements of a small business may be met by a single telephone, and probably a fax machine, or a few telephones served by a key system [1,3]. A larger business requires a private automatic branch exchange (PABX) [3,4] to serve a larger number of extension lines and provide connections between them and the public switched telephone network (PSTN). A business which operates on several sites will have a PABX at each. If extension users generate sufficient traffic between these PABXs, it may be economic to join the PABXs by a private network [5–8], rather than route traffic between them over the public network. Organisations also have a need for data communication [9,10] which may justify a private data network.

In order to understand about private networks it is first necessary to appreciate what is meant by the term 'private' in relation to tele-communication networks. When one considers the dictionary definition:

Private: not public, not open to or shared with or known to the public

the task may become marginally easier, but, when statutory regulations are also taken into account, further difficulties are introduced. Consequently, for simplicity's sake it is proposed to assume that genuinely private networks possess the following characteristics:

(a) The intersite circuits are leased from public telecommunications operators (PTOs).
(b) The nodal equipment which the intersite lines interconnect is all located on premises not belonging to the public-network authorities.

Provided that statutory regulations permit, such networks may interconnect with the public network and will therefore be capable of carrying traffic to and from the latter. They may also interconnect two or more public networks, but only to carry traffic related to the business of the private network operator. However, if the traffic is not related to the private network operator's business, the private network effectively becomes a third-party network and may therefore become subject to different regulations.

11.2 Voice networks

For many years organisations have had PABXs which provide connections between extension telephones, between these and the PSTN and to other PABXs over private networks. The private network of a large company may link many sites and this leads to the introduction of tandem switching centres [5], as shown in Figure 11.1.

A modern PABX [4] uses a digital switch with stored-program control (SPC), as shown in Figure 11.2. As a result of the introduction of digital PABXs, many private networks now also use digital transmission links, operating at 64 kbit/s and at 2 Mbit/s or 1.5 Mbit/s.

To meet business needs, a PABX is usually required to provide more facilities than a public exchange. The introduction of stored-program control to PABXs has extended the range of facilities available. Facilities provided by a typical PABX are listed in Table 11.1.

One important facility is *direct dialling in* (DDI). Callers are able to dial directly to extensions, without the intervention of the PBX operator. (Of course, there will also be a general directory number, giving access to the operator, for callers who do not know the number of the extension they

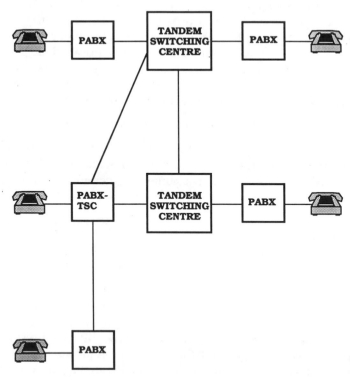

Figure 11.1   Private voice network

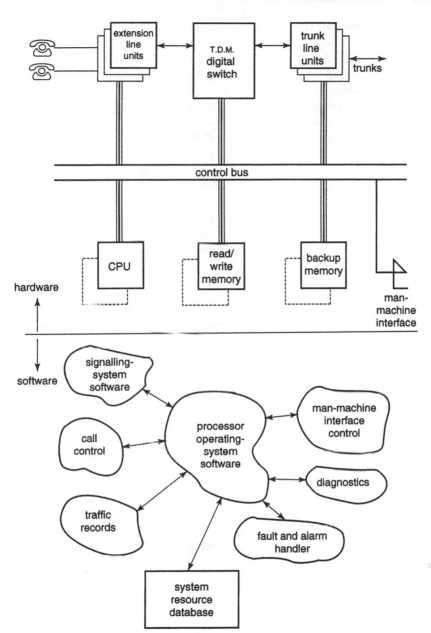

Figure 11.2 System architecture of SPC digital PABX

require.) The facility of *direct inward dialling* (DID) differs from DDI, because calls appearing on a particular exchange line are automatically routed by the PABX to a predetermined extension. This enables an executive to have a private exdirectory number on which to be called, without needing a separate

Table 11.1 Typical PABX user facilities and system facilities

Abbreviated dialling	Inter-PABX signalling (analogue)
Call back when free	Inter-PABX signalling (digital)
Call back when next used	Call logging
Call diversion (on busy)	Traffic recording
Call diversion (on no answer)	Data switching
Call diversion (follow me)	Modem pooling
Conference	Tandem switching
Direct dial in (DDI)	ISDN terminals
Direct dial out (selective)	Remote maintenance access
Call barring	Night service
Distinctive ringing	Music on hold
Group answering (hunt group)	Hotel facilities

telephone on another line. A further version, known as *direct inward system access* (DISA), enables a caller with a dual-tone multifrequency (DTMF) telephone to tone dial extensions after obtaining access to the PABX via the public exchange.

The application of SPC to PABXs has also led to the introduction of common-channel signalling. A widely used system is the *digital-private-network signalling system* (DPNSS). Common-channel signalling enables the facilities normally available on a single PABX to be extended over a complete private network. For example, a single operator at one site can answer all incoming calls for an organisation and route them to extension telephones at other sites. An extension user on one PABX can transfer calls to telephones on other PABXs at other locations.

Modern PABXs can also use automatic alternative routing, in order to achieve least-cost routing [11]. The price of a call over a private circuit is, of course, zero. However, when all circuits on a route in a private network are busy, the PABX can route a call via the PSTN instead. This is advantageous if the cost per annum of calls which overflow to the PSTN is less than the rental charge for the additional private circuits which would be needed to provide the required grade of service.

11.3 Data networks

Many organisations have data networks as an integral part of their data-processing systems [10,12]. A data network may have a star configuration based on a main-frame computer, as shown in Figure 11.3. Such networks may extend over several sites, by using modems on telephone lines or being connected by digital circuits. Local-area networks (LANs) are also employed extensively. LANs on different sites may be linked by a wide-area network

(WAN) and they may have gateways to a public packet-switched data network.

Cost savings can be obtained by combining networks for voice and data [13]. Figure 11.4 shows a private network which provides both voice and data services. These use common transmission links. However, separate circuits are provided for speech and data and the data traffic is not switched by the PABXs.

An integrated-services digital network (ISDN) can switch data traffic as well as speech traffic. This is discussed in Chapter 14. ISDN technology is applicable to private networks in addition to public networks. This has led to

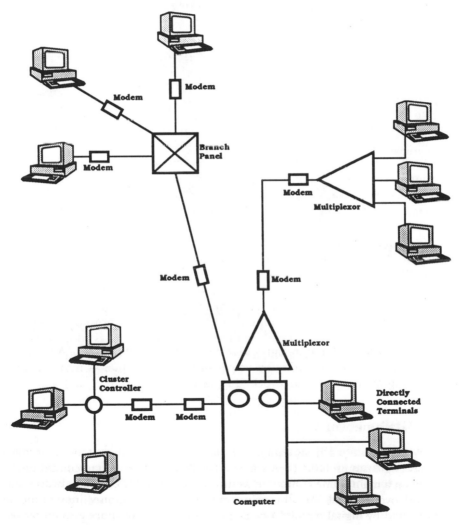

Figure 11.3   Private data network

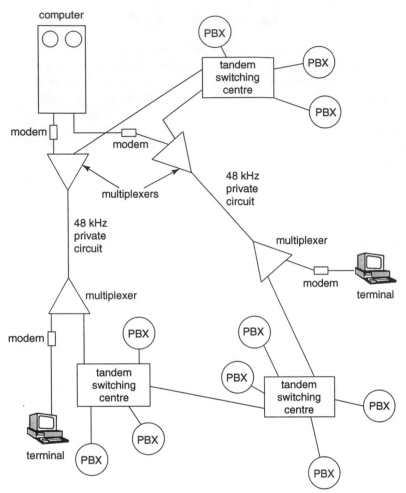

Figure 11.4　Combined voice-and-data analogue network

the introduction of the integrated-services PBX (ISPBX) [4]. As a consequence, separate networks for voice and data may merge into a single integrated-services private network [14], as shown in Figure 11.5.

11.4 Centrex

In *centrex* operation [8], equipment housed in a public telephone exchange provides a private switching service to a number of different customers. Each extension telephone is wired to the centrex switch instead of to a PABX on the customer's premises, as shown in Figure 11.6. Obviously, many more exchange lines are needed. However, the cost of this may not be excessive in the centre of a city where a concentration of high-rise office buildings enables

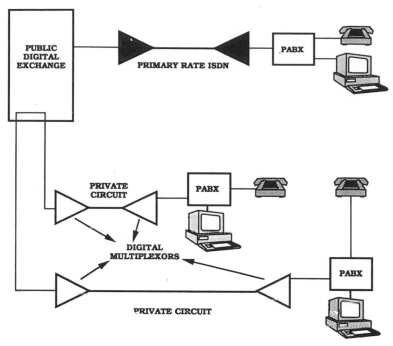

Figure 11.5   Digital network for voice and data

cable runs to be short. Multiplexers or concentrators may also be used to reduce cable costs.

Use of centrex enables an organisation to avoid the capital cost of a PABX, eliminates the need for space to accommodate it and takes care of maintenance. When moving to new premises, the customer does not have to move or replace PABX equipment. Thus, the use of centrex is attactive when accommodation is temporary. Centrex may also be attractive when a customer's staff have accommodation scattered over several buildings, each of which would otherwise need its own PABX.

As shown in Figure 11.6 the centrex equipment in the public exchange must cater for the individual numbering scheme of the customer, instead of the numbering scheme of the PSTN. It must also provide extensive facilities, equivalent to those of a PABX. This was not completely attained with public exchanges using electromechanical equipment, but it has been made possible with the introduction of SPC to main exchanges. As a result, the centrex service is now being offered more widely by PTOs.

11.5 Virtual private networks

There is another type of private network, the *virtual private network* (VPN). This is becoming more popular, especially in those countries, such as the UK

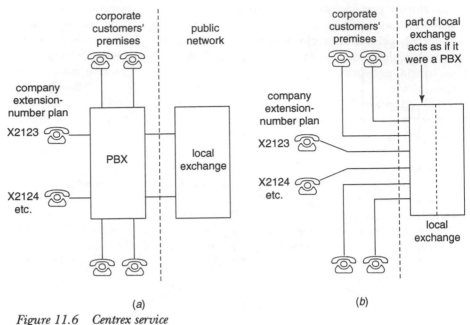

(a)   *(b)*

Figure 11.6   Centrex service

        a Use of PABX
        b Use of centrex

and the USA, which have adopted a very liberal policy as far as tele-communications-service provision is concerned. This new type of network has been given the title 'virtual' because it entails making channels, which are ostensibly part of a public network, exclusively available for a single user, through software partitioning.

In this era of digital transmission, it is difficult to appreciate the difference between this technique and that used to make permanent digital private circuits available. The main differentiator is probably that access to a virtual private network is made available via a large digital 'pipe' normally of 2 Mbit/ s capacity, although higher bandwidth is feasible because the feeder cable is usually fibre optic. Channels within this link are made available on an as-and-when-required basis. This differs from a permanent link, where channels are permanently allocated. The economic justification for a VPN is that fewer circuits are needed when private traffic is mixed with public traffic on a large group of circuits than when it is segregated onto a small group of circuits. This is illustrated by the example shown in Figure 11.7.

Because of this temporary provision, VPNs are charged on a usage basis, whereas fixed private circuits attract an annual rental. The tariffs for both types of links are within the control of public telecommunications-network operators. It is therefore probable that, in the future, the operators will attempt to dissuade users from adopting genuinely private networks by introducing competitive VPN tariffing. Despite these measures, it is unlikely that VPNs will ever be able to provide the specific functionality required of genuinely private networks; therefore the two are likely to coexist for some considerable time. This will almost certainly result in the extensive adoption of hybrid networks, consisting of a combination of genuine and virtual networks.

The adoption of VPNs in the USA has been accelerated by the fact that centrex exchanges have been available there for many years. It is a logical development to link these in virtual private networks.

In order to make VPN connections, public exchanges must recognise the numbering scheme of a private network and use its signalling protocols. Neither will be the same as those of the public network, so an additional database and different software are needed. If all exchanges in a PSTN are to be able to provide VPN facilities, it will be preferable to provide the database in a central location, rather than replicate it in every exchange. This is the basis on which an *intelligent network* operates, as described in Chapter 16.

Nationwide introduction of a VPN service will free organisations from

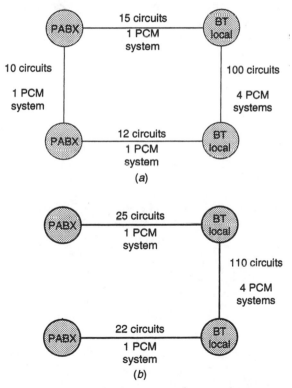

(a)

(b)

Figure 11.7   Principle of virtual private network

a Previous private network and part of PABX
b Replacement by VPN

constraints on where their staff can be located. For example, a member of staff can work at home and communicate with colleagues with the same facilities available as if they were all in the same building with a PABX.

11.6 Differences between public and private networks

In the earliest days of private networks, their functionality was at best the same as public networks, but generally less. Over the past 15 years, this situation has changed and most private networks have been developed to provide significantly more functionality than their public counterparts. The reason for this is the inherent inertia of the public networks due to geographic extent and universality. These factors considerably extend the period over which new technology can be economically adopted. In consequence, the very-much-smaller private networks can be modified to incorporate state-of-the-art techniques much quicker and at a lower cost than public networks. This means that the advanced features available in modern digital exchanges, such as *ring-back when free* or *call diversion* have been available in private networks for many years. It is only relatively recently that these features have started to become available in public networks.

Public networks provide some capability that private networks do not. The most notable of these is call charging. While, in some of the new digital public networks, the charging processes are software driven, in most of the old public networks the charging systems were hardware orientated. This meant that charging signals had to be passed over the network. This latter aspect is rarely necessary in a private network.

Other differences can apply because of the features available in PABXs. One of these stems from the ability in a PABX to 'recall' dial tone in order to establish another call, while the first is 'held'. The main reason for this facility in a PABX is to allow extensions, on receiving an external call, to make enquiries with another extension and then, if appropriate, transfer the external call to the other extension. This feature has never been provided in public networks, except when centrex is available.

Now that what constitutes a private network has been clarified, it is appropriate to consider the functionality of the network in order to understand its purpose fully. This can most easily be achieved by considering the International Standards Organisation (ISO) seven-layer model for Open Systems Interconnection (OSI) described in Section 1.7. Details of this model are shown in Figure 1.6.

Examination of Figure 1.6 reveals that layers 1–3 are concerned purely with interconnection. Layer 4 is concerned with both interconnection and interworking and layers 5–7 with interworking. This can be translated into protocol terms as follows:

layers 1–4: transportation
layers 5–7: application

When the model is applied to private telecommunication networks it can have differing connotations. For example, if the network is designed as a transport system only, i.e. to achieve interconnection between two peripheral devices and then to permit the free flow of information, it only needs to comply with layers 1–4. If, however, the network is designed to ensure that the information flowing between the two devices is comprehensible to both, compliance with all seven layers is necessary. It will now be appreciated that, for a telephone network, all that is necessary is compliance with layers 1–3. This is because compliance with higher layers is left in the hands of the two telephone users; once connected, they decide in what language they are going to converse and if they need to repeat anything (i.e. error correction in data-communications terms).

As opposed to this, for interworking when the two terminal devices are dissimilar data devices, the network, or some subsidiary device, must ensure compatibility. This is usually achieved by ensuring that both devices use the same application protocol or, alternatively, by arranging for the network or the subsidiary device to provide protocol conversion for one or other of the terminal devices.

The world's public networks have hitherto been essentially developed as transportation networks and are therefore incapable of providing the functionality which allows dissimilar data devices, or systems, to interwork with one another. It is for this reason that the need for private networks still exists and why they are different from the public networks. Alternatively, third-party value-added networks can be used to achieve the desired degree of interworking.

11.7 Network planning

11.7.1 General

Network planning is a logical process involving the following tasks:

(*a*) statistics gathering
(*b*) assessing usage in terms of both type and extent,
(*c*) predicting future growth and changes in usage patterns,
(*d*) deriving the most effective topology in functional and economic terms, and
(*e*) assessing which types of equipment will fully meet requirements.

11.7.2 Statistics gathering

Statistics which need to be collected include:

(i) type of potential usage,
(ii) quantity of potential usage, and
(iii) changes in use, in terms of type and quantity.

Although, hitherto, private networks have been used to carry traffic which previously was routed over the public network, this is less likely to be the case nowadays. The reason for this is that telephony private networks were employed to avoid the cost of public-network call charges. Modern digital private networks are more likely to be established to provide functionality not available in public networks.

As a result, the need to gather existing switched-network traffic details is less likely to be of paramount importance. Much more significant will be the necessity to assess the extent of textual and graphical information flows. Of course, if it is likely to result in a network which has spare capacity to accommodate telephony traffic, then statistics on this will also be required. Furthermore, with the advent of the public ISDN network, the likelihood that existing switched networks, both public and private, will be used for data transmission, is high. In this situation, information which is gathered should be in a form such that it can be readily applied to the planning of an ISDN private network.

All of this means that the main usage information required will be:

(*a*) information flows, in terms of maximum data flow in bits per second,

(*b*) demands, in time terms, when such flows will occur, and

(*c*) the duration when the flows will occur from any source.

In switched-network terms, this requires either the busy-hour traffic in erlangs or the quantity of calls and their average duration. Obviously, the type of usage of the network will also influence the type of information to be gathered. i.e. if it is to be used for an interactive data application, requiring a very short response time, a switched network is unlikely to be suitable. In this situation, it is then a matter of assessing just how large the data flow will be via a permanent link, which is also likely to be used for other applications. Quite often, the processor running the application can provide the required statistics, but if this is not possible then it will be necessary to use some other monitoring or counting device.

If it is necessary to gather switched-network traffic this is best done using call-information-logging equipment (CILE). This equipment can usually be connected to a V.24 output port available on most modern PABXs. The PABX will transmit call records to the output port which enables the CILE to collect and analyse them. Information obtained in this way will include:

(i)   date and time calls are originated,

(ii)   destinations of the calls, and

(iii) durations of the calls.

Once analysed, the call logger can present the information in the form of a report.

Some older PABXs may not be equipped with a V.24 port. In these cases a CIL scanner, connected to every external line, will be necesssary.

Whatever statistics are collected, it will also be necessary to attempt to estimate growth patterns. If it is possible to obtain statistics over a predetermined period of time, it is then feasible to undertake a forward projection in order to estimate growth [11].

As part of the statistics-gathering exercise, two other types of information will be needed. The first will be the result of an audit of all existing systems to identify the equipment installed on each site, i.e. its purpose, capacity, type and who supplied it. The second is assessing, in business terms, what systems and/or networks are required. This naturally entails intensive discussion with end users at all levels within the company.

Once all the appropriate statistics have been collected, the true planning process can commence.

11.7.3 Planning a network type and a topology

The main task involved in the planning process entails translating business needs into technological solutions. With advancing developments, this task has become much more difficult because a number of different solutions may be possible. In consequence, it may be necessary to derive several solutions and to choose the best in functionality terms which results in the lowest cost.

By and large, network topology will be determined by the location of end users. This will certainly dictate where each network node should be situated, but it will not reveal where strategic nodes should be placed, as this generally requires an intimate knowledge of the company's future plans. For example, there is no point in locating a strategic node on a site which is due to be dispensed with.

In the past, the initial network-planning process involved the production of a complete information-flow matrix. This would indicate the extent of each different type of information flow which was likely to prevail between any two sites. The next stage of the process entailed combining, as far as was feasible, all different types of information flow. Ideally this was carried out to indicate the peak flow of information which would prevail at any time. When the process was carried out manually, it was extremely time consuming and, as a result, a number of software packages have been produced to automate the task.

The network modelling tools which are available are unlikely to be able to undertake a complete design exercise. They need input from a person possessing both network expertise and a deep knowledge of the company that the network will serve. Another difficulty often materialises because it is rare for a network design to commence from a green-field situation. Usually, a number of networks already exist and will either have to be incorporated directly into the new network design, or at best, their functionality merged. Existing automated tools are unlikely to be able to achieve the direct incorporation process; consequently, an intensive period of interaction between the tool and the expert will be necessary.

So far only intersite information flows have been considered, but within-site networks cannot be ignored. To date, mainly because of statutory regulations, cables used for voice signals, which were ultimately destined to be routed over public networks, could not share the use of cables used for digital-data transmission. This is not now the case and, as a result, the opportunity now arises to utilise cabling schemes capable of conveying both voice and data.

Once all statistics have been gathered and the topology and functionality of the network decided, it will then be necessary to assess which equipment and systems will be capable of meeting the requirements. At this stage, although most of the planning will be complete, unless approval is gained to commit funds to the acquisition of the necessary systems, the exercise will be a waste of time. It is therefore important that the financial reasons for installing a private network, with its potentially high capital outlay, are clearly and unambiguously set out.

11.7.4 Financial considerations

Private networks will only be considered for adoption if they represent the most economic way of meeting a business need. Alternative methods of meeting these needs could be:

(*a*) use the public network,
(*b*) use an existing third-party network,
(*c*) persuade a service provider to install the required network on the basis that it can potentially be used by other users, or
(*d*) use a virtual private network.

Not all of these options may be available in every country. Also, a company may decide that the improved efficiency resulting from adoption of a private network may not be sufficient to justify the capital and ongoing expenditure. The business case for a network must be able to stand up to the rigorous application of any capital-investment rules that the company may seek to impose.

So far it has been assumed that the functionality of the required network is not already in existence and being operated by a third-party network operator. If this is the case, the private network required by the company concerned will have to be installed, commissioned and operated by that company. Other options do, however, exist. These are:

(*a*) have the network installed and commissioned and then handed over to a third party to operate; this is normally termed *facilities management*; or
(*b*) have a third party install the network at their cost and who will charge an ongoing fee for its usage; this is normally termed *outsourcing*.

Obviously, if some parts of the network already exist, for option (*b*) to apply, the assets forming part of the existing network, including its operational staff, will have to be taken over by the outsource company.

In any event, all of these options should have been considered before a business case is set out. Alternatively, all the options possible could be detailed in the business case. In this situation it will be necessary to undertake a net-present-value (NPV) comparison of all the options, as described in Chapter

19. On the assumption that formal approval for expenditure has been gained, it will then be possible to proceed with the actual provision of the network. This will involve two phases: implementation and commissioning.

11.8 Network implementation and commissioning

It is difficult to generalise on implementation and commissioning because each process is very much dependent on the technology, size and topology to be employed. It is, however, feasible to highlight some broad guidelines which would apply to any network.

(*a*) *Appoint a project manager*
It is likely that the vendor of the equipment to be installed will appoint a project manager. However, if trouble-free installation and commissioning are to be achieved, the ultimate operator of the network should also appoint its own project manager.

(*b*) *Use automated project-management tools*
The installation of a private network is usually an extremely complex process, with many separate but interrelated events. To avoid any of these causing slippage, it is advisable to track all events using a purpose-built automated project-management tool. Although these tools will not avoid delays, they will identify them early enough to allow remedial action to be taken, so that the overall project is not delayed.

(*c*) *Avoid disruption to existing services*
If the new network is designed to replace or augment an existing network, another important implementation guideline will be to ensure that existing services are not disrupted during the process. This requires extremely careful and detailed advanced planning. If it is fully appreciated exactly which services could be affected or changed by introduction of the new network, then all implications can be considered in advance.

(*d*) *Provide user training*
The process of providing an advanced digital network is not cheap. Under-utilisation should not be risked, just because its users have not received adequate training in its use. It should not be assumed that users have an intimate knowledge of telecommunications jargon and techniques. Instead, the benefits of the network to them personally should be explained in detail.

(*e*) *Train operational staff*
Although usage training is of paramount importance, training of operational staff should not be neglected. Preferably, the individuals concerned should have been deeply involved in the installation process so that, by the time commissioning has occurred, they will be completely familiar with all aspects of the network.

(f) Testing

Although testing will have proceeded throughout the installation period, it should not be assumed that once a particular aspect has been tested it can be forgotten until commissioning. Subsequent work can be disruptive; it is therefore important that final functional testing is carried out immediately prior to commissioning. If some actual users of the network can be involved in the process, so much the better, as they will be able to highlight potential problems resulting from the use of the network by nonexperts.

Once final testing has been completed, commissioning should begin without delay. Commissioning can be carried out in a number of different ways, depending upon the characteristics of the network. It can be phased in over a period of time, which will prolong the commissioning process and could entail additional planning. Alternatively a 'big bang' approach could be adopted, which will involve undertaking the changeover in a quiet traffic period, the length of which will depend upon the amount of work to be undertaken. Another important aspect of commissioning is to ensure that all operational procedures are in place and well rehearsed by operational staff.

11.9 Network operation

The need to ensure that the network is available, in a fully functional state, whenever required, is just as important as the network design itself. The procedures and systems concerned in network management and operational processes, should be planned and developed as carefully and thoroughly as the network itself.

There are two distinct aspects to operation of a network. The first is to develop a system capable of monitoring the network's performance and, ideally, of taking remedial action if a characteristic is observed which is outside a predetermined value. The second is to establish a facility whereby users can report difficulties and have them resolved quickly and efficiently. Both processes should be capable of storing historic information so that past trends can be observed and applied to future projections, thereby forestalling potential problems. In general, the information in question will be of most value when used to predict growth trends. Armed with such information, it becomes possible to forecast when additional capacity is needed and to undertake an appropriate augmentation exercise in a timely manner.

It is of particular importance to separate clearly the functions of planning, building and operation. Within a telecommunications project, close co-ordination is necessary between the plan and build tasks, whereas the operating task can be undertaken independently. However, the running of telecommunications and computer operations are also closely related and should preferably be undertaken as a single function.

A nebulous area exists where an existing system requires augmentation to provide increased capacity. In this situation, a good principle to adopt is to make the plan/build function responsible for all new developments, but to make the operating function responsible for generic operational development, e.g. increased capacity because of normal traffic growth. It is also important to give the operating function responsibility for all work carried out on a working system. Thus, although the plan/build function will have responsibility for planning and implementing the work, it should be carried out under the strict control of the operation function.

11.10 The future of private networks

Although the demise of private networks has been predicted for some time, especially by public-network operators, it seems likely that they will be around for some time, especially to serve large multinational users. However, it does need to be appreciated that they will not always be essential. As the public-network operators develop their networks, more and more of the required transport functionality will become available from the public domain. It is, however, unlikely that public networks will ever serve all applications required in business. Indeed, business interests would probably be best served if the public network operators confined their activities to transport of information only, i.e. layers 1–3 of the ISO model. General-service offerings for higher layers will be needed, but these will best be provided by specialist value-added service operators, using the transport facilities provided by the public-network operators. It also seems likely that the transport capability will more and more become a mixture of genuine and virtual private networks. This latter aspect will be greatly influenced by the adoption of outsourcing, because the outsourcing companies will inevitably seek the cheapest method of providing transport capability. This may render it extremely difficult for public-telecommunication-network operators to compete in the outsourcing business because they may find it difficult to justify one of their competitors' services, even if these are cheaper than their own.

For those companies which already operate their own private networks it is becoming more and more important regularly to review the reasons for the network's existence. In this regard, it is much easier to justify perpetuation of a network than to abandon it and use public facilities. It is equally significant to appreciate that a private network may not always be necessary to provide the perceived degree of functionality. This was undoubtedly the case in the past as far as data-communication networks are concerned. However, with the advent of digital public networks, and particularly the Integrated Services Digital Network (ISDN), this is unlikely to be the case for much longer. To exploit such services fully a significant amount of data-application development work needs to be carried out. For example, an application which allows the very fast set-up of ISDN calls from a personal computer could revolutionise data communication by permitting the economic use of a single digital circuit-switched network. Although the public ISDN can achieve this capability now, the required functionality is rarely available within personal computers.

It appears likely, assuming that current trends continue, that few businesses will operate their own exclusive private networks in the future. It is for this reason, more than any other, that the telecommunications manager should take steps to acquire business as opposed to technological expertise. In the past, the telecommunications manager has been fluent in technology and knowledgeable about business. The attributes required for the future will be to have fluency in business and be knowledgeable about technology.

11.11 References

1. MILLER, B.: 'Telecommunications for business' (Comm Ed Publishing, 1988)
2. CLARK, M.: 'Managing to communicate' (Wiley, 1994)
3. GRIFFITHS, J.M. (Ed.): 'Local telecommunications' (Peter Peregrinus, 1983)
4. GRIFFITHS, J.M. (Ed.): 'Local telecommunications 2' (Peter Peregrinus, 1986)
5. BELL, R.K.: 'Private telecommunications networks' (Comm Ed Publishing, 1988)
6. ELBERT, B.R.: 'Private telecommunication networks' (Artech House, 1989)
7. VALOVIC, T.S.: 'Corporate networks' (Artech House, 1993).
8. CLARK, M.P.: 'Networks and telecommunications: design and operation' (Wiley, 1991)
9. 'NCC handbook of data communication' (NCC Publications, 1982)
10. BREWSTER, R.L. (Ed.): 'Data communications and networks 2' (Peter Peregrinus, 1989)
11. HUNTER, J.M., LAWRIE, N., and PETERSON, M.: 'Tariffs, traffic and performance' (Comm Ed Publishing, 1988)
12. ELBERT, B.R.: 'Network strategies for information technology' (Artech House, 1992)
13. WALTERS, R.: 'Computer telephone integration' (Artech House, 1993)
14. BLACK, D.N., HARRISON, N.J., and HILL, R.J.: 'Migrating from the private digital network to the private ISDN'. *Proceedings of 2nd International Conference on Private Switching Systems and Networks*, 1992, *IEE Conf. Publ. 357*, pp. 8–118
15. TROUGHT, M.: 'Private networks: technology evolution to support the customer' Proceedings of the 2nd International Conference on *Private Switching Systems and Networks*, 1992, *IEE Conf Publ. 357*, pp. 24–30
16. SCOTT, J., and STANSELL, D.: 'Private networks: key considerations', Proceedings of the 2nd International Conference on *Private Switching Systems and Networks*, 1992, *IEE Conf. Publ. 357*, pp. 37–43
17. HARRIS, M.S.: 'Virtual private networks in Europe'. *Proceedings of 4th IEE conference on Telecommunications*, 1993, *IEE Conf. Publ. 371*

Chapter 12

Mobile communications

E.A.J. Boggis and C.I. Thomas

12.1 Introduction

Previous Chapters have concerned systems which are static. The customers' stations are in fixed locations. So are the nodes and links in the network, changing relatively slowly in response to customer demand. Although the network may contain radio systems, these are fixed point-to-point links. In contrast, when customers are on the move, the routes to them are constantly changing and communication with them must be by radio.

The simplest form of multiple-access radio system is called *net radio*. All users have transceivers tuned to the same frequency and hear all the traffic (propagation conditions permitting). A simplex channel is provided and a user switches from receive to send by means of a 'press to talk' button. Since only one conversation can take place at a time, a strict operating procedure must be observed for initiating and receiving calls. Such systems have been widely used for military purposes, by ships and aircraft and for local communication (e.g. on construction sites). Citizen's Band radio is another example.

In many systems, communication takes place between mobile units and a controlling base station. Examples are the private systems of taxi companies and fire, police and ambulance services. In this case, duplex operation is obtained by using two carrier frequencies; one is transmitted from the base station to the mobile units and the other used in the reverse direction.

Addition of a signalling channel to the speech channel enables a base station to call mobile stations selectively. This led to the introduction of public mobile-telephone services. The base station is connected to the PSTN in order to permit communication between its customers and the mobile-telephone users.

To accommodate a large number of mobile users, several radio channels must be provided and the mobile station is instructed which of these to use by a message sent over the signalling channel from the base station. The number of mobiles which the system can serve is determined by the number of radio channels and the amount of telephone traffic to be handled. The radio channels form a full-availability group of trunks accessed by the users. This system is therefore called *trunked mobile radio*.

A signalling channel can be provided without associated speech channels in order to obtain a paging system. The users have pocket receivers, but no transmitters. A paging system may provide only an audible 'bleep' to call the user, or it may transmit a short alphanumeric message which is displayed on the receiver. Various paging systems have ranges from a single site (e.g. for paging medical staff in a hospital) to nationwide coverage.

Early public mobile-radio-telephone services operated in the VHF band and so served wide areas. However, the available radio spectrum only accommodated a small number of channels and this restricted the number of customers that could be accommodated. Such systems are still used for private mobile systems (e.g. for road-transport fleets). In the UK, the former television Band III has been allocated for this purpose.

Public telecommunications operators now use *cellular-radio systems* [1–5]. These have provided an enormous increase in capacity and have attracted a corresponding growth in traffic. In a cellular network, a country is divided into a large number of small areas known as *cells*. Since the cells are small, low transmitter powers can be used and the same radio frequencies used in nonadjacent cells. Thus, for example, 1000 radio channels can serve about 1 000 000 users. To obtain sufficient radio spectrum for the large number of channels required, frequencies in the UHF band are used. In 1979, the World Administrative Radio Conference allocated the band from 862 MHz to 960 MHz for land-mobile-radio services.

Individual low-power short-range systems are used for cordless telephones. These also provide the basis of the *telepoint service*. Customers use cordless telephones to communicate with the PSTN via base stations situated in places where people congregate, such as airports, motorway service stations and shopping malls. Although the telepoint service is much cheaper than cellular radio, it is less useful because it cannot be used to receive incoming calls. In the UK, it was not a commercial success.

A geostationary satellite can provide a base station giving coverage over an extremely wide geographical area. For mobile communications, satellite systems have mainly been employed by specialist users, such as ocean-going ships and exploration teams in remote areas. The Inmarsat system [6] provides communications to ships at sea and its application is being extended to aircraft and other users.

Low-altitude satellites provide an alternative to geostationary satellites, but a large number are required to provide continuous coverage. A consortium led by Loral has proposed a system, called Globalstar, based on using 48 low-altitude satellites. In effect, this is a megacellular system providing global coverage. These systems will meet specific requirements, but they are unlikely to meet all the needs of the millions of customers for whom terrestrial cellular systems are designed.

12.2 Principle of cellular radio

The principle of cellular radio is illustrated in Figure 12.1, where each cell uses one of the available groups of frequencies (groups A–G). In this

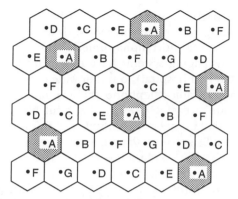

Figure 12.1 Principle of cellular-radio system (showing re-use of frequency groups A–G)

example, there are two other cells between every pair of cells using the same radio channels (e.g. group A), so cochannel interference between cells is negligible. Figure 12.1 shows an area tessellated with regular hexagonal cells. In practice, the effective boundaries of cells are irregular, because the distance over which effective communication is possible is determined by propagation conditions and these vary.

All cells normally have the same number of radio channels, but they are unlikely to have the same number of customers wishing to use them. This problem is solved, as shown in Figure 12.2, by having many small cells where there are many customers (e.g. in cities) and few large cells where there are few customers (e.g. in rural areas). Typically, cell radii may vary between about 1 km and 30 km. Originally, cells were served by a base station at a central site with an omnidirectional antenna. Nowadays, most masts carry directional antennas. As shown in Figure 12.2, three cells may be served by a base station situated where they meet and having antennae with 120° coverage.

12.3 Radio aspects

For a normal UHF radio link, the received signal is the sum of a direct wave and a reflected wave, as shown in Figure 3.21*a*, and the received signal strength varies with distance through a series of maxima and minima, as shown in Figure 3.21*b*. If the receiver is moving, the received signal will vary with time as its vehicle moves between these maxima and minima. In practice, the situation can be more complex, particularly in urban areas. There may be several reflected waves from different buildings. Moreover, the direct wave may be absent due to shielding by buildings. Propagation is then mainly by scattering and multiple reflections from surrounding obstacles. Hand-held mobile terminals are sometimes used inside buildings, where they may be

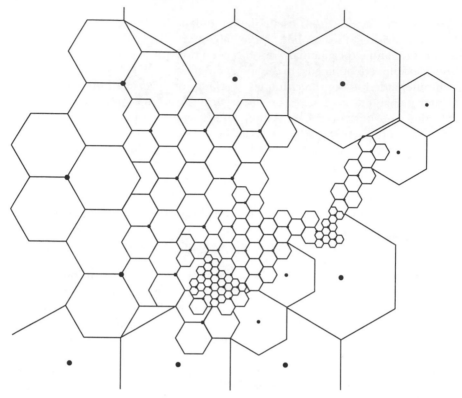

Figure 12.2   Use of cells of different sizes to cater for areas having many and few users

screened by metal in the structure of the building. If the receiver is moving, there is also a Doppler shift in the frequency components of the received signal.

It can be seen that the signal received by a vehicle moving in a cell is subject to frequent fading and interference results in a very poor carrier-to-interference ratio during a fade. Fades of up to 20 dB are frequent and fades exceeding 30 dB are not uncommon. Since fades occur at a distance of half a wavelength, and this is only 0.17 m at 900 MHz, a vehicle moving at 50 km/h can pass through several fades in a second. Since fading is a function of wavelength, it is selective, i.e. its effects are frequency dependent. If a signal has a relatively narrow bandwidth, the fading causes little distortion. However, for a broadband signal there is gross distortion. A high-bit-rate digital signal thus suffers dispersion of the pulses, which causes intersymbol interference.

Radiated power and receiver sensitivity must be adequate to maintain communication under these very adverse fading conditions. Also, the modulation method used must be able to cope with large changes in signal amplitude and with poor carrier-to-noise ratios. Analogue cellular systems therefore use frequency modulation. The second-generation cellular systems (described in Section 12.6) use digital modulation.

Characterising the mobile radio channel is not simple and much theoretical and experimental work has been done on it [1,3,5,7]. In practice, considerable effort needs to be devoted to cell planning, supported by computer-aided design techniques working on propagation models and digital maps and, of course, on-site optical and signal-strength surveys. By selecting the height of base-station antennas and using ground topology cleverly, good coverage of the cells can be obtained.

The concept of cellular radio depends on avoiding cochannel interference to the nearest cell where a frequency is re-used. One of the key aspects of obtaining good-quality cellular communication is therefore to use the lowest power levels which give a satisfactory carrier-to-noise ratio. The transmitter power of a mobile can be switched to several different levels and these are selected by the base station over a control channel. Using the minimum power level also extends the battery life of hand-held mobile telephones.

12.4 Operation of a cellular system

12.4.1 General

Each cell has a *radio base station* (RBS) for communicating with the users' *mobile stations* (MS). Duplex communication between the MS and the RBS is by a pair of radio channels. The *downlink* transmits from the RBS to the MS and the *uplink* from the MS to the RBS. Both voice channels and control channels are provided in each direction.

As shown in Figure 12.3, the RBSs in a group of cells are connected to a *mobile-service switching centre* (MSC), which controls their functions. The MSCs are connected by fixed links and have interfaces to the PSTN. Thus, calls can be made between mobile users and customers of the PSTN and between mobile users in different areas.

12.4.2 Subscriber identification and validation

In the fixed network, billing is by connection and is easily controlled. It is unlikely that someone will break into your premises and use your phone, thus making you liable for the cost of that call. For a mobile system, the dedicated connection does not exist; the only information available to identify a mobile subscriber comes from the identity provided to the system by the terminal equipment used by the subscriber.

As the directory number allocated to a mobile telephone is relatively easy to change, the telephone is also tied to a unique equipment-identity number. The network then requires the telephone to identify itself with both numbers, the number pairs having previously been stored by the network.

For further security, a PIN code can be used which is allocated to or derived by the subscriber but again linked to the identity of the terminal equipment. This area is known generally as *authentication*.

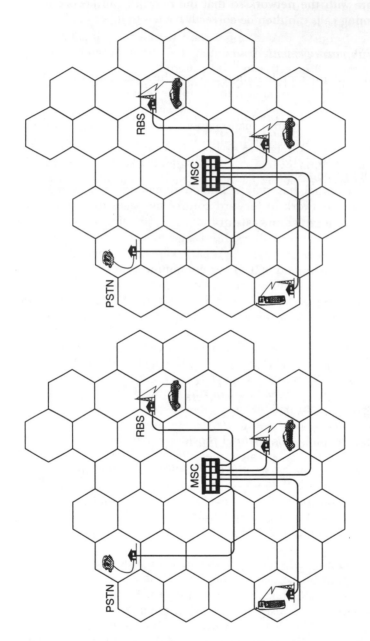

Figure 12.3   Cellular mobile-radio network

RBS=radio base station
MSC=mobile-service switching centre
PSTN=fixed network

All this information must be stored by the network and compared with that received from the telephone. The telephone must be identified for validation purposes every time it communicates with the network. The telephone must also communicate with the network so that the network can keep track of where it is; incoming calls can then be correctly routed to it.

12.4.3 Mobility management

For an originating call, the mobile station accesses the system, thus identifying its location. However, this is not known for a mobile terminating call. In fact, for both categories of call it is necessary to know which mobiles are active in a particular area and to store information about them locally, i.e. identification numbers and level of service. If this does not happen, every access would produce a request for data from the home data-storage facility, with possible delays and signalling congestion.

As a broadcast channel has only a limited signalling rate, it becomes impossible in a large network to forward control messages to all currently active telephones in all areas. In a large network incoming calls occur in excess of 100 per second at peak times. If it is known where the telephone is located, signalling can be selective, i.e. by area. An area is, in general, a group of cells connected to a common MSC or a subset of the cells connected to an MSC.

12.4.4 Registration

The process of notifying the network that a telephone is active on the system, i.e. switched on, is known as *registration*. This requires the telephone to signal to the network and can be used as a method for keeping track of the telephone's movement. The simplest form requires the telephone to announce itself on a regular basis. This is known as *periodic registration*.

The telephone continuously monitors system-overhead messages by listening to a broadcast channel. This channel provides system information to the telephone, part of which is an area identification. The telephone will automatically select a broadcast channel from a different cell if the quality of the signal received falls below a certain threshold. If, on selecting a new broadcast channel, a different cell identification is received the telephone will use this change of identifier to trigger a registration, letting the network know that the telephone is now active in the new cell. This type of registration is known as *forced registration*.

A further category of registration is *switch-on registration* and occurs when the telephone is first switched on.

12.4.5 Home data storage

The network must store data related to the mobile station, e.g. directory number, equipment serial number, class of service. To avoid requesting this information from the data store every time a system action occurs, such as registration or call origination, the data are stored temporarily locally in the MSC of the area where the customer is active.

When a registration occurs, the switch which has received that registration requests the customer data from the *home switch* or *home-location register*. At the same time, the home switch records where the request has come from, uses this information to identify geographically where the mobile station now is and records that it is, in fact, active. Calls to the customer can now be routed to the correct switch and subsequently to the correct base stations.

All the time a customer is active in an area, there is no need to communicate continually with the home database. A *visitor location register* has been created locally, thus minimising the interswitch signalling requirements.

12.4.6 Call set up

For a call originated by a mobile station, the following sequence of operations takes place:

(*a*) The user enters the number to be called.
(*b*) The call is initiated by pressing the 'send' button.
(*c*) The mobile checks the system-overhead data on the broadcast channel and performs a system access.
(*d*) This message contains information such as the subscriber identity, the mobile serial number and the dialled number.
(*e*) The system confirms reception.
(*f*) The MSC performs a validation on the identities offered and the dialled number.
(*g*) The mobile is assigned a voice channel. If one is not free, a release message is sent to the mobile and a 'call failure' tone is presented to the user.
(*h*) The mobile tunes to the allocated voice channel and, provided that certain signalling conditions are fulfilled confirming that two-way communication has been established, the call proceeds with ringing followed by conversation.

For a call to a mobile station, the following sequence occurs:

(*a*) An incoming call to the network is routed to the switch where the mobile is known to be active.
(*b*) If the customer is not marked as active, a message can be sent to the call originator giving notification of this.
(*c*) The RBS sends a 'page' message on the broadcast channel.
(*d*) The mobile detects the page and responds with a system access containing the validation information.
(*e*) The system confirms receipt of the mobile access.

(*f*) Following a successful validation, the mobile is instructed to tune to a voice channel.

(*g*) The system sends a message to the mobile to instigate ringing.

(*h*) The user lifts the handset and the audio path is completed between the customers.

12.4.7 Handoff

The mechanism by which a call is maintained even though the subscriber may have moved several tens of kilometres since the call was set up is known as *handoff* or *handover.* This involves passing the call from cell to cell as follows:

(*a*) The quality of the call is continuously monitored both for the absolute signal level and relative interference level.

(*b*) The cells surrounding the cell on which the call is currently in progress periodically determine if they are receiving a better signal from the mobile. In addition, the mobile itself may provide information to the base station on the performance of the downlink.

(*c*) If, as a result of these monitoring functions, the system decides that the call would be better maintained on one of these surrounding cells, then the system assigns a voice channel in the new cell, sets up a new voice path to that channel and instructs the mobile to retune accordingly.

(*d*) Once the system has confirmed that the mobile has appeared on the new channel, it releases the old channel in the original cell.

12.5 Analogue systems

12.5.1 Cellular radio

The first-generation cellular systems use analogue transmission. They employ frequency-division multiple access (FDMA). Each cell has a group of carrier frequencies and two of these (one for the uplink and one for the downlink) are allocated for each connection.

There are a number of cellular-radio standards in use. These include: the Advanced Mobile Phone System (AMPS) used in North America, the Nordic Mobile Telephone Service (NMT) used in Scandinavia, the Total Access Communication System (TACS) used in the UK, Network C used in Germany and Portugal, Radiocom 2000 used in France, RTMS used in Italy, the Nippon Automatic Mobile Telephone System (NAMTS) used in Japan and UNITAX used in China and Hong Kong.

All these systems operate on similar principles, but they use incompatible standards. For example, AMPS uses carriers spaced at 30 kHz, whereas TACS uses 25 kHz spacing. Thus, a mobile station designed for one system cannot be used on another. This makes it impossible for a mobile to operate while

Table 12.1   Comparison of key system parameters

System	TACS	GSM	DCS-1800	CT2	DECT
Forward band, MHz	935–950	935–960	1805–1880	864–868	1880–1900
Reverse band, MHz	890–905	890–915	1710–1785	864–868	1880–1900
Multiple access	FDMA	TDMA	TDMA	FDMA	TDMA
Duplex	FDD	FDD	FDD	TDD	TDD
Carrier spacing, kHz	25	200	200	100	1728
Channels/carrier	1	8	8	1	12
Bandwidth/channel, kHz	50	50	50	100	144
Modulation	FM	GMSK	GMSK	FSK	GMSK
Modulation rate, kbit/s	N/A	271	271	72	1152
Voice rate, kbit/s	N/A	22.8	22.8	32	32
Speech codec	N/A	RPE-LTP	RPE-LTP	ADPCM	ADPCM
Uncoded voice rate, kbit/s	N/A	13	13	32	32
Control-channel name	–	SACCH	SACCH	D	C
Control-channel rate, bit/s	–	967	967	2000	6400
Control-message size, bits	–	184	184	64	64
Control delay, ms	–	480	480	32	10
Peak power (mobile), W	0.6–10	2–20	0.25–1	10mW	250 mW
Mean power (mobile), W	0.6–10	0.25–2.5	0.03–0.25	5mW	10mW
Power control	Yes	Yes	Yes	Yes	No
Voice-activity detection	Yes	Yes	?	No	No
Handover	Yes	Yes	Yes	No	Yes
Dynamic channel allocation	No	No	No	Yes	Yes
Minimum cluster size	7	3	3	N/A	N/A
Capacity per cell, duplex channels/cell/MHz	2.8	6.7	6.7	N/A	N/A

N/A means not applicable

roaming across international borders.

Some parameters of the TACS system are listed in Table 12.1.

12.5.2 Cordless telephones

The first-generation cordless telephones also use analogue transmission. A European specification known as CT1 has been adopted in several countries [8]. However, there is no interoperability between the products of different manufacturers. This is satisfactory for cordless telephony, but it is an impediment to use in a telepoint service.

The UK version of CT1 provides eight HF carrier frequencies, in the range 1.642–1.782 MHz, from the base unit to the telephone and eight VHF frequencies, in the range 47.456–47.544 MHz, in the reverse direction. These can be used in 40 different combinations.

12.6 Digital systems

First-generation cellular-radio systems are already showing signs of being overloaded. The number of users has grown rapidly since their introduction and this rapid growth is still continuing. A new approach to system design is therefore required to accommodate more users within the existing available radio spectrum. This suggests trying to obtain more channels per cell. However, this means an increase in signalling activity as more handoffs occur and base stations handle more access requests and registrations. Thus, improved efficiency in access and handoff processing is required in addition to greater spectrum efficiency.

Digital transmission offers the possibility of reliable communication in the presence of high levels of noise and cochannel interference and this is of prime importance for cellular systems. However, if a voice signal is converted to a digital signal by conventional PCM, it requires a much greater bandwidth than for analogue transmission. The application of digital transmission in cellular-radio systems therefore depends on the use of more sophisticated methods of voice coding to give a lower digit rate than PCM.

Digital transmission offers alternatives to the use of frequency-division multiple access (FDMA). One possibility is to use code-division multiple access (CDMA) [3,5]. Each channel modulates a different pseudorandom sequence of binary pulses and these are combined on a common carrier. CDMA can operate successfully with almost zero signal-to-noise ratio. However, the receiver requires complicated digital signal processing to extract the signal of its particular channel. In time-division multiple access (TDMA), samples of the signal of a particular channel are transmitted in allocated time slots in repetitive frames containing a time slot for each channel. The same carrier frequency can thus be shared by all channels and it can be used for both uplink and downlink by sending in each direction at a different time in the frame. In both methods, each channel uses the complete bandwidth provided, instead of it being partitioned among the channels as in FDMA. This has the incidental advantage that a base station can use a single transmitter and receiver for a group of channels instead of one for each channel. CDMA is used for the Qualcom system [5] in the USA, but other systems use TDMA.

12.7 The GSM system

12.7.1 Principles

In Europe, a second-generation cellular standard called GSM has been adopted for a pan-European system. The specification for this was developed by a group set up by the Conference Européene des Postes et Télécommunications (CEPT) which was called the Groupe Spéciale Mobile. The initials GSM are now used to indicate 'Global System for Mobile Tele-communications' because, although it was designed primarily as a European system, it is now deployed in many parts of the world. Some parameters of the GSM system are shown in Table 12.1.

The GSM system [4,5] has been allocated the frequency bands from 935 MHz to 960 MHz for its downlinks and 890 MHz to 915 MHz for its uplinks. It uses TDMA with eight full-rate channels per carrier. The carrier spacing is 200 kHz and this gives the same spectral occupancy as an FDMA system with 25 kHz channel spacing. It is also expected that half-rate speech channels will be deployed, thereby doubling the spectral efficiency.

Since the TDMA access has eight time slots per radio carrier and the duration of each time slot is 0.58 ms, the frame duration is 4.6 ms. At the data rate of 271 kbit/s, multipath propagation leads to deep fades and to dispersion which causes intersymbol interference. Transmission errors are combated by channel equalisation and by forward error-correcting coding. In addition, frequency hopping at 217 hops per second is used to provide a frequency-diversity effect.

The form of speech encoding used is called *regular-pulse-excited linear-prediction coding* (RPE-LPC) [5]. This generates a pulse train at only 13 kbit/s (compared with 64 kbit/s for PCM). The addition of parity bits to give a forward error-correcting code increases the digit rate to 22.8 kbit/s. Since fading causes bit errors to occur in bursts, interleaving is also used. The information bits of a channel are dispersed among a number of code words. Thus, an error burst which corrupts a large part of a complete code word is dispersed over a number of information blocks and affects only a small part of the rearranged output of each at the receiver.

The GSM system is intended to allow mobiles to operate at distances of up to 35 km from a base station and the time taken for a signal to travel 70 km from a base station and back is then 233.3 μs. A guard period of 252 μs is therefore provided in the access burst. Since the data sent back by a mobile must fall into the correct time slot at the base station, the send timing at each MS must be individually adjusted. This is done by the BS measuring the round-trip delay at regular intervals and sending timing-advance signals to the MS throughout a call.

The GSM system supports other services besides telephony. These include fax and data at a maximum bit rate of 9.6 kbit/s. In addition, a 'short-message' service provides the facility to send or receive alphanumeric messages of up to 160 characters. Unlike conventional pagers, this is a confirmed-delivery service. The system must therefore store messages until the mobile station has been switched on.

12.7.2 System architecture

The architecture of the GSM system is shown in Figure 12.4. The functions of its principal entities are described below. Other network entities, which are not described here, are the equipment-identity register (EIR), the network-management centre (NMC), the operations and maintenance centre (OMC) and the short-message-service centre (SMS–SC).

The mobile station (MS) has two components. The first is the mobile equipment (ME), which contains all the audio and radio functions. There are five classes of GSM mobile equipment, ranging from the 20 W class 1 vehicle-mounted to the 0.8 W class 5 hand-held terminal.

The second component of the MS is the subscriber-identity module (SIM), which contains all the subscriber's unique information, such as the subscriber identity, PIN codes, authentication key and any abbreviated dial lists or subscriber preferences and defaults. The GSM SIM is a 'smart' card, the same size as a credit card, and typically contains an 8-bit processor and up to 8 kbyte E²PROM/RAM. The SIM can be moved by its owner from one MS to another MS and is actually the first real deployment of 'personal telecommunications' as well as of 'mobile telecomms'. To verify the authenticity of the mobile

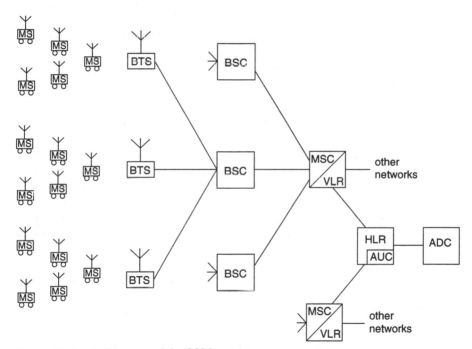

Figure 12.4   Architecture of the GSM system

ADC=administrative data centre
AUC=authentication centre (located at HLR)
BSC=base-station controller
BTS=base-transceiver controller
HLR=home-location register
MS=mobile station
MSC=mobile-service switching centre
VLR=visitor-location register (located at MSC)

station (i.e. ensuring that the network not only knows where it is, but can check that it is who it says it is!), it is periodically challenged with a random number. The mobile station's subscriber-identity module (SIM) uses a secure algorithm and a hidden individual identity key to calculate a response. If the response received back from the mobile station is as expected by the network, then the mobile is deemed to be authenticated. As a byproduct of the authentication activity, it is also possible to derive a ciphering key which is then used to encrypt calls between that mobile and the base station.

A base transceiver station (BTS) is the physical equipment used to provide the radio coverage within a cell. It contains the cellular-radio transmitters and receivers, a mast and either a small building or large cabinet. A mast may provide the coverage for more than one cell by using directional aerials, and it may support a microwave dish for the link to the main network.

The base stations have two types of channels:

(*a*) Traffic channels: These carry user data or speech. They are configurable to the type of use.
(*b*) Signalling channels: These control the behaviour of mobile stations or carry signalling to and from mobile stations. Different channels are used for broadcast control, common control (paging, random access and access grant), individual dedicated stand-alone MS-network signalling and individual traffic-channel-associated MS-network signalling.

The BTS also controls the transcoders required to convert the speech encoding used over the air interface (13 kbit/s regular-pulse-excitation–long-term-prediction) to 64 kbit/s A-law PCM.

To reduce system costs, a large proportion of the radio control functions are centralised in a base-station controller (BSC), which can support tens or possibly hundreds of base stations. The functions of traffic-channel allocation and handover are two which benefit from this centralisation. The BSC also has an important role in concentrating the BTSs' traffic from their lower-capacity links on to a single link to the MSC. Indeed, physically positioning BSCs in a network at their 'least-cost' location is an intriguingly complex problem.

The predominant functions of the mobile-service switching centre (MSC) are identical to an ordinary PSTN local exchange. For example, the MSC receives routing digits, selects routes, signals to other exchanges, generates call records for billing and provides the network-management system with statistics and alarms. However, the MSC is usually closely coupled to a visitor-location register (VLR) which performs additional functions specific to a mobile network. For example, the VLR holds the subscription information for every mobile station located within its coverage area (i.e. the coverage area of all the BTSs served by the BSCs which are in turn served by the MSC), and so it is able to process calls for its subscribers. In addition, the VLR stores the current location area of each mobile station so that it is able to page correctly. When a mobile station enters a new MSC/VLR coverage area, the

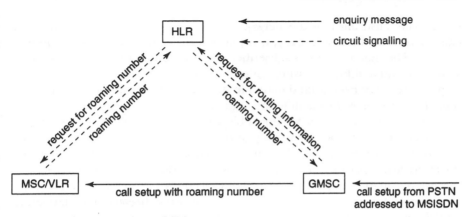

Figure 12.5   Call setup in GSM system

information in the old MSC/VLR is deleted.

The home-location register (HLR) is a database which holds the permanent record of the subscription information. Each customer is assigned to an HLR, and only one HLR. Alterations to HLR subscription information, e.g. change of a 'call-forwarded-to' number, can be made either from the mobile station or from the home-network operations and administration centre. Whenever the mobile station roams into the coverage area of an MSC/VLR, that MSC/VLR receives a temporary copy of most of the database. In return, the HLR stores the identity of the MSC/VLR to which the mobile station has roamed.

The HLR has a role in directing mobile-terminated calls to the correct MSC, as is shown in Figure 12.5. When a mobile-terminated call attempt is made into a mobile network, the receiving gateway switch (GMSC) initiates an enquiry to the HLR. This, in turn, sends the mobile identity to the appropriate MSC/VLR and requests in return a temporary *roaming number*. The roaming number is then used to set up the connection to the correct VLR; when the call arrives, the MSC/VLR, having allocated the number, knows which mobile station is being called and hence can page it.

The administrative data centre (ADC) controls the configuration of individual subscribers' service profiles within the HLR. These service profiles contain the list of the basic and supplementary services as well as the MSISDN (mobile station international ISDN number) of each subscriber.

12.7.3 GSM signalling

Because all the interfaces in a GSM have been standardised to be 'open', it is possible for the HLR, gateway MSC and MSC/VLR to be in different countries. Thus, a mobile station can roam abroad. It is also possible to construct networks with equipment from different manufacturers. This is due to the effort that manufacturers and network operators have put into the

ETSI-GSM standards working groups. Part of the reason for their success has been a rigid adherence to the layered protocol model. The remainder of this Section develops the concepts of multilayered protocols, and the services delivered by the different layers.

The signalling systems used between the different GSM network entities are derived from existing telecommunication standards. There are clear commercial reasons for doing this. So, for example, all circuit control between MSCs in most networks is by CCITT Signalling System 7 [9] and the call-control signalling between the mobile station and MSC is similar to Q.931 used for ISDN basic-rate access [10]. However, there are other signalling needs caused by mobility management and radio-resource control.

The mobility-management signalling requirements between the MSC/VLR and HLR are met by a specially designed CCITT no.7 *mobile application part* (MAP). The capabilities which are supported by MAP include:

(*a*) transfer of subscribers' service data from HLR to VLR;
(*b*) location updating from MSC/VLR to HLR;
(*c*) supplementary service management from MS via MSC/VLR to HLR;
(*d*) requests for routing information for mobile-terminated calls from gateway MSC to HLR to MSC/VLR; and
(*e*) requests for authentication-checking data from MSC/VLR to HLR.

Given that the CCITT Signalling System 7 network is being used to transfer MAP, then the message-transfer part (MTP) is used to provide error-free transfer. However, additional protocol layers are required to convert the service from MTP into one which is suitable for MAP usage. The full protocol stack specified for MAP is shown in Table 12.2.

The signalling between the mobile station and the network falls into three

Table 12.2 MAP protocols

Protocol layer	Function
Mobile-application part	Defines mobility-management operations
Transaction-capability-application part component sublayer	Provides remote operations 'framework'
Transaction-capability-application part transaction sublayer	Provides transaction management, association binding and version control
Signalling-connection-control part	Provides extended addressing within a point-code and/or beyond a point-code domain
Message-transfer part (layer 3)	Provides addressing within a CCITT Signalling System 7 network's point-code domain
Message-transfer part (layer 2)	Provides error detection and recovery on a link
Message-transfer part (layer 1)	Provides 64 kbit/s links

categories:

(*a*) broadcast system information,
(*b*) mobility management, and
(*c*) call control.

There is a series of broadcast messages from the BTS to all mobiles within the cell. The information contained in these broadcast-system-information messages includes the identity of the network, location area and cell, frequencies of channels and general information upon how an MS should behave. The only requirement from the underlying signalling protocol is error detection. If the message is detected as corrupted and cannot be reconstituted using its forward error correction then it is discarded. This is acceptable, because another copy will arrive a few seconds later. This error detection is achieved by a cyclic-redundancy checksum on the message between the BTS and MS. Random-access, paging and access-grant messages use a similar technique, except that a preset number of retransmissions is requested by the application at the outset of the procedure.

The mobility-management signalling between the mobile and network terminates within the MSC/VLR. Consequently, there are additional signalling-protocol requirements. First, across the air interface, error detection and recovery are performed using a modified version of LAPD, the link-layer protocol for the basic-rate ISDN access. Between the BTS and BSC, a more standard variant of LAPD is employed. Fortunately, the addressing capability of LAPD is adequate to identify which channel at a BTS is involved. However, the addressing problem between the BSC and MSC is far greater. This addressing issue is solved by establishing a different signalling-connection-control-part (SCCP) connection for each mobile-station-to-MSC signalling session. Consequently, the BSC can transfer the mobility-management message transparently between the BTS link and the appropriate SCCP connection. The SCCP connection uses the services of MTP to provide error-free delivery between the BSC and MSC. The call-control signalling is handled in an identical fashion to mobility-management signalling.

During a call, there is a continual stream of messages between the MS and BSC reporting the power levels of adjacent cells' broadcast-control channels, the performance of the current traffic channel, and from the BTS to MS requesting changes to the MS's transmitted power level. The measurements are averaged at the receiving entity; error detection is therefore required but error recovery at the link layer is unnecessary, indeed it is actually a nuisance because it will delay the reporting of more up-to-date information!

12.8 Personal communication networks

There has been rapid growth in demand for cellular radio and it is forecast that demand will grow still more in future, particularly if costs can be reduced. For example, there are now over three million cellular customers in the UK and it has been estimated that this number could increase to eight or ten million! There is insufficient frequency spectrum available in the 900 MHz band to cater for this.

These forecasts led to proposals to introduce systems operating at higher frequencies. This allows the re-use of RF carriers at shorter distance, in order to use smaller cells, known as *microcells*. Thus, many more cells can be provided and this enables many more customers to be accommodated. In the UK, frequencies in the 1800 MHz band have been allocated for this service, which is known as a *personal communication network* (PCN) [5] A standard for such systems, known as DSC 1800, has been developed by ETSI. Some parameters are listed in Table 12.1.

PCN systems [5] operate on the same principles as GSM. However, the GSM system caters for car-portable terminals transmitting up to 20 W, whereas PCN caters only for hand-portable terminals transmitting at 0.25 W or 1 W. Base stations are also of low power and so should be small and cheap compared with those for conventional cellular systems.

Microcells have a radius of only a few hundred metres and propagation conditions can cause gaps between them in radio coverage. This can be combated by providing an *umbrella cell* covering several microcells. To make a call, a mobile first tries to find a free channel in a microcell. If this is unsuccessful, a second attempt is made in the umbrella cell. This not only improves reliability, but also enables the network operator to avoid the cost of equipping microcells in areas where traffic is very light.

If it can be sufficiently cheap, a PCN network may eventually become an economic competitor to the fixed-cable network for normal customer access. This will enable the same telephone, with the same directory number, to be used at home or away from it.

12.9 Digital cordless telephones

The first digital cordless telephone is the CT2 system [5,8], used in the UK and several other countries. This operates in the band 864.1–868.1 MHz. A channel spacing of 100 kHz enables a choice of 40 channels to be provided. The common air interface specified for CT2 telephones [11] enables them to be used in telepoint applications. Some parameters of the CT2 system are listed in Table 12.1.

The same carrier frequency is used for transmission in both directions by using time-division-duplex operation. A frame of 2 ms length contains 144 bits sent at 72 kbit/s. The first half of the frame is used for transmission from the fixed station to the portable and the second half for transmission from the portable to the fixed station. Each contains 66 or 68 bits: 64 for speech and two or four for signalling. This caters for speech transmission at 32 kbit/s (using ADPCM) and 1 kbit/s or 2 kbit/s for signalling. Since distances are short, a guard interval of only 7.6 μs is provided between the two halves of the frame to accommodate propagation delay.

A European standard, known as Digital European Cordless Telecommunications (DECT), has now been produced [5,8]. This also uses time-division-duplex operation, but with 12 channels per carrier and a carrier spacing of 1.782 MHz in the band from 1880 to 1900 MHz. Some parameters of the DECT system are also listed in Table 12.1.

The DECT system is more complex than CT2, but it is intended for a wider range of applications [12]. These involve data as well as voice communication, including use for access to LANs and X.25 packet-switched networks. The additional cost, due to the comprehensive facilities of DECT, make it unlikely to replace CT2 for basic cordless telephony. The two technologies will probably prove complementary over the range of possible applications.

12.10 Conclusion

Personal mobile services already exist. However, present cellular-radio services are expensive and lower-priced telepoint services, based on cordless telephones, are less useful. New digital systems are being introduced, including the pan-European GSM system and PCN networks.

Designers and standards bodies are already considering possible third-generation systems which could lead to a universal personal mobile service [13, 14]. In Europe, this is called the *Universal Mobile Telecommunications Service* (UMTS). Internationally, it is called the *Future Land Mobile Telecommunications Service* (FLMTS). The objective is a universal service capability supported by a series of complementary radio interfaces.

If this is linked to an intelligent fixed network, it could give customers identical services whether on a wired or a radio connection. The use of a telephone number associated with a person, rather than a place, should achieve the goal of being able to use the same number to call somebody whether they are at home, at work, in the same country or travelling abroad.

12.11 References

1. JAKES, W.C.: 'Microwave mobile communications' (Wiley, 1974)
2. LEE, W.Y.C.: 'Mobile cellular communications' (McGraw-Hill, 1989)
3. PARSONS, J.D. and GARDINER, J.G.: 'Mobile communication systems' (Blackie, 1989)
4. MACARIO, R.C.V. (Ed.): 'Personal and mobile radio systems' (Peter Peregrinus, 1991)
5. STEELE, R. (Ed.): 'Mobile radio communications' (Pentech Press, 1992)
6. DALGLEISH, D.I.: 'An introduction to satellite communications' (Peter Peregrinus, 1989)
7. PARSONS, J.D.: 'The mobile radio propagation channel' (Pentech Press, 1992)
8. TUTTLEBEE, W.H.W. (Ed.): 'Cordless telecommunication in Europe' (Springer-Verlag, 1990)
9. MANTERFIELD, R.J.: 'Common-channel signalling' (Peter Peregrinus, 1991)
10. CCITT Recommendation Q.931: 'ISDN user-network interface Layer 3 specification for basic call control'

11. GARDINER, J.G.: 'Second generation (CT2) telephony in the UK: telepoint services and the common air interface', *Electron. Commun. Eng. J.* 1990, **2**, pp. 71–78
12. HOWETT, F.: 'DECT: beyond CT2', *IEE Rev.* 1992, **38**, pp. 263–267
13. CHIA, S.T.S., and GRILLO, D.: 'UMTS: mobile communications beyond the year 2000', *Electron. Commun. Eng. J.*, 1992, **4**, pp. 331–340
14. MACARIO, R.C.V. (Ed.): 'Modern personal radio systems' (IEE, 1996)

Chapter 13
Packet switching
J. Atkins

13.1 Introduction

Early demand for switched data services, during the 1960s and early 1970s, arose largely from the high cost of computing equipment. To make the most effective use of the expensive data-processing equipment, time sharing by remote terminals became an attractive choice for many users. The ubiquitous public switched telephone network (PSTN) was pressed into service to provide the necessary switched access. However, as computer technology developed, it became clear that the PSTN would not be adequate for many of the new applications and that public switched networks designed specifically for data communication would be needed [1].

Packet switching, first proposed in the 1970s, initially for military applications, was identified as the best way of implementing these networks and the CCITT (now the ITU-T) began work to develop the necessary standards. The most important of these was Recommendation X.25. First published in 1976, X.25 defined a robust data-transfer service which provided the foundation for the rapid growth of both public and private packet-switched networks, and these networks are now in widespread use throughout the world.

During the 1980s, as personal computers became established as an essential part of the workplace, *local-area networks* (LANs) were introduced to interconnect computer equipment within offices, buildings or sites. These LANs use 'shared-medium' network topologies for which packet switching was the natural choice. The term *wide-area network* (WAN) is often used to distinguish data networks (such as X.25 networks) which do not have the LAN's restrictions on geographical scale. Although the general principles of packet switching remain the same, the differences in technology between LANs and WANs are substantial and warrant separate treatment [2]. The reader should also be aware of the term *metropolitan-area network* (MAN) which is used to refer to packet-switched networks based on a development of shared-medium LAN technology (usually known as the IEEE802.6 MAN) but which is capable of covering a 'metropolitan' area with a radius of up to several tens of kilometres.

The X.25 standard was designed to use analogue transmission systems which are noisier and more prone to transmission errors than their modern digital counterparts. Analogue circuits also tend to have substantially lower transmission speeds than digital circuits of comparable price and therefore introduce more delay to data crossing the network. (We will see later that 'round-trip' delay is important in packet switching.) To overcome these impairments, X.25 has a complex data-transfer protocol which can detect and correct transmission errors. This complexity absorbs a lot of processing power in packet switches and terminal equipment, which limits throughput. (X.25 services tend to be limited to data rates up to 64 kbit/s, though one or two network operators offer high-speed X.25 services at rates up to 10 Mbit/s.) It can also seriously interfere with higher-layer protocols which may be running end-to-end between the communicating terminals, i.e. it is not very 'transparent' to higher-layer protocols.

To take advantage of the higher transmission speeds and much lower error rates offered by modern digital transmission systems, new types of packet-switched services have recently been developed which use very much simpler data-transfer protocols than that of X.25. Of particular importance are *frame relay* and *switched multimegabit data service* (SMDS). Both of these offer higher throughput than X.25, together with a high level of transparency to higher-layer protocols [3].

After introducing the basic ideas of packet switching, this Chapter describes X.25, the standard on which today's 'mature' packet-switched public data networks are based. It then gives an outline of Frame Relay and SMDS, which are the emerging packet-switched services designed for the digital era.

13.2 Packet-switching principles

13.2.1 Store-and-forward switching

Computers handle information in blocks. They store data in 8-bit bytes or multiples of this, and these are organised into larger blocks or 'files', usually with a hierarchical naming or 'directory' structure, so it is not surprising that data communication is really about transferring blocks of information rather than continuous bit streams. These blocks may be very short, such as individual 8-bit 'characters' representing letters of the alphabet or similar symbols, or they may be complete files. Usually they are somewhere in between.

If we wish to transfer continuous bit streams through a network we will set up a continuous connection between the communicating parties at the start of the transaction and clear it when the transaction has been completed. This, of course, is circuit switching as used in the familiar telephone network. However, as Figure 13.1 shows, blocks of data can be switched through a network without setting up a continuous end-to-end connection. They can be stored temporarily at each switch *en route* before being forwarded on an appropriate outgoing circuit. This method is known as *store-and-forward*

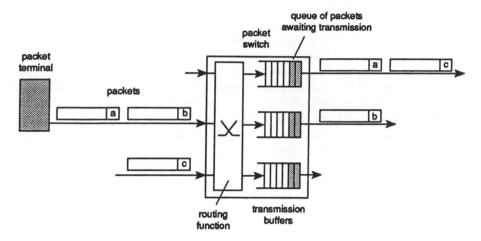

Figure 13.1 Store-and-forward switching

switching or *message switching*.

To avoid the transmission of short messages being unduly delayed by the sending of a long message, the long message can be broken down into short blocks of data known as *packets*. Short messages (consisting of single packets) can then be interleaved between the packets of the long message. This is known as *packet switching*.

Message switching or packet switching requires that each data block carries control information which the switches use to decide how to route the data block. As Figure 13.2 shows, the term 'packet' refers to the data block plus the control information in the *header* which is attached to the front of the data block (in order of transmission). For reasons which will become clear, the header also carries other control information in addition to that used for routing.

Note that Figure 13.1 gives a highly simplified representation of a packet

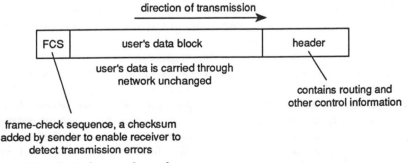

Figure 13.2 Basic format of a packet

switch. For example, in a real packet switch the routing function (together with other control functions such as flow control and error control) would tend to be implemented in software and the packets would also queue at the input for the available header processing power.

In practice, moving packets from an input (processor) queue to an output (transmission) queue may be accomplished simply by passing pointers between input and output software processes rather than by physically moving packets between storage areas. Such details are not covered here. Our purpose in this Section is to illustrate the basic ideas involved.

Packet switching then is based on store-and-forward switching and its characteristics are different from those of circuit switching. For data communication it has the following advantages:

(*a*) efficient transmission,

(*b*) low end-to-end error rates, and

(*c*) speed matching.

Efficient transmission is obtained on the transmission links between switches on a route because packets associated with different transactions are interleaved so that the available bandwidth is shared between users. Thus, users only occupy transmission capacity when they have information to send (unlike a connection through the telephone network which has transmission capacity dedicated to it even when the communicants are not speaking). Packets routed to the same outgoing transmission link compete for the available transmission capacity. They are queued for transmission in order of their arrival (usually), and, when their turn comes, they are transmitted at the full bit rate of the link. This can be contrasted with time-division multiplexing (TDM) as used in circuit-switched networks in which each user gets only a fraction of the bandwidth.

In the same way, packets belonging to different transactions can be interleaved on the local access link connecting the user to the serving switch, permitting simultaneous communication with a number of remote users. This is of great value where many users want to access the same remote computer at the same time, in an electronic-messaging system for example. Before packet switching was available, anyone wishing to access a remote multi-user mainframe computer had to set up a separate connection through the telephone network with a modem at each end. This involved a large number of access circuits at the mainframe with a modem for each one; the modem room for a time-sharing bureau could be very large. The introduction of packet switching enabled time-sharing bureaux to provide the same multiple-access capability with only a few access circuits and the large modem rooms disappeared.

Low end-to-end error rates can be obtained because each packet is stored temporarily at each switch. Thus, any errors introduced in transmission can be corrected by arranging for corrupted packets to be retransmitted. This error correction can be performed on a link-by-link basis. By this means, the resulting end-to-end error rate seen by the user can be made very small indeed. To facilitate error detection, each packet has to carry an error-checking code which is used at the receiving end of the link to detect the presence of transmission errors. Usually this is achieved by adding a frame-check sequence (FCS) to the end of the packet, often referred to as the 'trailer', as shown in Figure 13.2.

Another benefit arising from store-and-forward switching is that the network effectively buffers the users' terminals from each other so that they can use access circuits of different speeds. For example, an electronic-mail server which would typically serve several hundred remote users simultaneously would probably use 64 kbit/s access circuits, while the individual user might have only a 2.4 kbit/s access circuit. The danger arises, of course, that a faster terminal may swamp a slower terminal by sending data faster than it can be digested. It is therefore necessary to operate flow control between the terminals to prevent this. The description of X.25 in Section 13.3 explains the commonest way of achieving this, using what is known as a 'sliding-window' mechanism.

13.2.2 Connection-oriented and connectionless services

There are two distinct types of packet service: *connection-oriented* and *connectionless*. In the connectionless service there is no concept of setting up a connection between the communicating parties. The network treats every packet as a self-contained entity. It has no sense of packet sequence; successive packets from one terminal to another are treated as independent and may even take different routes through the network. Every connectionless packet, sometimes known as a *datagram*, has to carry the full address of the destination user.

However, many transactions involve the exchange of large numbers of data packets, for example transferring a large file. A more efficient use of network resources can then be achieved using a connection-oriented service in which a 'connection' is set up between communicating parties before dialogue can begin. This is not a continuous physical connection as in the circuit-switched case; rather it is a logical route established through the network over which all packets belonging to the connection pass using store-and-forward switching at each switch. To distinguish it from a 'real' connection this type of connection is known as a *virtual connection* or *virtual circuit*. The aim is to convince users that they have a real connection whilst reaping the benefits of efficiency and flexibility that store-and-forward switching bring. Part of this trick, of course, is to make sure that packets belonging to a virtual connection are delivered in the same order as they are sent, and we will see how this is achieved in practice when we look at X.25.

Like real circuits, virtual circuits may be either permanent, i.e. set up at subscription time by the network operator, or set up by the user when required. From the user's point of view a permanent virtual circuit (PVC) is always in the data-transfer phase. A switched virtual circuit (SVC) involves signalling between the user and network.

It is important not to confuse 'connection oriented' and 'circuit switched'. While circuit switching is necessarily connection oriented, they are not the same thing. Connection oriented means that there is a connection-setup phase, followed by a data-transfer or 'conversation' phase, and eventually there is a connection-clear phase. However, the 'connection' may be either circuit switched or packet switched.

13.2.3 Characteristics and performance

With circuit switching, a connection has well defined requirements in terms of network resources (transmission and switch capacity). If the network has the necessary resources available when a user attempts to set up a new connection, the attempt will succeed. If it has not, the attempt will fail and the call will be lost. For this reason circuit-switched networks are often referred to as 'loss' systems. However, once a connection has been established, the user has sole use of the associated resources and other network traffic, whether established connections or new connection attempts, does not interfere with it.

The characteristics of circuit-switched connections are also well defined and consistent. End-to-end delay is short and constant, consisting mainly of propagation and switching delays. Also, the end-to-end performance reflects that of the underlying transmission systems. Furthermore, any failures within the network which affect the resources used by a connection generally result in loss of the connection.

Packet-switched networks, however, have completely different characteristics. A virtual connection does not have well-defined requirements. Even characterising packet traffic is somewhat problematical! To characterise fully the traffic a particular virtual connection will present to the network, we need to know the probability distributions of packet-arrival times (or interarrival times) and packet lengths, and different virtual connections will generally have different traffic profiles. In practice this information is not known.

As illustrated in Figure 13.1, once the packets enter the network, at each switch *en route* they queue for transmission on the appropriate outgoing link, the capacity of which is shared between many virtual connections on a packet-interleaved basis. Characterising this aggregated traffic on the interswitch links is somewhat easier than for individual virtual connections, since the law of large numbers comes into play and a Poisson model is commonly used for (comparative) ease of calculation. This model assumes that packet arrivals are random and that packet lengths have a negative exponential distribution. (The Poisson model is a favourite in queueing theory because of the resulting simplicity of the mathematics, but it also has the merit of giving a reasonable approximation to what actually happens in practice.)

The ideal switched network allows the user to set up a connection whenever required, with no transmission errors, no transit delay and no interruptions caused by equipment failure or inadequate provision. In practice this ideal is not attained, largely because of the need to dimension networks on an economic basis of provision rather than the maximum possible demand. In addition, practical transmission systems cannot avoid introducing propagation delay and occasional errors.

The main impairment introduced by a circuit-switched network is 'blocking', i.e. a call attempt failing because the network resources needed are not available at the desired time. There is also always the possibility, usually very small, that an established connection might fail.

The main impairments introduced by a packet switched network are:

(*a*) cross-network delay, which can vary for successive packets belonging to the same virtual connection because of queueing. In contrast with circuit-switched networks, packet-switched networks are referred to as 'delay' systems; and

(*b*) loss of packets and possible reduction of throughput, usually caused by congestion.

Because users compete dynamically for the network's resources on a packet-by-packet basis, a packet-switched network accepts each packet from the user without knowing whether it will actually have the resources available to deliver it. There will therefore inevitably be times, hopefully rare, when a packet-switched network will accept packets which it does not have the resources to handle. Typically, a packet switch somewhere *en route* would run out of transmission buffers. As we will see, X.25 networks operate a data-transfer protocol which enables such a congested network temporarily to stop a user sending further packets into the network until the congestion has cleared. This is *network flow control*. Frame-relay and SMDS networks, on the other hand, do not operate explicit network-flow control. They leave it to the higher-layer protocols running directly between the communicating parties to relieve the load on the network if they detect the onset of congestion (for example by a sudden increase in the cross-network delay).

13.2.4 A packet-switched network as a queueing system

A packet network can be viewed as a 'queue of queues'. The behaviour and performance of such queueing systems are notoriously difficult to model in detail, even when the traffic is well characterised [4]. In practice, a combination of analytical and simulation methods is often used. However, one can obtain a flavour of how such queueing systems behave by looking at an isolated queue, as shown in Figure 13.3 which represents an individual packet switch.

The queueing-theory parlance is that 'customers' arrive, queue (if necessary), are 'served' by a 'server' and exit the system. The time the server takes to deal with a customer, when that customer reaches the head of the queue, is the 'service time'. In our packet-switching example, packets are

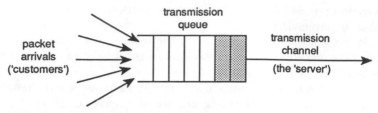

Figure 13.3   Simple single-server queueing system

customers, the transmission channel is the server, and the service time is the time it takes to transmit the packet (which is proportional to the packet length).

The system is fully characterised if we know the arrival statistics of the packets, the service-time statistics, and the 'queueing discipline' (i.e. the order in which customers in the queue are dealt with by the server). In this simple case we will assume first-come-first-served, but in a practical example different packets in the queue may have different priorities or there may be a number of queues served by a transmission link, one queue for each of several priority classes.

The behaviour of even a single queue such as this may be difficult to analyse, especially if the dynamic behaviour of the system is of interest. It all depends on how amenable the arrival and service-time statistics are to mathematical treatment. For simplicity we restrict ourselves here to a system in equilibrium, i.e. the traffic statistics are constant and a steady state has been reached. We will also assume that packets arrive at random (i.e. a packet is as likely to arrive in any given short interval as in any other interval of the same duration) and at a constant average rate (i.e. the same number of packets arrive in any long interval as in any other interval of the same duration). Packet lengths have an equivalent distribution which is of random length with a constant mean length. This is the best-known simple queueing model and is known as the $M/M/1$ queue [4].

Figure 13.4 shows how delay and queue length (strictly speaking the number of packets in the system including any being transmitted) vary with increasing link utilisation. Remember that this represents a single packet switch. The end-to-end performance will not be exactly the same, but the general pattern will be similar. Remember also that any protocols operated between the switches (e.g. network-flow control and link-by-link error correction) would need to be incorporated into the network model, since they may significantly modify the behaviour of this simple example.

At low levels of link utilisation, the queueing delay increases fairly slowly as the offered load increases, but beyond a utilisation of about 50% the performance deteriorates rapidly. This dramatic sensitivity of packet switches to overload is illustrated very clearly. In practice, this means that packet switches have to keep a close eye on visible indicators of incipient congestion, such as transmission-queue size, and take protective action when preset thresholds are reached. They should not wait until all buffers are full, or even nearly full, before taking corrective action.

This very simple model provides useful guidance in practical design situations. For example, a transmission utilisation of 50% means that on average a packet spends as long in the queue waiting to be transmitted as it does actually being transmitted. Also, at 50% link utilisation there is on average less than one packet in the transmission queue. In fact, though it cannot be deduced from Figure 13.4, if only 10 packets-worth of storage was provided in the transmission buffer, fewer than 1 packet in 1000 would be lost for lack of buffer space at a link utilisation of 50%. In practice, a transmission utilisation of 50% is commonly used in network specifications.

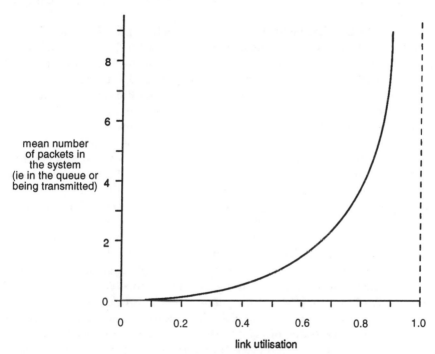

Figure 13.4   Behaviour of a transmission queue in a packet switch

13.2.5 Congestion and network flow control

Even in well-designed packet networks, the unpredictable nature of the packet traffic means that short-term peaks will inevitably occur from time to time somewhere in the network. We have seen how sensitive the transmission queues are to even small increases in link utilisation if the utilisation is high, as it could be during traffic peaks. So there are likely to be occasions when a switch runs out of buffers and becomes congested. Because this congestion loses packets, which probably triggers requests for their retransmission from adjacent switches, the congestion tends to spread. Practical packet-switched networks incorporate *network-flow control* and *congestion management* which are designed to reduce the likelihood of congestion arising, contain any congestion which does occur and promote graceful recovery.

Network-flow control can be thought of as an element of congestion management. It involves detecting the onset of congestion, preferably before it reaches a level which calls for drastic action such as discarding packets. To achieve this, the switches may simply monitor key traffic indicators, such as queue lengths, but it may also involve switches keeping track of transit delays through the network, as these rise rapidly as congestion sets in (see Figure 13.4), and switches keeping their neighbours informed of approaching or actual congestion.

When congestion has been detected, the switches should operate mechanisms which effectively throttle overload traffic at source, i.e. from the users, both to prevent the situation from getting worse and to assist graceful recovery. There are numerous ways of doing this. To a large extent the designer's choice depends on the data-transfer protocol operating between the user and the network. The X.25 protocol includes explicit features which enable the network temporarily to stop the user from sending any more packets into the network, but frame relay does not have this capability and depends to some extent on the willing co-operation of users in reducing their demands. An important consideration is how to apply this throttling 'fairly' over users. (We will see later how service 'contracts' help in this.)

If, despite applying network flow control, a switch actually runs out of buffer space for packets, it will of course discard them. This may lead to the demand on the network actually increasing because of the greater incidence of packet retransmissions. There is therefore a danger that the congestion will grow. Congestion management is concerned with controlling this and recovering the situation as rapidly and gracefully as possible.

Figure 13.5 shows the effect of congestion on a network's throughput (i.e. its capacity to carry traffic). Initially, the network throughput increases linearly with the total traffic offered to the network, since it has no difficulty carrying all the traffic it is offered. As the offered traffic continues to increase, however, the transmission queues get longer as the utilisation of the interswitch circuits increases (see Figure 13.4), and packets take longer to get through the network. Each packet will therefore consume more of the network's resources (especially buffer storage) and the network's ability to carry additional traffic diminishes. Eventually, as severe congestion sets in, further increases in offered traffic will actually reduce the capacity of the network to carry traffic, as shown in Figure 13.5*a*. Introduction of network-flow control reduces this deterioration, as shown in Figure 13.5*b*.

The effectiveness of network-flow control will determine how frequently the network creeps into the 'onset-of-congestion' zone. The effectiveness of

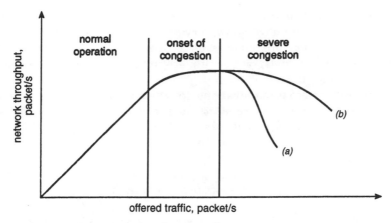

Figure 13.5   *The effect of network flow control*

    a Without flow control
    b With flow control

the congestion management will determine how well the network sustains its throughput in the 'severe-congestion' zone.

13.3 Network-design choices

13.3.1 Network topology

The choice of network topology will depend mainly on the number and geographical distribution of terminals, the cross-network delay targets, the speeds of the links between switches, the performance of the available switches and, of course, the nature of the application. For a tactical military network, where survivability is important, a highly-coupled mesh would obviously be appropriate.

A public switched network, on the other hand, is probably best organised as a hierarchy, an arrangement which permits large-scale growth while effectively constraining cross-network delay. Interswitch links in current public packet-switched networks generally have a capacity of 48–64 kbit/s. With typical cross-network delay targets of about 300 ms, this means at most a three-level hierarchy.

Packet-switched corporate-data networks would be designed against cost/performance criteria which reflect the company's business and geographical locations (international leased lines are particularly expensive!) and would probably have an *ad hoc* structure.

13.3.2 Interswitch protocol

The distinction has already been made between virtual-call and datagram operation in relation to the service seen by the user, i.e. the user's interface with the network. The same options arise for the interfaces between the switches. Of course, for a datagram service the question of virtual circuits in the network does not arise. However, a virtual-call service may be implemented using either a datagram or vitual-circuit protocol within the network, as shown in Figure 13.6.

Despite the numerous ramifications, for example on the choice of routing strategy, neither approach seems to have a decisive advantage. Thus, in practice, both methods are used.

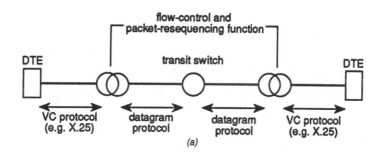

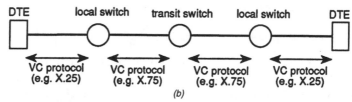

Figure 13.6   Network protocols

    a Datagram network protocol
    b Virtual-circuit network protocol

13.3.3 Routing

Routing is concerned with how the switches choose the 'correct' outgoing route given the identity of the destination terminal. In the early work on military networks, survivability was a prime requirement and routing strategies were favoured which did not need knowledge of the network's topology. Generally, these made very poor use of network resources, caused long transit delays, or both. Practical routing strategies for public packet-switched networks exploit full knowledge of network topology.

The simplest approach is to use a fixed routing look-up table at each switch, with one entry for every possible destination switch. These tables would be calculated using well known methods on the basis of known link capacities and an assumed traffic matrix for the network. If the traffic was smooth and accurately reflected the assumed traffic flows, and the network elements were reliable, this simple scheme would be fairly effective. In practice, these conditions are not usually found and a large variety of adaptive-routing algorithms has been devised in which the routing tables are automatically updated from time to time in an attempt to improve the utilisation and performance of the network.

These adaptive routing strategies may be classified according to whether the table updates are based on purely local usage information (e.g. queue lengths within the switch), on information received from other switches, or on both, and whether the tables are recalculated locally by each switch, centrally at a network routing centre, or a combination of these. It is obvious that adaptive routing interacts with flow and congestion control, and some networks take the logical step of combining them in a single unified strategy.

The scope for adaptive routing is clearly greater in a datagram transit subnet where a separate routing decision may be made for every packet. In a virtual-circuit network the routing decision is made only once, at virtual-call set-up time.

13.4 The X.25 standard

13.4.1 Structure of the standard

This Section presents a brief overview of X.25, currently the commonest packet-switched service used in wide-area data networks. It is intended only as an introduction; a complete description would run to several hundred pages! But it serves to put some flesh on the skeleton of packet switching given in Section 13.2. However, a CCITT/ITU Recommendation is not a complete guide to implementation, and differences inevitably exist between X.25 packet networks and in the detailed services and features they support. Moreover, the standards do not stand still; they evolve to meet new requirements and satisfy the ever-increasing expectations of customers. Thus, different networks will not necessarily support the same version of the standard.

In standards parlance the customer's terminal is known as a *data-terminal equipment* (DTE) and the network is usually referred to as the *data-circuit-terminating equipment* (DCE). Strictly speaking, X.25 defines the procedures, rules and data formats (i.e. the protocol) applying at the interface between the DTE and the DCE [5], as shown in Figure 13.7, but the service interface as seen by the user, i.e. the interface over which data are sent and received, is actually buried inside the terminal.

The DTE may be any device which supports the X.25 protocol, e.g. a desktop personal computer or a mainframe host computer, or it may consist of one or more 'dumb' terminals, such as low-speed asynchronous character-oriented VDUs together with a *packet assembler–disassembler* (PAD) which

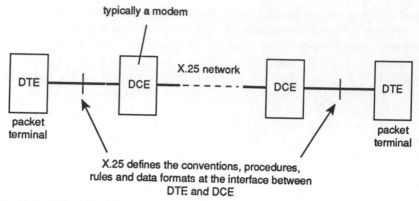

Figure 13.7   What X.25 defines

converts between the character-mode protocol and X.25, as described in Section 13.4.5. If access to the packet network is provided over an analogue line, the DCE will be the modem.

X.25 is structured in three distinct layers as shown in Figure 13.8, reflecting the lowest three layers of the OSI seven-layer reference model [6]. The

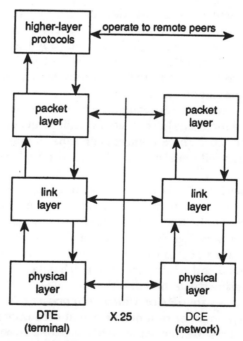

Figure 13.8   The structure of X.25

principle of the OSI reference model is that each layer provides a defined service to the layer above, and that this service is achieved by means of a 'peer-to-peer' protocol. For X.25, from the bottom up the layers are:

(i) *The physical layer.* This provides a circuit between the link-layer process in the terminal and the link-layer process in the serving packet switch so that they can interchange information.

(ii) *The link layer.* This provides for the error-free transfer of packets over the access circuit. The link-layer protocol (usually referred to as LAPB) includes procedures for setting up and disconnecting the link layer, for transferring packets (including flow control, sequence control, correction of transmission errors and error recovery).

(iii) *The packet layer.* This is concerned with setting up and clearing virtual calls and managing the exchange of data packets during the data-transfer phase, again including flow control and sequence control.

13.4.2 The packet-layer protocol

13.4.2.1 Virtual circuits

X.25 is a virtual-circuit protocol which supports multiple simultaneous virtual connections to different remote terminals. A user can have *permanent virtual circuits* (PVCs), or *switched virtual connections* (SVCs), or any combination of PVCs and SVCs. Permanent virtual circuits are set up by the network provider at subscription time and may be considered to be permanently in the data-transfer phase.

13.4.2.2 Virtual-call setup and clear

Switched virtual circuits are set up by an initial exchange of signalling packets, as shown in Figure 13.9.

A terminal wishing to set up a new SVC creates a CALL REQUEST packet and passes it (via the link layer and physical layer) to the serving packet switch. This packet is then routed by the network to the called terminal where it is delivered as an INCOMING CALL packet. The called terminal accepts the call by returning a CALL ACCEPTED packet, which the network delivers to the calling terminal as a CALL CONNECTED packet. The virtual connection then enters the data-transfer phase. These four 'signalling' packets contain full addresses for both calling and called terminals. The addresses used in public data networks [7] conform with CCITT Recommendation X.121, as described in Section 9.2.5.

When a virtual connection is set up it is allocated a *logical channel identifier* (LCI). This is contained in all packets in the data-transfer phase to identify to which of the user's virtual connections each packet belongs. Remember that the user may have a large number of virtual connections set up simultaneously to different remote terminals. When a virtual connection is cleared, its logical channel number is returned to a 'pool' for subsequent reuse.

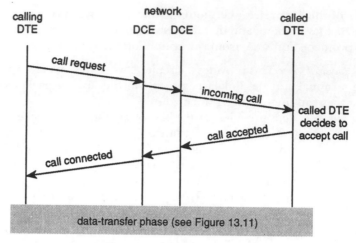

Figure 13.9   X.25 virtual-connection setup

Either terminal may initiate clearance of the call at any time, using the packet sequence shown in Figure 13.10.

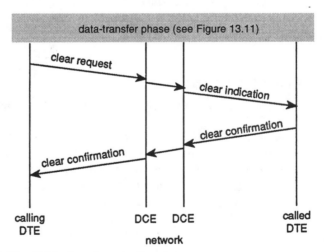

Figure 13.10   X.25 virtual-connection clear

13.4.2.3 The data-transfer phase

Several different types of packet may be used during the data-transfer phase. The most important is the DATA packet, the type normally used to carry user data. To facilitate sequence control, every DATA packet carries a *send sequence number P(S)*, which is a modulo-8 (or optionally modulo-128) count beginning at '0' for the first packet to be transmitted after call set-up.

X.25 uses what is known as a 'sliding window' flow-control mechanism to control the rate at which a user may send DATA packets into the network for each virtual connection. This permits the user to send up to an agreed maximum number of packets *w* into the network, after which no more will be accepted until further permission is given. This permission takes the form of acknowledgments returned by the receiver to indicate that previous packets have been received. As long as there are fewer than *w* DATA packets awaiting acknowledgment, the transmit window is open and further packets may be sent on the virtual connection. As soon as the number of DATA packets awaiting acknowledgment reaches *w*, the window closes and the network will temporarily refuse to accept any more. The receipt by the user of outstanding acknowledgments for that virtual connection will reopen the window and further DATA packets may then be sent.

w is known as the maximum window size. It is an important parameter for the user to set because it has a major influence on the throughput (i.e. the packets per second) which may be achieved on a virtual connection. Usually it is set in the range 2–7. For each virtual connection this flow-control mechanism operates independently for each direction of transmission.

To facilitate acknowledgment, packets contain a receive sequence number *P(R)* which is set equal to the send sequence number expected in the next packet to be received, that is, one more than the send sequence number of the last packet that has been acknowledged.

Figure 13.11 gives an example of packet-layer data transfer. We assume that

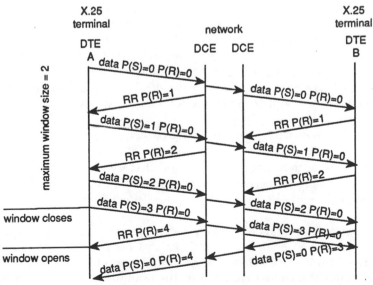

Figure 13.11   Example of X.25 packet-layer data transfer

data transfer has just started so that all sequence numbers start from '0'. Also, for simplicity, we assume that the maximum window size is 2.

Terminal A begins by sending a DATA packet with send sequence number $P(S)$ set to '0'. The receive sequence number $P(R)$ is also set to '0', indicating that the next DATA packet terminal A expects to receive on this virtual connection should have send sequence number set to '0'.

If the acknowledgment process depended solely on DATA packets to carry acknowledgments, then the acknowledgment to this first DATA packet would have to wait until the switch serving terminal A had received a DATA packet from terminal B for delivery on this virtual connection, which could cause long delays. So two other types of packet are used to carry acknowledgments during data transfer: the receive-ready (RR) packet and the receive-not-ready (RNR) packet. They are short, only three octets, and carry receive sequence numbers but not send sequence numbers.

As shown in Figure 13.11, a receive-ready packet is used to acknowledge receipt of a DATA packet from terminal A with send sequence number '0'. The receive sequence number in this RR packet is set to '1', indicating that the next DATA packet from terminal A is expected to have a send sequence number of '1'.

This acknowledgment is generated by the network, not the remote terminal, so it indicates only that the first DATA packet has been received by the serving packet switch. It does not mean that it has been received by the remote terminal DTE B. This is typical; acknowledgments usually have 'local' significance, not end-to-end significance. There is a mechanism which can be used to force end-to-end rather than local acknowledgment (the so-called D-bit facility, described below). However, with the cross-network delays commonly found in X.25 networks (typically several hundred milliseconds) this can have a devastating effect on throughput. End-to-end acknowledgment therefore tends to be used only in applications which involve small amounts of data and where it is important to have assurance of delivery, such as EFTPoS transactions (electronic funds transfer at the point of sale).

This first DATA packet is delivered by the network to terminal B, which acknowledges its receipt by returning a receive-ready packet to the network with $P(R)$ set to '1'.

Terminal A then sends another DATA packet, with $P(S)$ set to '1'. $P(R)$ is still set to '0', since it has not yet received any DATA packets from terminal B that need to be acknowledged. This second DATA packet is acknowledged by the network using a receive-ready packet with $P(R)$ set to '2', and it is delivered to terminal B which also acknowledges it using a receive-ready packet.

Terminal A then sends two DATA packets in rapid succession, with send sequence numbers set to '2' and '3'. At this point the window closes for terminal A, since it now has the maximum permitted number of DATA packets awaiting acknowledgment (in this example, the maximum window size is set to '2') and no more DATA packets may be sent until the window is re-opened. In this example DATA packets numbered '2' and '3' are acknowledged by a single receive-ready packet with *P(R)* set to '4', illustrating the important point that a single acknowledgment can be used to acknowledge receipt of a number of DATA packets. This re-opens the window and terminal A can then send further DATA packets.

After receiving the DATA packet with *P(S)* set to '2', terminal B sends its first DATA packet to terminal A. This DATA packet has *P(S)* set to '0' since it is the first one terminal B has sent. It has *P(R)* set to '3', acknowledging the receipt of DATA packet number 2 from terminal A. However, when this DATA packet from terminal B reaches the packet switch serving terminal A this *P(R)* is changed by the switch from '3' to '4' because the next DATA packet the switch expects from terminal A should have the send-sequence number set to '4'. This example emphasises that X.25 defines what happens at the interface between a terminal and the network, rather than what is happening between the two terminals.

This simple example of packet-layer data transfer illustrates some of the basic ideas. It illustrates sequence control: a terminal would immediately know if a packet had been lost or misordered in transit since the send-sequence numbers would be out of sequence. It illustrates flow control: a terminal cannot simply send DATA packets whenever it feels like it. It may only send them when the transmission window is open. However, there are numerous subtleties in this flow-control procedure that it is not possible to cover here. The reader should refer to the complete CCITT X.25 recommendation for a full description.

Figure 13.12 shows the format of a DATA packet. In addition to the send and receive sequence numbers *P(S)* and *P(R)* and the logical channel identifier *LCI*, there are several other bits of interest. They are briefly outlined

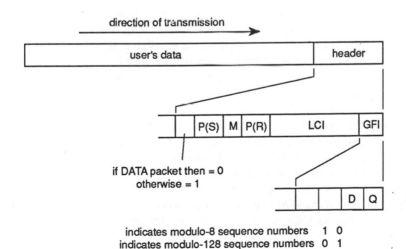

Figure 13.12   Format of X.25 DATA packet

below.

(*a*) *The D-bit*

We have already seen that one of the issues which often causes difficulties with X.25 is the question of whether acknowledgments apply only over the local DTE-DCE interface, or whether they have end-to-end significance and indicate that the remote terminal has actually received the acknowledged packets. This issue was clarified in the 1980 revision of X.25, which provided for an explicit indication of whether acknowledgments are local or end-to-end. This is accomplished by setting a bit in the packet header (the 'delivery confirmation' bit, or more commonly the D-bit) to '1' if acknowledgments are to have end-to-end significance, and to '0' if acknowledgment is local. It should be recognised, however, that some X.25 networks may not implement the D-bit option.

(*b*) *The M-bit*

In many cases the user wants to send a data block which is too long to fit into a single X.25 packet. This data block has to be segmented (i.e. broken up) into a sequence of smaller blocks which will each fit into an X.25 packet. There is a one-bit field in the header of a DATA packet which can be used to indicate to the remote user that the packet is part of such a sequence. This is the 'more-data' bit, or M-bit. For example, if the user's data block required a sequence of 10 X.25 packets to carry it, then the M-bit would be set in the first nine. The M-bit is passed transparently (i.e. without modification) through the X.25 network from one user to the other.

(*c*) *The Q-bit*

The 'qualifier' bit, or Q-bit, is another single-bit field in the header of a DATA packet which is passed transparently through the X.25 network from one user to the other. It is used to indicate the nature of the information in the packet's user-information field. For example, it could be set to '1' to indicate that the packet's user information is control information, and set to '0' to indicate that it is data. In Section 13.4.5 we will see an example of how it is used to indicate that a DATA packet is carrying X.29 control messages between a packet terminal and a packet assembler/disassembler (PAD).

13.4.2.4 The interrupt procedure

This feature enables a packet terminal to send information to the remote terminal without following the normal flow-control procedure. To do this, the terminal sends a DTE INTERRUPT packet, which is delivered to the remote terminal as a DCE INTERRUPT packet carrying the same data field as the original DTE INTERRUPT packet. Acknowledgment is by means of an INTERRUPT CONFIRMATION packet, which has end-to-end significance. Only one INTERRUPT packet may be outstanding (i.e. unacknowledged) at any time, and the interrupt procedure does not affect the transfer of DATA packets.

Until 1984 the interrupt packet was limited to a data field of one octet. This was increased to 32 octets in the 1984 revision. INTERRUPT CONFIRMA-TION packets do not carry user data.

13.4.2.5 Receive-ready (RR) and receive-not-ready (RNR) packets

We have already seen how receive-ready (RR) packets may be used for efficient acknowledgment of DATA packets, but they also have a more direct role to play in flow control.

RNR packets are used to indicate a temporary inability to accept any more DATA packets on the associated virtual call. A terminal (DTE) or network (DCE) receiving an RNR packet should not send any more DATA packets on that logical channel until given permission to resume. An RR packet is used by a terminal or the network to indicate that it is ready to receive data packets, and in effect cancels an RNR condition.

13.4.2.6 Virtual-call-reset procedure

During the data-transfer phase a virtual call may be reset to effect recovery from a minor error condition. The reset procedure resets send sequence numbers to '0', and removes any data or interrupt packets which may be queued by a terminal for transmission, in transit in the network, or received by a terminal but not yet acknowledged. The packet sequence for effecting reset is shown in Figure 13.13.

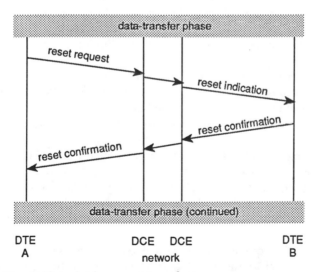

Figure 13.13   X.25 packet-layer reset procedure

13.4.2.7 Restart procedure

In the event of a serious error condition, it may be necessary completely to reinitialise the DTE–DCE interface covering all virtual connections on that interface. This is achieved using the restart procedure, which uses a similar message sequence to that for reset (see Figure 13.13), but with RESTART REQUEST, RESTART INDICATION and RESTART CONFIRMATION packets.

13.4.3 The link-layer protocol

13.4.3.1 Frame format

The X.25 packet-layer protocol depends on a reliable means of transferring packets between terminal and network. The link-layer protocol provides this. It includes procedures for setting up the link (i.e. establishing the data-transfer state), data transfer and clearing the link down. During data transfer, the protocol operates sequence-control, error-control and flow-control procedures. However, these are completely separate from those operated at the packet layer.

Because of overlap in the development of HDLC by ISO and X.25 by CCITT, the original link-layer procedure was based on draft HDLC procedures which were subsequently revised before publication. This is known as LAP (Link Access Procedure). However, in 1977 a fully-standardised link-access procedure was agreed, known as LAPB, based on the HDLC asynchronous balanced mode of operation. Both are still included in X.25, but LAPB is the preferred version and all new implementations follow this. Only LAPB will be covered here.

Packets are carried in LAPB frames having the general format shown in Figure 8.15. Several types of frame are used. The INFORMATION frame (often simply I-frame) is used to carry packets. Other types of frame are used for control and supervision of the data link; in most of these the information field is absent.

Each frame begins and ends with a flag sequence of '01111110' which is used by the receiver to identify the frame boundaries. It is necessary to prevent the flag sequence (01111110) from being simulated within the frame between the beginning and end flags. This is achieved by zero-bit insertion and deletion (otherwise known as bit-stuffing and destuffing), as described in Section 8.5.2.

The frame-check sequence (FCS) is a two-octet checksum added to the frame immediately before the end flag. It is calculated by the sender from the bit pattern between the last bit of the start flag and the first bit of the FCS field by applying a specific coding algorithm. At the receiving end, a complementary algorithm is performed which identifies the presence of transmission errors. This process is not completely foolproof in that there are combinations of errors which cannot be detected; however, these are rare and the likelihood of transmission errors remaining undetected is very small.

13.4.3.2 Link-layer setup and disconnection

Before the link layer can transport packets over the DTE-DCE interface it has to be set-up in the data-transfer state. This is achieved by the exchange of frames shown in Figure 13.14. The network indicates that it is able to set up the data link by transmitting contiguous flags. The terminal initiates link setup by sending a *set-asynchronous-balanced-mode frame* (SABM). Basically this is a subset of HDLC. The network confirms setup by returning an *unnumbered-acknowledgment frame* (UA) and the link enters the data-transfer state.

To disconnect the link, the terminal sends a *disconnect frame* (DISC). The network responds with a UA frame and the link enters the disconnected state.

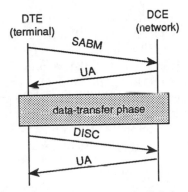

Figure 13.14   Setting up and disconnecting at the link layer

13.4.3.3 Link-layer data transfer

X.25 packets are transferred over the link in INFORMATION frames or I-frames. The link layer operates a sequence-control and flow-control scheme which is basically the same as that used by the packet layer for a virtual connection. All I-frames are numbered (usually modulo-8, optionally modulo-128), using a send sequence number $N(S)$. All I-frames contain a receive sequence number $N(R)$ which indicates that all I-frames numbered up to and including $N(R)$-1 have been received correctly. These are analogous to $P(S)$ and $P(R)$ as used by the packet layer.

These sequence numbers are used by the link layer to ensure correct sequencing of frames and to operate a sliding-window flow-control algorithm exactly like that already described for the packet layer. Again, flow control operates independently in each direction of transmission.

Like the packet layer, the link layer has receive-ready (RR) and receive-not-ready (RNR) frames. RNR frames are used to indicate a busy condition, i.e. a temporary inability to deal with any more I-frames. RR frames are used to clear the busy condition reported by an earlier RNR frame. They efficiently acknowledge the receipt of I-frames if there are no I-frames ready to be transmitted that could carry the acknowledgments, exactly like their packet-layer counterparts.

13.4.3.4 Error recovery

Received frames which contain errors are simply discarded. They may have failed the FCS check, indicating transmission errors, they may be too short, or they may contain invalid addresses. A sequence error, where the $N(S)$ is different from that expected (probably owing to a frame having been discarded), requires the missing frame or frames to be retransmitted. This is achieved by returning a REJECT frame (REJ). On receiving a REJ frame, the terminal or network initiates sequential retransmission of I-frames starting with the one indicated by the $N(R)$ contained in the REJ frame.

A FRAME REJECT RESPONSE frame (FRMR) is used by the terminal or network to report an error condition not recoverable by retransmission of the identical frame. This initiates the data-link-resetting procedure, in which the terminal or network sends a SABM command. On receipt of a UA response, data transfer can resume with all sequence numbers beginning from '0'. If, on correct receipt of a SABM, the terminal or network determines that it cannot recover from the error condition, it will return a DISCONNECT MODE frame (DM) and the link will enter the disconnected state.

13.4.3.5 Summary of LAPB frames

The above very brief overview of LAPB has introduced all nine types of frame and outlined their function. These frames fall into three distinct classes: information frames (which carry X.25 packets), supervisory frames (which are involved in sequence and flow control and error control), and unnumbered frames (so-called because they do not contain send or receive sequence numbers). The control field for each class has a distinct format, as shown in Figure 13.15.

LAPB frames are defined as commands or responses. The I, DISC and SABM frames are always commands. DM, UA and FRMR frames are responses. REJ, RNR and RR frames may be commands or responses. The P/F bit is referred to as the poll bit in commands and the final bit in responses. If a command is sent with the P-bit set then the associated response must have the F-bit set. It is used as a handshaking mechanism to resolve ambiguities where there are two commands awaiting acknowledgment. It may also be used to force an acknowledgment to an outstanding I-frame. For this brief introduction to LAPB the distinction between commands and responses is not important, but it is useful to be aware of the terminology. In fact, the reader will look in vain for an explanation of these terms in CCITT Recommendation X.25. Those interested in this detail should refer to the ISO standard for HDLC from which LAPB is derived [8].

bit	8	7	6	5	4	3	2	1	
I	-----N(R)-----			P	-----N(S)-----			0	information frames
REJ	-----N(R)-----			P/F	1	0	0	1	supervisory frames
RNR	-----N(R)-----			P/F	0	1	0	1	
RR	-----N(R)-----			P/F	0	0	0	1	
DISC	0	1	0	P	0	0	1	1	
DM	0	0	0	F	1	1	1	1	
FRMR	1	0	0	F	0	1	1	1	unnumbered frames
SABM	0	0	1	P	1	1	1	1	
UA	0	1	1	F	0	0	1	1	

Figure 13.15   Control-field formats of LAPB frames

13.4.4 The physical layer

The physical layer is concerned with the electrical and physical aspects of the interface and the procedures needed to effect data transfer at the bit level. Typically, the physical layer would embrace modems and equivalent devices which convert data signals into a form suitable for transmission over the access circuit.

13.4.5 X.25 support for character-mode terminals

Many of the terminals in daily use are simple character-mode devices which do not have X.25 capability. However, it is often important for these terminals to interwork with X.25 packet-mode terminals. For example, a character-mode terminal may wish to access a remote electronic-mail server connected to an X.25 network. This interworking is achieved by means of *packet assembler/disassembler* equipment, invariably known simply as a PAD, as shown in Figure 13.16.

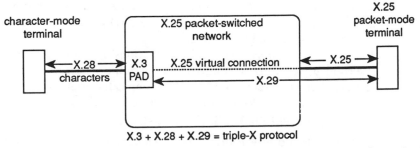

Figure 13.16   Interworking of X.25 packet terminals and character-mode terminals

Character-mode terminals send and receive characters, which are usually IA5 characters [9], (where IA5 stands for International Alphabet number 5). The PAD assembles the character stream from the terminal into X.25 packets which are then switched through the packet-switched network and delivered to the remote X.25 terminal. In the other direction of transmission, the PAD disassembles the packets and transmits the characters they contain one at a time to the character-mode terminal.

Three CCITT/ITU standards [10,11,12] define what the PAD should do and how the two types of terminal (character-mode and X.25) should communicate with it (and thus with each other). These are recommendations X.3, X.28 and X.29, often known simply as the Triple-X protocol.

X.3 defines the characteristics of the PAD in terms of parameters. For each character-mode terminal it serves, the PAD holds a set of parameters (a profile) reflecting the capabilities of the terminal and the functions which are to be performed by the PAD.

A character-mode terminal sends information to the PAD a character at a time. When a pre-agreed 'data-forwarding' character (such as a carriage return or control character) is sent by the terminal, the PAD assembles the string of characters it has just received into the information field of an X.25 DATA packet and sends it to the remote X.25 terminal. The identity of the data-forwarding character is one of the X.3 parameters in each terminal's profile. Alternatively, the X.25 packet will be assembled and sent on expiry of the 'data forwarding timeout', which is another parameter in the profile, ranging from zero to 12.8s in steps of 50 ms.

The complete set of X.3 parameters is too extensive to cover here in detail, but the following examples give a flavour of what is involved:

Parameter number	Description
1	*PAD recall using a character.* This enables a character-mode terminal to escape from the data-transfer state in order to send PAD command signals.
2	*Echo:* This, when set, provides for characters received from the character-mode terminal to be echoed back to the terminal as well as being interpreted by the PAD.
3	*Selection of data forwarding characters.* This allows selection of one or more characters to be recognised by the PAD as an indication to complete assembly and forward a packet.
4	*Padding after carriage return.* This provides for the automatic insertion by the PAD of nonprinting 'padding' characters in the character stream sent to the character-mode terminal after a carriage return to allow a printer time to complete the carriage return before printing the next character.

X.28 defines commands and procedures used by character-mode terminals to control the PAD parameters and to set up and control a virtual call. It also defines the service and control signals sent back by the PAD in response to those commands.

X.29 defines the commands and indications which may be sent between the PAD and the remote packet terminal, enabling the latter to read PAD parameters and set them remotely (if permitted), to request clearance of the virtual call and to indicate command errors (in either direction). These commands and indications are carried in X.25 DATA packets, which are distinguished from user DATA packets by having the Q-bit in the header set to '1', as described in Section 13.4.2.3.

The PAD is shown in Figure 13.16 as an intrinsic part of the packet-switched network, but it may alternatively be part of the user's equipment. For example, a multi-user PAD at the customer's premises can support several character-mode terminals simultaneously. However, as far as the network is concerned, such customer premise-based PADs are simply X.25 packet terminals.

13.5 Frame relay

13.5.1 Principles

Like X.25, frame relay [3,13–19] is a virtual-call protocol; there is no provision for a connectionless service. But it differs from X.25 in two important ways. First, the call-control signalling is 'out-of-band', i.e. it is carried in a different logical channel to the user's information. Secondly, it employs a very much simpler protocol in the data-transfer phase. These key features of frame relay reflect the fact that it was first developed as an ISDN packet service, being based on the ISDN service architecture in which user information and signalling are kept logically separate from one end of a connection to the other, as shown in Figure 13.17.

The ISDN can be viewed as two closely related 'subnets': a signalling subnet, shown below the broken line in Figure 13.17, and a switched-information subnet. The signalling subnet provides the call-connection control capability needed to set up, control and clear connections in the switched-information subnet. Much of the power and flexibility of the ISDN accrues from this logical separation of signalling and user information.

From the user's point of view a frame-relay virtual connection would be set up in exactly the same way as a circuit-switched connection (such as a telephone call), details of which can be found in Chapter 14. The same out-band signalling protocol (but enhanced to support the additional features needed for packet services [20]) is used to create the connection. However, instead of a circuit-switched channel through the switched information subnet, the user would get a frame-relay virtual connection.

In a practical implementation, the frame-relay switches would be accessed (without the user's involvement or knowledge) by setting up circuit-switched connections through the main ISDN switch block. If the user signalled a

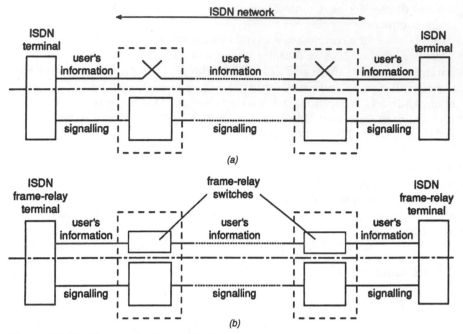

Figure 13.17   Frame relay in the ISDN

    a Circuit-switched service
    b Frame-relay service

request for a frame-relay virtual connection, a path would be set up automatically to the nearest frame-relay switch which would then set up the virtual connection in conjunction with other frame-relay switches in the network.

13.5.2 Protocols

During the data-transfer phase a very simple frame-relay protocol [21] operates between the terminal and the network. Information is carried in variable-length packets, invariably referred to as 'frames', with the simple format shown in Figure 13.18.

As with X.25, the user may have a number (potentially a large number) of virtual connections set up simultaneously to different destinations. The header of each frame contains a label agreed between the user and network at connection-setup time, known as the *data-link-connection identifier* (DLCI), which uniquely identifies each virtual connection. Also, like X.25, these virtual connections may be SVCs or PVCs, or a combination of both.

The frame-relay data-transfer protocol corresponds to a very thin 'slice' of the link-layer protocol operated by X.25. As illustrated in Figure 13.19, this is referred to as the data-link core protocol (or in standards parlance the data-

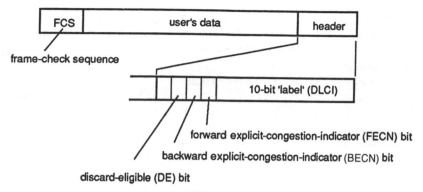

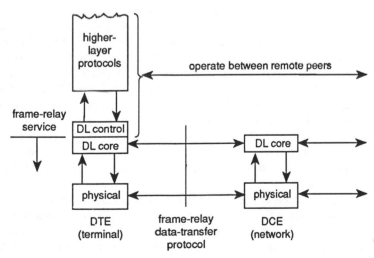

Figure 13.18 Frame-relay frame format

link core sublayer). Together with the *data-link control sublayer,* this forms the link-layer protocol as defined by the OSI Reference Model (ISO 7498). Unlike X.25, there is no explicit flow-control procedure operating as part of the frame-relay data-transfer protocol. Specifically, the core protocol covers the following:

(*a*) Frame delimiting, alignment and transparency. This is exactly the same as for X.25, where the standard flag sequence is used to indicate the beginning and end of each frame, together with zero-bit insertion and extraction.

(*b*) Frame multiplexing and demultiplexing using the 10-bit DLCI (see Figure 13.18).

(*c*) Checking that the frame is neither too long nor too short and that it consists of an integral number of octets (before zero-bit insertion).

(*d*) Detection of transmission errors using the frame-check sequence at the

Figure 13.19 Frame-relay protocol stack for data transfer

end of each frame. If transmission errors are indicated, the network simply discards the frame. Unlike the X.25 link layer, no attempt is made to correct these errors. It is left to protocols operating end-to-end between the terminals to take care of error correction.

(*e*) Congestion control indications. These are carried by the forward and backward explicit-congestion-notification bits (FECN and BECN, respectively) shown in Figure 13.18. The FECN bit would be set by the network to indicate to the receiving terminal that the frame had encountered congestion. Similarly, the BECN bit would be set by the network to indicate to the sending terminal that frames it is sending into the network may encounter congestion. These indications would be used by higher-layer protocols operating end-to-end between the terminals to adjust the rate at which they send information.

(*f*) Priority. The discard eligible (DE) bit (see Figure 13.18) is used by the network as a simple indicator of priority. Frames with the DE bit set would, in the event of congestion, be discarded by the network in preference to frames in which the DE bit was not set.

The simplicity of the frame-relay data-transfer protocol can be gauged from the fact that it can be described in so few words. Basically, a terminal sends frames into the network, and the network does its best to deliver them to the intended recipient. If congested resources are encountered *en route* (such as a full transmission buffer), this is indicated to the terminals, but frames may be lost. Also, if a frame contains errors, the network simply throws it away and leaves it to higher-layer protocols operating end-to-end between the terminals to correct it.

13.5.3 Applications

Its simplicity makes frame relay useful at higher data rates than X.25 and gives it a high degree of 'transparency' to higher-layer protocols. These merits were quickly recognised for application outside the ISDN and the earliest implementations are not ISDN based; they are stand-alone data networks [3]. Indeed, the standards specifying frame relay as an ISDN bearer service now have their counterparts specifying frame relay as a non-ISDN data-transmission service [22,23]. The basic approach has been to preserve everything in the standards which is not ISDN-dependent, while freeing the remainder from the ISDN framework. For example, the signalling protocol used to set up and clear SVCs is the same, but in the non-ISDN case it is carried in frames with a specific DLCI reserved for signalling rather than in the ISDN D-channel.

The main application driving the introduction of frame relay has been the rapidly growing market for LAN interconnection. One of the features of this application is that it does not really require random interconnection. What is needed is the packet equivalent of leased lines between a company's LANs, so the earliest implementations of frame relay provide only PVC services; there is no provision for setting virtual connections up by means of user signalling. In effect the signalling subnet of Figure 13.17 is missing.

A description of frame relay would not be complete without mention of the parameters which determine the service the user will get. While the protocol described above specifies the nature of the data-transfer service, it does not say anything quantitative such as how much data the user may transfer in a given time. In practice, a user would not buy a service without knowing what is obtained for the money, and frame relay is no exception.

The three key parameters which are used to define the service contract between the user and service provider are the *committed information rate (CIR)*, the *committed burst size* (B_c) and the *excess burst size* (B_e), as follows:

(*a*) *CIR* specifies the maximum rate, averaged over a period T_c, at which the network will accept data with a 'guarantee' that it will be delivered. The nature of this 'guarantee' is a feature of the specific service subscribed to and is likely to be different from different service providers. In practice, the guarantee usually means 'best endeavours'. It is in the nature of packet services generally that guarantees are not absolute! During the period T_c a terminal would typically transmit a number of frames separated by contiguous flags as interframe time fill.

(*b*) B_c specifies the amount of data (in bits) that the network will accept in time interval T_c, with the 'guaranteed' likelihood of delivery. T_c is normally calculated as B_c/CIR.

(*c*) B_e specifies the amount of data in bits, over and above B_c, that the network is prepared to accept in time interval T_c for delivery to the recipient.

For any frames transmitted during a period T_c that take the number of data bits transmitted above B_c but below B_c+B_e, the network will set the discard eligible bit (DE) and route the frames to the recipient. If these frames encounter congestion in transit through the network, they will be discarded in preference to frames which do not have the DE bit set. Any frames which would take the number of data bits transmitted during the period T_c above B_c+B_e are discarded at the point of entry; the network makes no attempt to deliver them.

13.6 Switched multimegabit data service

13.6.1 General

A different way of achieving simplicity is to opt for a datagram or connectionless service (see Section 13.2.2). The basic feature of connectionless working is that each packet sent by a terminal is treated by the network as a self-contained transaction. There is no sense of setting up a virtual call between communicating terminals and issues of sequence and error correction do not arise within the network.

The switched multimegabit data service [3,24–27] (SMDS) provides a public wide-area connectionless high-speed packet-switched data service. As

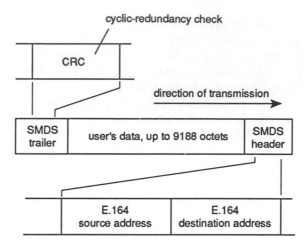

Figure 13.20   SDMS packet format

with frame relay, one of the prime applications for SMDS is LAN interconnection. The current generation of LANs, the 'shared-medium' LANs (i.e. Ethernet, token ring and token bus), are intrinsically connection-less in operation. The idea is that, when used for LAN interconnection, SMDS will give the user the impression of one large LAN rather than a number of interconnected smaller ones.

SMDS packets carry an information field of up to 9188 octets. As shown in Figure 13.20, the information field is preceded by a 36-octet header containing the complete addresses of the sender (the source address) and the intended recipient (the destination address). There are also 10 other fields in the SMDS header, the functions of which are not covered in this very brief introduction. Following the information field is a trailer of up to 11 octets, which contains an optional cyclic-redundancy check (CRC) equivalent to the frame-check sequence field already encountered in X.25 and frame relay. This checksum would typically be omitted if a higher-layer protocol handled error detection. Whether it is present or not is indicated by a bit in the header.

The source and destination addresses have the format and structure specified in ITU/TS recommendation E.164, the standard for public-network numbers in the ISDN era, described in Section 9.2.4. Two types of destination address may be specified, known as individual addresses and group addresses. An individual address is exactly what it suggests; it identifies a single recipient. The group address, however, is used when it is desired to send the same SMDS packet to a number of recipients simultaneously. A group address identifies a group of individual addresses and the serving local switch copies the SMDS packet to every recipient identified by that particular group address. The destination address in the resulting packets actually delivered to each recipient is therefore the group address rather than the recipient's individual address. This tells the recipient who else has received the packet, a feature which can be very useful in transactions where security and accountability are important.

13.6.2 The interface protocol

13.6.2.1 The protocol stack

The protocol stack for SMDS is illustrated in Figure 13.21. The terminology here is that the SMDS interface protocol (SIP) operates across the subscriber-network interface (SNI). It should be no surprise to find that SIP is a layered protocol. Hence, the similarity between Figure 13.21 and Figure 13.8, the corresponding illustration for X.25. However, they are not strictly comparable. Unlike the X.25 case, the three levels of the SMDS protocol stack do not correspond with layers 1–3 of the OSI reference model. In fact, the three levels of the SIP correspond to the medium-access-control (MAC) sublayer which, together with the logical-link-control (LLC) sublayer, forms the data-link layer of the OSI reference model in respect of LANs.

13.6.2.2 SIP Level 3

In the transmit direction, SIP level 3 forms the SMDS packet, including the user information to be transferred (typically a LAN packet) in the user data field. In the receive direction, it checks that the received packet contains no errors and then extracts the user data and passes it up to the higher-layer

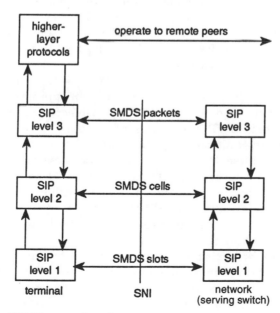

Figure 13.21   SDMS protocol stack

protocols. The service provided to the user is essentially technology independent.

By using address screens contained in the SMDS switches, the network can restrict the delivery of SMDS packets to one or more sets of specific destinations. By comparing the destination address contained in an SMDS packet with the list of permitted destination addresses contained in the address screens, the switch serving the sender can filter out and discard packets addressed to unauthorised destinations. In a similar way, the switch serving the recipient can filter out packets by performing address screening on the source address contained in the packet. This feature permits the SMDS network to support virtual private networking (VPN) whereby the customer is given the illusion of having a completely private network, though it is actually provided on a public network shared with many other customers.

13.6.2.3 SIP Level 2

SIP level 2 is the means of transporting the SMDS packets across the SNI and is technology dependent. The initial definition of SMDS has borrowed an existing protocol rather than develop a new one. Specifically it uses a subset of the IEE 802.6 metropolitan-area-network (MAN) protocol [28], also known as the DQDB (distributed-queue dual bus) protocol (ISO/IEC 8802-6). This subset is often known as the access DQDB.

The use of DQDB in this application permits a number of terminals to be attached to a dual bus as shown in Figure 13.22. However, in a multiterminal configuration, all the terminals attached to an SNI must belong to the same customer. This permits point-of-entry policing. Since every SNI is dedicated to a specific customer, the network can check that the source address in each SMDS packet transmitted has actually been allocated to that customer and prevent fraudulent misrepresentation.

DQDB also permits a number of SMDS packets to be received or transmitted across an SNI 'simultaneously'. It is not appropriate in this brief outline to give a detailed description of the DQDB access protocol. However, the general picture is that DQDB carries information in fixed-length units of 44 octets. In the transmit direction therefore SIP level 2 has to break the SMDS packet into these 44-octet units for transmission across the SNI. To each 44-octet unit is attached a 9-octet header and a 2-octet trailer to form a 53-octet data unit. This is called a 'cell', by analogy with the 53-octet cell used in ATM. In the receive direction, SIP level 2 has to reassemble the SMDS packets from the sequence of cells received. The information in the segment headers is used to aid this reassembly process.

SIP level 2 operates a slot-reservation scheme by which the terminal equipment gains access to the dual bus for transmission of these cells. The full DQDB protocol permits information to be transferred directly between different terminals attached to the access DQDB in multiple-terminal configurations, as well as across the SNI between the terminals and the serving switch in the SMDS network. By interleaving cells belonging to different packets, a number of SMDS packets can be sent and received 'simultaneously'.

13.6.2.4 SIP Level 1

SIP level 1 is concerned with the bit-level transport of information across the SNI.

13.6.3 Access classes

To enable customers to buy the SMDS service which best suits their particular needs, it is available in a number of 'access classes', each of which provides for different levels of traffic as indicated below. Each is defined by a *sustained information rate* (SIR), which is the long-term average rate at which user information may be sent across the SNI.

The standard access classes and their SIRs are as follows:

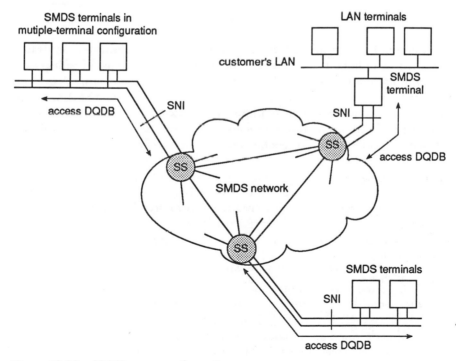

Figure 13.22 SDMS access configurations

SNI = subscriber network interface
SS = switching system

Access class	SIR (Mbit/s)
1	4
2	10
3	16
4	25
5	34

SMDS has so far been defined for use in Europe with access circuits of 2.048 Mbit/s (E1) and 34.368 Mbit/s (E3). The corresponding rates for North America are 1.544 Mbit/s (DS1) and 44.736 Mbit/s (DS3), respectively. Clearly, the above access classes are available only with E3 or DS3 access circuits. For E1 and DS1 access, the maximum effective user data rates that can be achieved are approximately 1.5 Mbit/s and 1.17 Mbit/s, respectively, when the overheads have been accounted for.

The above access classes are enforced by the switch serving the sender. (There is no access-class enforcement in respect of packets received from the network.) For this purpose, the switch uses a mechanism known as the credit manager to regulate the rate at which the network will accept packets for delivery [3].

13.7 Conclusion

This Chapter has introduced the principles of packet switching and illustrated them with reference to X.25, the most widespread packet-switched service currently used for wide-area data communications. The picture has been brought up to date with a brief description of frame relay and SMDS, two of the most recent developments in public packet-switched data networks. More complete descriptions of frame relay and SMDS can be found in the defining documents, though more accessible descriptions are available [3].

13.8 References

1. DAVIES, D. W. *et al.*: 'Computer networks and their protocols' (Wiley, 1979)
2. HALSALL, F.: 'Data communications and open systems' (Addison-Wesley, 1996), 4th edn.
3. ATKINS, J., and NORRIS, M.: 'Total area networking: ATM, frame relay and SMDS explained' (Wiley, 1995)
4. KLEINROCK, L.: 'Queueing systems' (Wiley, 1975)
5. CCITT Recommendation X.25: 'Interface between data terminal equipment (DTE) and data circuit terminating equipment (DCE) for terminals operating in the packet mode and connected to public networks by dedicated circuits'
6. ISO 7498: 'Information processing systems. Open Systems Interconnection. Basic Reference Model'. Part 1: 1994 'The basic model'; Part 2: 1989 'Security architecture'; Part 3: 1989 'Naming and addressing'
7. CCITT Recommendation X.121: 'International numbering plan for public data networks'
8. ISO 7776: 1986: 'Data communications. High-level data link control procedures. X.25 LAPB-compatible DTE link procedures'
9. CCITT Recommendation T.50: 'International reference alphabet'

10. CCITT Recommendation X.3: 'Packet assembly/disassembly (PAD) facility in a public data network'
11. CCITT Recommendation X.28: 'DTE/DCE interface for a start-stop mode DTE accessing the PAD in a public data network situated in the same country'
12. CCITT Recommendation X.29: 'Procedures for the exchange of control information and user data between a PAD facility and a packet-mode DTE or another PAD'
13. HOPKINS, H. H.: 'The frame relay guide' (Comm Ed, 1993)
14. CCITT Recommendation I.122: 'Framework for frame mode bearer services'
15. CCITT Recommendation I.233.1: 'ISDN frame relaying bearer service'
16. CCITT Recommendation I.233.2: 'ISDN frame switching bearer service'
17. CCITT Recommendation I.370: 'Congestion management for the ISDN frame relaying bearer service'
18. CCITT Recommendation I.372: 'Frame mode bearer service network-to-network interface requirements'
19. CCITT Recommendation I.155: 'Frame mode bearer service interworking'
20. CCITT Recommendation Q.933: 'DSS1 (digital signalling system no 1) - signalling specification for frame mode basic call control'
21. CCITT Recommendation Q.922: 'ISDN data link layer specification for frame mode bearer services'
22. CCITT Recommendation X.36: 'Interface between DTE and DCE for public data networks providing frame relay data transmission service by dedicated circuit'
23. CCITT Recommendation X.76: 'Network-to-network interface between public data networks providing the frame relay data transmission service'
24. Bell Technical Reference TR-TSV-000772: 'Generic systems requirements in support of SMDS'
25. Bell Technical Reference TR-TSV-000773: 'Local access system generic requirements, objectives and interfaces in support of SMDS'
26. Bell Technical Reference TR-TSV-000774: 'SMDS operations technology network element generic requirements'
27. Bell Technical Reference TR-TSV-000775: 'Usage measurement generic measurements in support of billing for SMDS'
28. ISO/IEC 8802-6: 1994: 'Information technology. Telecommunications and information exchange between systems. Local and metropolitan area networks. Specific requirements. Distributed queue dual bus (DQDB) access method and physical layer specifications'

Chapter 14

Integrated services digital networks

J.M. Griffiths

14.1 Introduction

Throughout the history of telecommunications the variety of services provided to customers has been increasing. To cater for this, two different strategies have been employed:

(*a*) to offer a single network of universal application which would support all services. The advantages of this strategy include the economies of scale and the flexibility with which service may be provided; and

(*b*) to offer networks specially tailored to the needs of the user. Examples are the packet and telex networks. The advantages are obvious, but their small size does increase costs. In fact, the ubiquitous telephone network has carried considerably more data than the specialist networks.

In practice, there has long been a differentiator between data and telephony. The telephone network has been analogue and data transmission is essentially digital. The conversion of the telephone network to digital operation through the use of PCM has removed this distinction. The modern telephone network is essentially a circuit-switched 64 kbit/s data network. Note that the move to digital operation had nothing to do with the needs of data users; it was simply the cheapest way of providing telephony using modern technology. Until the early 1990s modems operating on the PSTN offered the highest easily-available bit rates for data transmission. However, the PSTN has an inherent upper bound of information rate set by Shannon's law which relates the maximum information rate I to the bandwidth F and signal to noise ratio S/N as follows:

$$I = F \log_2(1 + S/N)$$

With a bandwidth of 3.4 kHz and a typical signal-to-noise ratio of 30 dB, this gives a maximum information rate of about 35 kbit/s. Modems operating at 28.8 kbit/s are near the limit of developments in that area. Thus, the next step to 64 kbit/s is entirely natural. This is illustrated in Figure 14.1.

A digital network of local and trunk exchanges and interconnecting links is known as an integrated digital network (IDN). An *integrated services digital network* (ISDN) [1] extends digital working to the customer, but this needs to

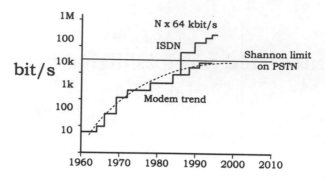

Figure 14.1   Growth of data rates on switched connections

overcome one further obstacle and that is the local-access link from customer
to local exchange which, even in the digital-switching and trunk-transmission
era, remained analogue.

14.2 Digital transmission over customers' copper pairs

The transmission of 64 kbit/s over existing copper pairs in the customer-
access network, which were originally designed only for 3.4 kHz speech, was
regarded as impossible until the 1970s. The problem is not only the resistance
(and consequent associated losses) of the pairs, shown in Figure 14.2, but also
the crosstalk between pairs. There is no screening between the pairs, and
reliance is placed on twisting of the pairs to avoid coupling between them.
Near-end crosstalk of the order of 55 dB is often regarded as a working figure

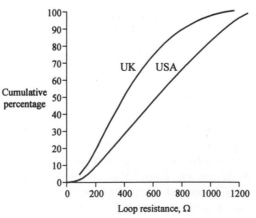

Figure 14.2   Distribution of resistances of customers' lines

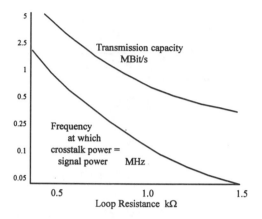

Figure 14.3   Transmission capacities of customers' lines

at the frequencies in question. Once again Shannon's law can be consulted to find the potential bandwidth of the network, and this is perhaps surprisingly large. Figure 14.3 gives some indication of the value. The upper curve shows that the potential bandwidth, when limited by crosstalk from other similar systems, is of the order of 1 Mbit/s on typical local-network pairs. Shannon's law does not, however, indicate how the transmission should be achieved. The lower curve does give some guidance. It indicates the frequency at which the signal power equals the interference power; in consequence, any use of spectrum above this frequency will result in more noise than signal being added, which is obviously counterproductive. Thus, steps must be taken to ensure that the signal spectrum lies well below this frequency.

The system chosen [2] encodes two binary symbols into one 4-level symbol and is hence known as '2-binary - 1-quaternary', or 2B1Q for short. As this reduces the modulation rate, and hence the bandwidth required, by a factor of two, it has proved possible to accommodate two 64 kbit/s channels plus a 16 kbit/s signalling channel on such a bearer. A further 16 kbit/s is used for monitoring of the transmission system, giving a total digit rate of 160 kbit/s. As two bits are encoded as 1 'quat', the resulting modulation rate is 80 kbaud.

Figure 14.4 shows how the losses of typical cable pairs vary with frequency. The increasing loss with frequency means that pulses spread as they travel along the cable. To compensate for this, some form of equalisation is required, which must be made adaptive to compensate for the wide range of mixes of cable type that can be encountered. The usual form of this equaliser is the decision-feedback equaliser shown in Figure 14.5. This has the advantage of removing noise from the compensating signal. Obviously it only has this advantage if a correct decision has been made as to the previous symbols.

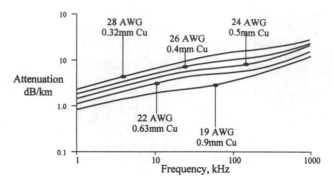

Figure 14.4   Cable attenuation/frequency responses

The system as described so far only provides for one direction of transmission. As only one pair is normally available between exchange and customer, some means is required to provide full-duplex operation. Early attempts revolved around using different frequency bands for each direction of transmission, or alternately transmitting in one direction then the other direction. Both of these techniques used at least twice and, more realistically, three times, the signal bandwidth. We have seen above how restricted the bandwidth needs to be due to the crosstalk limitation; hence these techniques are now only of historical interest.

The technique that is now generally in use involves cancelling the echo of the transmitted signal. This is demonstrated in Figure 14.6. The data to be transmitted arrive at the top left-hand corner. The box labelled TX drives the cable pair indicated to the right. To this pair is also connected the receive path to the receiver RX. The problem obviously is that the signal incoming from the right will be swamped by the transmitted signal, since the received-signal level may be 40 dB (voltage ratio 1/100) below the transmitted-signal level. The device in the centre of the diagram cancels the transmitted signal. One signal to be cancelled is obviously that travelling round the direct path from transmitter to receiver, but further echoes arise from signals which travel along the pair until they are reflected at discontinuities (e.g. changes of wire gauge) in the path, or the far-end termination. The echo canceller operates by storing the past data in a shift register. The contents of the shift register are then multiplied by coefficients which model the echo path and are summed to give an echo signal which is subtracted from the incoming signal to leave only the received signal from the distant end.

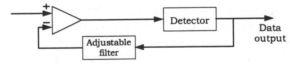

Figure 14.5   Decision feedback equaliser

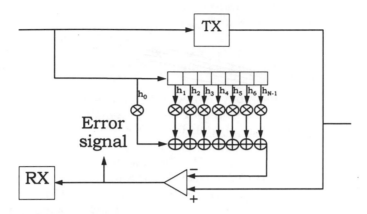

Figure 14.6 Echo canceller

Every pair in the local network is different, so the coefficients have to adapt automatically to the line. Let us consider the case where there is no received signal. In that case, any incoming signal must be due to echoes. In Figure 14.6 the current data bit is labelled h_0 and historic data bits are labelled h_1, h_2, ... The corresponding multiplying coefficients are c_0, c_1, c_2 etc. The adaptation process is quite simple in principle. Let us consider the case where h_0, h_1, h_2, h_3, h_4, h_5, h_6 are $+1$, -1, $+1$, -1, -1, $+1$, $+1$, where -1 is used rather than zero to keep the mathematics symmetrical. If the error signal is positive, then it is necessary to increase the value of the echo signal, and this can be achieved by increasing the multiplying coefficients which are multiplying an h containing a $+1$ (i.e. c_0, c_2, c_5, c_6) and decreasing those coefficients multiplying a -1 (i.e. c_1, c_3, c_4). Thus, with a fairly random string of data, the coefficients will converge to the perfect values. If a received signal now arrives, it will cause some incorrect modification to the coefficients; however, if the coefficient-incrementing factors are small, this is of little effect. The incoming signal cannot be itself cancelled, as the canceller has no knowledge of it.

Of course, despite the simple description given above, echo cancellation is not quite as easy as it sounds. Additional complication arises from:

(*a*) the fact that the actual signal is quaternary;
(*b*) the need to have a large coefficient increment to obtain rapid initial setup of the coefficients, but a small increment once convergence has been achieved;
(*c*) the assumption above implied that echoes add linearly; if the transmitter is in any way nonlinear, then the echo cannot fully be cancelled;
(*d*) the need to extract a clock from the received signal;
(*e*) the need to avoid any correlation between the two directions of transmission, as otherwise the canceller will try to cancel the correlated part of the received signal; and

(*f*) the sheer magnitude and precision required; typically, echo cancellers have to cancel 20–30 symbols and have coefficient precisions of 18 bits.

Notwithstanding these difficulties, single chips which implement these functions are available and essentially solve the problem of ISDN local transmission.

The method described above is used to provide, over a single cable pair, two 64 kbit/s communication channels, called B channels, together with a shared 16 kbit/s signalling channel, called the D channel. In addition, network providers also offer an interface to the ISDN with 30 B channels and a common 64 kbit/s D channel. This borrows the technology of the trunk network by using a 2.048 Mbit/s bearer on two metallic pairs or an optical fibre. In North America, the equivalent service offers 23 B channels. The standard 24-channel 1.544 Mbit/s format is modified to allocate one channel to signalling. The 2B+D service is known as *basic-rate access* and the 30B+D and 23B+D services are known as *primary-rate access*.

14.3 The customer interface

14.3.1 Reference configuration

The objective of the ISDN designers was to facilitate the connection of a wide range of terminal equipment, including data and nondata equipment. The analogue telephone was taken as a primary model, in which terminals are simply connected in parallel. This is extended in the ISDN so that terminals connected in parallel can be addressed individually, either on the basis of their type so that calls can automatically be directed to a compatible terminal, or on the basis of location so that, for example, a particular telephone of several will be caused to ring. One inadvertent feature of analogue telephony service, that of being able to intercommunicate between terminals connected in parallel, is not provided; this omission is widely believed to have been unfortunate!

The reference configuration of the customer's interface is shown in Figure 14.7. This shows reference points and components:

(*a*) The U reference point
This is not now actually acknowledged in the ITU recommendations, but is widely understood to be the line interface described in Section 14.2.
(*b*) The NT1 (Network Termination)
This terminates the line transmission system and interfaces to the customer.
(*c*) The T reference point
This is the interface to the customer. For political reasons in the USA the U reference point is legally defined as the customer interface. However, this

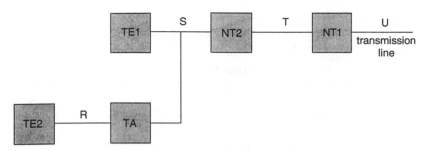

Figure 14.7   Reference configuration at customer's premises

does give rise to operational difficulties, such as the absence of an NT1 and its maintenance features, the difficulty of introducing new forms of access network and the indeterminacy of the maximum line length when the customer's premises wiring has to be included.

(*d*) The NT2
This may be present as a switching function, such as a PABX or a LAN, or it may be omitted.

(*e*) The S reference point
This is electrically identical to the T interface.

(*f*) The TE1 (Terminal Equipment)
This is an ISDN terminal.

(*g*) The TA (Terminal Adapter)
This is used to adapt from ISDN standards to older standards, such as RS232 or similar.

(*h*) The R reference point, which is the non-ISDN interface.

(*i*) The TE2, which is a non-ISDN terminal.

It is expected that the commonest configuration [3] will have no NT2 and terminals will simply be connected in parallel on a four-wire bus as shown in Figure 14.8. Up to eight terminals can be connected as shown over a distance of 200 m, limited by the transmission delays between them. If only one terminal is connected, then this can be up to about 1 km from the NT1, limited by the transmission loss. The bus is terminated by 100 Ω resistors at each end. Power is fed along the four wires to operate one terminal in emergency (normally a telephone); all other terminals are locally powered. In fact, eight wires are specified, allowing one extra pair to feed power to terminals from the NT1 and another pair to feed power from a terminal to the NT1 or another terminal. Neither extra pair is commonly used and so will not be discussed further. Nevertheless, the connector has to be provided with the extra pins [4].

14.3.2 Layer 1

The Layer 1 signal on the bus [5] is a three-level signal at 192 kbit/s. Zero volts represents a logical 1, and a logical 0 is represented by either +0.75 V or

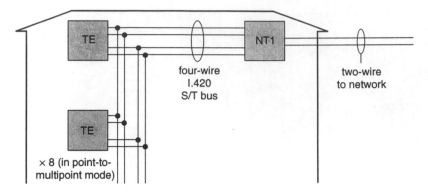

Figure 14.8   Network configuration at customer's premises

> *TE = terminal equipment*
> *NT = network termination*

− 0.75 V. The two polarities are used alternately so that the overall signal is balanced and hence not distorted by passage through a transformer. Figure 14.9 shows the structure of the frame. The top line is the waveform from NT to terminal equipment. The two B channels are shown shaded. The signalling is carried in the D bits. The bits marked F are framing bits used to synchronise the system. The bits labelled L are chosen so that each frame is entirely balanced. The frame is balanced between the vertical dotted lines. Other bits, F_A, N (auxiliary framing), M (multiframing), A (activation) and S (for future standards), are beyond the scope of this short Chapter.

The lower frame is from the terminal equipment to the NT1. It is very similar in structure to the opposite direction. Two features distinguish it:

(*a*) Each B channel and the D channel may come from a different terminal and so there are more balance bits, one associated with each, to ensure that the whole is balanced.

(*b*) Any terminal may send information to the D channel. To avoid corruption of information due to two terminals sending simultaneously, the D channel is echoed back to the terminals in the E bit. The terminal sending impedances are arranged so that a logical 0 will overwrite a logical 1. A transmitting terminal monitors the E bit and if this does not reflect what it is sending then there must have been another transmitting and so the terminal backs off.

14.3.3 Layer 2

Layer 2 in the signalling channel [6] is designed to give a secure link between the terminal and the exchange processor. The protocol is a version of HDLC known as Lap D. Its structure is shown in Figure 14.10. After any run of five '1's, a '0' is always inserted. This means that the flags containing six '1's uniquely delimit the frames. The service-access point identifier (SAPI) can be

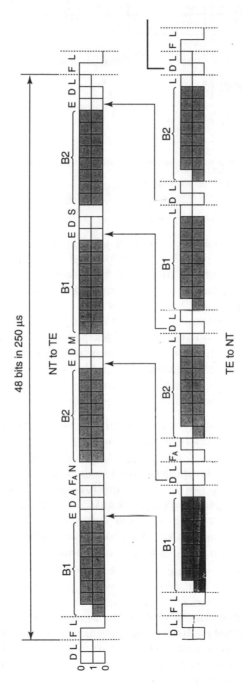

Figure 14.9 Layer-1 frame structure

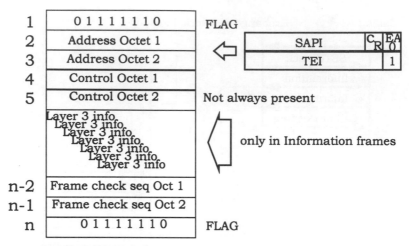

Figure 14.10   Layer-2 frame structure

used to separate different services (e.g. telephony, packet) and the terminal-endpoint identifier (TEI) identifies a particular physical terminal. The command/response (CR) bit identifies the direction of transmission. The EA0, EA1 bits allow extended addresses to be implemented should that ever prove necessary. The control fields include sequence numbers so that it can be determined that no frames have been lost. The frame-check sequence (FCS) ensures that there has been no error in the body of the frame.

14.3.4 Layer 3

The purpose of layer 2 is to carry the information of layer 3 which controls the call [7]. There are about 25 messages defined at this level, the lengths of which can vary between a few octets and over 100 according to their type. The longest is the initial setup message. It may contain information on the bearer-channel requirements, routing and display information, compatibility-checking information and user–user information. Figure 14.11 indicates the messages involved in a simple call establishment. The initial setup message from terminal to exchange is acknowledged by the exchange. This is followed by further, probably routing, information. When the exchange has all the information it needs, it advises the terminal that the call is proceeding and sends a setup message to the far end. In this case, it is a telephony call and there are two telephony terminals at the far end with TEIs of 5 and 8. They both ring and advise the exchange in alerting messages. They are both answered, as indicated by the connect messages. The exchange only connects the first telephone to be answered and releases the other telephone. It should be noted that every action is controlled by the exchange and terminals only act when they are instructed.

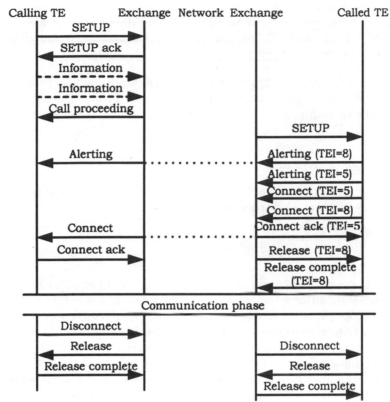

Figure 14.11 Messages transmitted for making a basic call

14.4 ISDN services

14.4.1 General

The most obvious difference between the ISDN and the PSTN is the bandwidth it offers as a data service. The availability of 64 kbit/s has an impact at three levels:

(*a*) The higher bit rate may be exploited to give greater throughput. Examples are electronic mail and file transfer. In these cases there is no fundamental change to the service provided.

(*b*) The higher bit rate gives opportunities to upgrade services with new opportunities. An example is facsimile. Group 3 fax is not well suited to the transmission of more than, say, 10 pages. At 64 kbit/s each page can be sent in a few seconds and transmission of tens of pages becomes a real possibility.

(*c*) Fundamentally new services emerge. For example, high-quality speech and music can be encoded at 64 kbit/s and 128 kbit/s, respectively, using rather complex algorithms. They can then be transmitted as a direct service to the customer, or for use by broadcasters. Videophones can also be put in this class. Although analogue videophones and digital phones operating at 14.4 kbit/s have been offered, the quality of the image has generally been such as to discourage buyers.

Services benefit not only from the greater bandwidth but also from the fact that, for the first time, terminal apparatus has a proper signalling system which allows communication between the exchange processor and the terminal.

14.4.2 Bearer services

A *bearer service* indicates only the characteristics of the connection required by the calling terminal. Examples are:

(*a*) 64 kbit/s unrestricted
In this case a transparent channel is required.
(*b*) Telephony
In this case the interface at a European terminal would be A-law PCM and at a North American terminal would be μ-law PCM. In a connection between them there would be an A- to μ-law conversion, possibly circuit-multiplication equipment on intercontinental links, and echo suppressors. Part of the link could even be analogue.
(*c*) 3.4 kHz audio
This might be used for a modem connection and would be as for telephony, but full-duplex operation would be allowed with no circuit multiplication and no echo suppressors.
(*d*) Packet service
Instead of offering a circuit-switched service, a packet service may be offered. The format may be X.25 or frame relay. A particularly interesting possibility is the use of the D channel for this purpose. It is clear that 16 kbit/s is a gross overprovision for signalling and the excess capacity could provide access to a packet service [8].

When primary-rate access is used, it is also possible to use fewer than 30 or 23 channels in order to obtain channels with greater digit rates for broadband applications such as video communications. These are:

(*a*) the H_0 channel at 384 kbit/s
(*b*) the H_{11} channel at 1536 kbit/s (for use with 1.544 Mbit/s access)
(*c*) the H_{12} channel at 1920 kbit/s (for use with 2.048 Mbit/s access).

Both H_{11} and H_{12} may be used as single channels, or they may carry multiplexed H_0 channels. In order to switch these channels, a digital exchange must be able to make connections using n time slots (where $n=6$ for H_0, $n=24$ for H_{11} and $n=30$ for H_{12}).

14.4.3 Supplementary services

The *supplementary services* [9] offered by ISDN are similar to the many services widely available on PABXs. By using the signalling system and the intelligence of the exchange processor, a whole range of public services can be implemented.

Examples are:
(i)   Calling/called-line identification presentation/restriction
(ii)   Closed user group
(iii)   Call waiting
(iv)   Direct dialling in
(v)   Malicious-call identification
(vi)   Multiple subscriber number
(vii)   Subaddressing
(viii) Terminal portability
(ix)   User-user signalling
(x)   Advice of charge
(xi)   Completion of call to busy subscriber
(xii)   Add-on conference call
(xiii) Meet-me conference
(xiv)   Call-forwarding unconditional
(xv)   Call-forwarding busy
(xvi)   Call-forwarding no reply
(xvii) Call deflection
(xviii) Freephone
(xix)   Three-party service

These services can be implemented in several ways. The simplest way is to define sequences of keypad depressions using the * and # keys as delimiters in the same way that supplementary services are set up on multifrequency telephones. However, these tend to be rather user unfriendly, and a protocol called remote-operations service element (ROSE) has been defined [10]. This allows a direct interaction between terminal and exchange processor.

14.4.4 Teleservices

Teleservices are those which require particular terminal equipment [11]. Bearer services only indicate the capability of the channel. However, it is necessary to know a lot more before two terminals of even moderate complexity can interwork. For example, there are several ways of implementing a videophone service. A teleservice includes the information needed to ensure interworking. Examples include the definition of a videophone and the group 4 facsimile service. In the early days of such a service, the only way of operating is to have terminals of the same manufacture at the two ends. Standard teleservices are the route by which the market is opened up, so that terminals of any manufacture can be used and purchased locally.

14.5 ISDN evolution

The ISDN has often been criticised either for being unnecessary (Integration that Subscribers Don't Need) or inadequate. In fact the ISDN should be seen as a natural evolution, not to be oversold by excessive hyperbole or undersold because of the novelty of some of its features [12]. Those with the biggest difficulty in the roll out of the ISDN were the classic marketing people. They like to be able to offer a customer a 'killer application', while all the technologists were emphasising the wide range of services it could support. Marketeers also like to start with a small geographical area in which to try out their wares with a small number of skilled technical people to support the initial service. The technologists, however, were emphasising the ubiquity of the service based on the fact that it could be supported by any digital local exchange, of which there are now many. Marketing people were also concerned that in many cases ISDN would simply cause a switch from existing products (e.g. leased lines) rather than generate new business. In the early days it is also true that only a limited range of products were available (e.g. digital telephones, computer cards). Figure 14.12 shows how the basic-rate service has rolled out in various countries. The general slopes indicate that the increase is by a factor of 2–3 per annum. The benefits of the earlier starts in Japan, the USA, Germany and France are somewhat offset by the fact that their services are not to international standards and hence some retrospective modification will be needed.

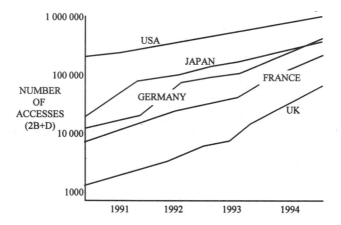

Figure 14.12   Growth of basic-rate ISDN

Figure 14.1 shows a continuing upward trend, implying a move to broadband services. An approximate formula indicates that customers require a bit rate of the order of $10^{(Y-1940)/12}$ where Y is the year in question. Initially, some additional capacity can be obtained by concatenating 64 kbit/s channels. Unfortunately, complexity is introduced because there is no way of ensuring that every channel has a similar delay. The only practical way of doing this would be to ensure that every channel remains in the same 30-channel (24-channel in North America) group. Unfortunately, even if the exchange processor could accommodate this, problems would arise from the dimensioning of the network. A network is designed to offer an agreed grade of service to users of single channels. If someone then chooses to take several channels in a single group, not only is the probability low that the number of channels will be free, but taking those channels reduces the grade of service to other users. By way of example, a 30-channel system can carry about 19 erlangs of traffic at 1-in-200 grade of service. If a videoconferencing unit wishes to take six channels, the probability that these are available is only about 1 in 10, and if that is successful the resulting grade of service for single channels will be only 1 in 20. This is shown in Figure 14.13. In the core network this problem may be less serious, as many 30-channel systems may be available to be selected. However, a relatively-small link to a PABX will not have that advantage.

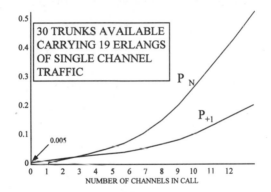

Figure 14.13   Grades of service with N× 64 kbit/s traffic

30 trunks available carrying 19 erlangs of single-channel traffic
P_N=GOS for N× 64 kbit/s call
P_{+1}=GOS for subsequent single-channel call

An alternative solution is to select channels at random and to install buffers which will build all channels out to the same delay [13]. However, the difference between a terrestrial link and a satellite link is about 14 000 bits at 64 kbit/s. Many items of terminal equipment are now available, not only to provide the buffering but also to measure the delays automatically and to adjust the number of channels employed as demand for bandwidth changes.

Although the use of multiple 64 kbit/s channels offers an immediate resource, it is clear that the true future of broadband service lies in the use of asynchronous-transfer-mode (ATM) techniques, as described in Chapter 15.

The use of a very simple cell with a 5-octet header and 48-octet information allows the very-high-speed processing needed to offer services at aggregate rates of 150 or 600 Mbit/s. Early offerings are only on the basis of permanent circuits (actually virtual circuits as no fixed connection is in place), but the eventual move to a public switched service is the next evolutionary step to a broadband ISDN.

14.6 References

1. GRIFFITHS, J. M.: 'ISDN explained', (Wiley, 1992), 2nd ed.
2. ANSI T1.601: 1988: 'ISDN basic access interface for use on metallic loops', American National Standards Institute
3. CCITT Recommendation I.420: 'Basic user-network', ITU
4. ISO 8877: 'Interface connector and contact assignments for ISDN basic access at reference points S and T', International Standards Organisation
5. CCITT Recommendation I.430: 'Basic user-network - layer 1 specification', ITU
6. CCITT Recommendation Q.920: 'ISDN user-network interface data link layer — general aspects', ITU
7. CCITT Recommendation Q.930: 'ISDN user-network interface layer 3 — general aspects', ITU
8. CCITT Recommendation I.122: 'Framework for providing frame mode bearer services', ITU
9. CCITT Recommendation Q.932: 'Generic procedures for the control of ISDN supplementary procedures', ITU
10. CCITT Recommendation X.219: 'Remote operations: model, notation and service definition', ITU
11. CCITT Recommendation I.241: 'Teleservices supported by ISDN', ITU
12. GRIFFITHS, J. M.: 'The integrated services digital network', Institution of British Telecommunications Engineers, Structured Information Programme, section 9.4
13. CCITT Recommendation H.221: 'Frame structure for a 64 kbit/s channel in audiovisual teleservices', ITU

Chapter 15

Broadband networks

G. N. Lawrence

15.1 Introduction

Existing national telecommunication networks have developed mainly for telephony, since this service generates the major proportion of the traffic to be carried. The evolution of integrated digital networks (IDN) enables digital channels at 64 kbit/s and 2 Mbit/s or 1.5 Mbit/s to be extended to customers' premises and this has led to the development of integrated services digital networks (ISDN), as described in Chapter 14.

There are services, for example high-definition colour television, which cannot be accommodated within the bandwidth of the existing ISDN. This network has therefore been called *narrowband ISDN* (N-ISDN). The demand for broadband services is expected to increase, so plans are being considered [1] for the introduction of a future *broadband ISDN* (B-ISDN). This is also referred to as the *Integrated Broadband Communications Network* (IBCN).

The terms 'narrowband' and 'broadband' are essentially relative, and no clear definition exists as to where one ends and the other begins. In the context of telephony, the basic 300 Hz–3.4 kHz band is clearly narrowband. In transmission systems, the multiplexing together of many such channels onto analogue or digital carrier systems leads to these carrier systems being defined as broadband. Alternatively, a broadband carrier system may carry fewer channels of a high-bandwidth service such as colour television.

In relation to switching systems, most analogue crosspoints switch only one speech channel, normally in the audio band. Digital switching systems have the capability of switching 64 kbit/s channels. The N-ISDN has a requirement to be able to switch and transport from end to end basic channels operating at 64 kbit/s. In addition, higher bit rates can be switched from source to destination using the multislot technique, where $n \times 64$ kbit/s is a requirement of the service [2]. However, most N-ISDN networks are unable to guarantee integrity of time-slot sequence; thus, users of multiple 64 kbit/s slots may find that the sequence of slots delivered has been rearranged during the call. As there are no network signals provided by the ISDN to cater for this problem, it is necessary to provide additional terminal equipment, whose function is to request a connection for $n \times 64$ kbit/s and multiplex and synchronise them for presentation to the N-ISDN. Similar equipment at the receiving end will detect a rearrangement of time slots as a loss of synchronism. An exchange of end-to-end signals must then occur to resynchronise and establish the correct sequence of time slots.

The N-ISDN offers to users basic rate (2B+D channels) at 144 kbit/s and primary rate (30 B+D) at 2.048 Mbit/s. The former used to be the highest bit rate that could be carried over ordinary subscriber copper pairs. For primary-rate access, special pairs can be employed, or systems such as ADSL (Asymmetrical Digital Subscriber's Line) and HDSL (High Speed Digital Subscribers Line), and now increasing use is made of optical fibre. In the latter case, higher-order multiplexing can be offered between the customer and the local exchange. However, for switching purposes this will usually be demultiplexed to 2.048 Mbit/s.

Thus, it is generally accepted that the *narrowband* ISDN network is for customer connections and switching up to 2.048 Mbit/s, and the *broadband* ISDN will be for services which require switching for bit rates in excess of 2.048 Mbit/s. However, two points should be noted:

(*a*) The N-ISDN will be multiplexed and demultiplexed to and from higher bit rates, to achieve economies in the transmission network.
(*b*) The B-ISDN will be required to carry services requiring less than 2.048 Mbit/s for compatibility reasons.

The characteristics of services required are summarised in Figure 15.1. They are considered in more detail in Section 15.2.

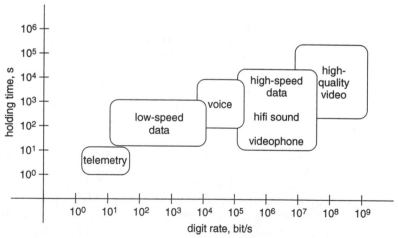

Figure 15.1   Services expected in broadband networks

15.2 Broadband services

15.2.1 *General*

When people communicate face to face, they generally make use of the fact that they can hear the other person, and see them with all their movements. They may also use documents containing text, graphics or diagrams during their discussion. However, when they use the telecommunication network that currently exists, they are strictly limited. Ways have been found to get round some of these limitations, for example, using facsimile and slow-scan TV. Ideally, the B-ISDN should offer users what they really want. This section examines these fundamental requirements, and the bandwidths needed to realise them.

15.2.2 *Audio*

The current 64 kbit/s PCM is proving to be more than adequate for good-quality voice, particularly in the all-digital environment of the N-ISDN. Indeed standards have been prepared to use 64 kbit/s to provide high-quality 7 kHz audio bandwidth [3]. Conversely, in the interests of conserving bandwidth in the radio environment, hand-held mobile units are exploiting 32 and 16 kbit/s codecs for speech [4]. It is unlikely, therefore, that audio will require broadband techniques.

15.2.3 *Video*

Slow-scan television in black and white, suitable for video-monitoring purposes, is well suited to the 64 kbit/s N-ISDN, and the two B channels will easily accommodate a black-and-white videophone service. It is when the user demands colour, good definition and natural movement that the bandwidth can rise. Good-quality colour videoconferencing [5] is currently available using from 384 kbit/s to 2 Mbit/s, thanks to the skill of the video codec designers. The latest international standard [6] (H.261) is now being used to provide even-better picture quality at all rates between 128 kbit/s and 2 Mbit/s. The higher the bit rate chosen, the better the picture quality, but at 384 kbit/s the picture quality will be as good as the current H.120 standard at 2 Mbit/s. Therefore, videotelephony and videoconferencing are unlikely to require broadband, because of the good quality achieved by the video-compression techniques used in the coding algorithms. For transmission of documents with detail too fine to be resolved by an H.261 codec, the user may use colour fax transmission, perhaps assigning all 2 Mbit/s for a short period for this purpose.

In videotelephony and videoconferencing the cameras do not normally move to any great extent, either panning or zooming, and the motion by the subject is not extensive. However, applications such as the relaying of a surgical operation for teaching purposes require fine resolution and good movement definition, equivalent to current colour-television broadcast standards. For these, bit rates of up to 33 Mbit/s may be required [7].

This bit rate would also be required for the general distribution of current colour-television services. Thus, the B-ISDN is designed to accommodate telecommunications *and* entertainment distribution services carried by the same network operator, using the same local-loop plant (coaxial cable or fibre). However, even 33 Mbit/s is not adequate for television contribution services, where the high-quality picture generated in the studio is switched to different locations before being broadcast. For this, 135 Mbit/s is required [8].

Networks can also be used for video-retrieval purposes. Examples of applications would be video-on-demand, distance learning, where the user chooses which course material to view, or teleshopping in which the mail-order catalogue is replaced by a dial-up video service. These would be catered for by networks offering switching and transmission at rates between 128 kbit/s and 135 Mbit/s depending on the quality required by the customer.

Finally, the needs of the approaching high-definition television (HDTV) service should also be considered. With approximately twice as many lines as the current television service, and a wider picture format giving 50% more picture area, the minimum bit rate for ordinary distribution of HDTV is expected to be about 150 Mbit/s. This would also be the bit rate for HDTV medical applications and video-retrieval services. The HDTV contribution services can be expected to require bit rates of the order of 1 Gbit/s.

The uncertainty surrounding the values of these higher bit rates is that video-coding techniques are expected to make use of variable-bit-rate codecs [4]. These are variable in two senses. In the first, the codecs adapt to suit the needs of the terminal at the far end, thus giving compatibility between two video terminals designed for different levels of quality of picture. Clearly, the cost of this adaptation must lie with the more-expensive higher-grade terminal, or the network service provided to it, and not with the cheaper lower-grade terminal. But it will be important for the future take-up of interactive video services that a range of terminals can be offered to the potential user, which differ in cost and therefore quality, and that they can interwork with each other, albeit at the lowest common level.

In the second sense, variable-bit-rate codecs adapt to suit the content of the image being transmitted. Thus, although 150 Mbit/s is required for the satisfactory distribution of HDTV moving colour images, a lower bit rate would be adequate for those periods when the picture is relatively static. This has the advantage of dramatically lowering the mean bit rate required over a period of time, thus reducing the network resources required, and (presumably) the cost.

Thus, it can be seen that although the N-ISDN can cope with the current terminals and services associated with video, new services planned and already in the pipeline will need to be supported by a broadband network.

15.2.4 Data

There is already a clear distinction between narrowband and broadband data. For the user of a VDU or PC, working with a modem over the PSTN or a leased line to a computer, a bit rate of 19.2 kbit/s yields a reasonably-low screen-refresh time, and is more than adequate for the keyboard-generated data in the upstream direction. Most of the delays in this application are caused by insufficient central-processing-unit power at the mainframe computer. Thus, a basic-rate ISDN channel of 64 kbit/s is more than ample for this type of application. However, a PC operating on a local-area network with a file server needs a much higher bit rate for downloading software and data files. This is currently 10–100 Mbit/s, although it is of a very bursty nature. Companies on more than one site will want to exchange data at these rates between their LANs, to create a metropolitan-area network (MAN) or wide-area network (WAN). For this, a broadband public or private network will be required, and techniques are being developed to satisfy the need. The bit rates required will lie in the range 10–150 Mbit/s, with the lower end being used for text interchange and the upper end for graphics and images (still pictures).

Another application which requires a high bit rate between terminal and computer is the computer-aided-design (CAD) process. Users currently have to site the terminal and the computer within reach of the same LAN if they want fast response times and short screen-refresh times. A broadband network offering 150 Mbit/s connections will enable the user of a CAD system to work from home.

In the financial sector, we are already seeing an increase in the use of electronic point-of-sale terminals and electronic funds transfer, and the 2 Mbit/s N-ISDN will assist in this growth. However, with the increase in the movement of goods and people that is expected, the amount of data exchange generated by this source will grow, and will almost certainly require broadband techniques to cope with the volume. Similarly, a unified European air-traffic-control system will require a broadband network, even though it may be private and separate from other users. The amount of data generated by a single radar screen is already considerable, and is growing as techniques become more sophisticated. Currently, each air-traffic-control zone has its own processing system, and aircraft-handover techniques are relatively primitive, but the application of broadband switching and transmission will lead to better and safer air-traffic control for Europe.

15.2.5 Mobility

The growing popularity of mobile communications has demonstrated that there is a market for this type of service, even though it is currently required more by the business sector than the residential sector. If the technical and frequency-spectrum-allocation problems can be resolved such that the cost of full mobile communications falls dramatically, then market predictions are for a widespread takeup of the service. This would lead to an increase in the data being exchanged between switching centres catering for mobile users as the users roam, in order that the location registers in the network keep their databases up to date. In other words, a large number of mobile users who are moving about would generate a large amount of infrastructure data transmission *even if they did not make or receive any calls*. The current method of providing mobile communications is to have separate overlay networks linked into the fixed telecommunication networks at certain well defined access points. This would not be satisfactory in the future; therefore mobile communications must be an integral part of broadband networks.

It has been stated in Section 12.7.1 that mobile terminals are using very low bit rates for their audio connections. There is a growing demand to be able to connect mobile fax machines and laptop PCs into the mobile network, but these too are currently constrained by radio bandwidth to offering low bit rates. If these constraints were to be removed, then it can be envisaged that bit rates up to 2 Mbit/s would be required. Although some users may wish to have their portable PC form part of a cordless LAN running at up to 100 Mbit/s, this would be very costly in radio bandwidth and is unlikely to become a widespread public service. Thus, the majority of users are unlikely to require B-ISDN for the terminal-access network.

It could be envisaged that, for the purpose of instant electronic news gathering, portable video terminals would want broadband access into a B-ISDN network. Although bit rates of at least 33 Mbit/s would be required, coverage would not be extensive and therefore the service could not offer true mobility outside a very restricted area.

One aspect of mobility which is being given attention is the issue of the mobility of the user around fixed terminals, in a network where the call is addressed *to the user*, not the terminal. The first prerequisite of such a system is that a personal-numbering scheme must be available; the second is that the user must inform the network which terminal is his nearest. This would be achieved by logging-in at the terminal, using a card with or without a cordless capability (e.g. infra-red). The network would then update its location registers and route incoming calls to the correct terminal. It would be possible to achieve automatic logging-in with no action from the user, but this would generate vast amounts of infrastructure data unnecessarily, as the user passed intermediate terminals between start and finish points. However, the proper use of such a system would require a broadband network to handle the large number of log-in data packets generated.

15.2.6 Network management

As telecommunication networks become more complex, so does their management, and sophisticated techniques are being applied to the monitoring and control of the networks from central service points. This in turn generates more data, and it can be expected that the *management* of a broadband network will itself require broadband switching and transmission techniques in order for it to be realised.

15.2.7 Flexibility of services

The above Sections have shown that, while audio will not require anything other than narrowband techniques, all other services will sooner or later require a broadband network to be available. Because of the additional cost of broadband services over narrowband, what the user will really require is *bandwidth on demand*. Users will probably accept that they must pay for an access network that has the ability to carry broadband services up to say 150 Mbit/s. However, if they make an ordinary narrowband audio call using a multiservice terminal, they would not expect to have to pay the same as for a video call. They should be able to switch from lower bandwidth to higher bandwidth during the call, ideally only paying for the bandwidth which is actually used. This will make the network extremely flexible in use, so that audio, video, data, text, graphics and images can be transmitted as and when required. The switching technique to enable this to happen is the *asynchronous transfer mode* (ATM) and this is described in Section 15.4.

15.3 Broadband transmission

Transmission equipment has for many years used analogue and digital multiplexing techniques to realise point-to-point broadband carrier systems. However, these have employed hierarchical and asynchronous techniques for multiplexing. While being perfectly adequate for point-to-point bearers, they have not given the ability to provide flexible allocation and management of the bandwidth available. These systems are designed to provide the economic transport of many channels between two end points, but they do not allow easy access to these channels at any intermediate points. The only way access can be gained to a single 64 kbit/s channel is to demultiplex the whole system down to its individual channels, extract or insert the desired information and remultiplex to a higher order again for onward transmission.

The synchronous digital hierarchy (SDH), which is also known in the USA as SONET (synchronous optical network), provides flexibility in broadband transport and flexibility in accessing the bandwidth available. The multiplexing techniques are fully synchronous and allow add–drop multiplexers or crossconnect equipment to be placed anywhere in the transmission network. Such equipment needs only to recognise the framing pattern and establish synchronism with it in order to be able to locate and provide direct access to any multiple of 64 kbit/s up to the full rate of the bearer employed. As the ITU-T is currently specifying bearer rates which include 155 Mbit/s, 622 Mbit/s and 2.5 Gbit/s, it can be seen that high degrees of flexibility are available, with fine resolution. These features will prove invaluable in a broadband network.

15.4 Switching

15.4.1 Circuit and packet switching

The two techniques of circuit switching and packing switching are well established in today's telecommunication networks. Circuit switching is used for voice traffic, and provides a unique path from user to user for the duration of the call. Packet switching is available for data traffic which is generated in bursts; it enables more efficient use to be made of transmission facilities, although at an increase in cost and complexity at the switching nodes.

Digital circuit switching can cater for several different digit rates [2] at multiples of 64 kbit/s, up to a maximum of 2 Mbit/s. However, we have seen that some future services will require higher digit rates. Moreover, this solution lacks flexibility. Segregated switches operating at different bit rates are needed. Thus, during congestion, the system may be unable to accept a low-bit-rate call even if it has free high-bit-rate channels. We have also seen that some services produce 'bursty' traffic (e.g. data) or variable-bit-rate traffic (e.g. from video coders). Circuit switching must cater for the maximum required bit rate throughout the duration of each call, thus leaving resources underutilised for much of the time.

In packet switching, the paths taken by successive packets may not be the same. Overhead information is added to the data to enable the network to route it correctly and the recipient has to assemble the packets in the correct sequence. In addition, delays are incurred when packets queue at switching nodes. Only small delays can be tolerated in telephone connections, so packet switching is not normally used for speech. Even if packets are re-assembled in the correct order and the mean overall delay is tolerable, there is the problem of varying delay. It is necessary to store received packets in a buffer in order to emit the received speech samples at a constant rate. Use of such a buffer increases the overall delay and speech samples which arrive when the buffer is full must be discarded, causing a further degradation of the transmission quality [9].

These considerations led to the adoption of a modified form of packet switching for broadband services. This has been called *fast packet switching* [10] (FPS), *asynchronous time division* [11] (ATD) and the *asynchronous transfer mode* [12,13] (ATM). The term ATM has been adopted by the ITU-T. The word 'asynchronous' is used because the method allows asynchronous operation of the clocks at a sending and a receiving terminal. Any timing difference between these clocks is removed by inserting or removing empty packets in the information stream. It does not imply that transmission links or switches operate asynchronously.

15.4.2 ATM principles

The asynchronous transfer mode uses a high-speed bit stream on which are interleaved the packets for different connections. The packets are of a fixed size, known as *cells*. Thus, for example, a 34 Mbit/s video coder will transmit cells much more frequently than a 64 kbit/s voice coder.

ATM must meet the requirements for switching all services, including both speech and data. For data, a very low error rate is essential, but delay can be tolerated. For speech, quite a high error rate can be tolerated, but delay (and jitter on it) must be minimal. These conflicting requirements have been called *semantic transparency* and *time transparency*, respectively. The requirement for time transparency also applies to video information if it accompanies a sound signal. For example, the lip movements in a picture of a speaker must occur in time with the speech.

When existing packet networks for data transmission, such as X.25, were designed, the available links in the telephone network had a relatively poor error performance. (A bit-error rate of 10^{-6} was considered excellent at the time.) Complex protocols were therefore necessary to perform error control at every link in a connection. Packets were of variable length, so rather complex buffer management was required inside the network. However, this was possible because operating speeds were low. Delays were relatively long, but time transparency was not a requirement.

Subsequent developments have permitted simplifications to be introduced for ATM, which enable it to obtain sufficient time transparency for use in a B-ISDN, while retaining adequate semantic transparency. The resulting features of ATM are as follows:

(*a*) Packets are of fixed length, known as cells.

(*b*) ATM operates in a connection-oriented mode. Before information is transmitted, a virtual connection is set up. The necessary resources for this are then reserved. However, if insufficient resources are available, the call is refused. This enables the network to guarantee connections with small packet losses (typically 10^{-12} to 10^{-8}).

(*c*) There is no error protection or flow control on a link-by-link basis. Error control for the information in cells (but not for their headers) can be omitted because of the very low error rates of modern transmission systems. If error control is needed to obtain a still lower error rate, it can be done on an end-to-end basis by the terminal equipments.

(*d*) The header is short. The functions of the header are limited to identifying the virtual connection to ensure correct routing of the cell. Other header functions provided in conventional packet switching, such as sequence numbering and flow control, are not provided. However, since an error in a header would cause misrouting of a packet, error-control bits are included. Because the header is short and simple, its processing in an ATM switch is rapid and queuing delays are therefore small.

(*e*) The information field is short. Use of a short information field also reduces the length of the cell, and thus the packetising delay at a sending terminal. In addition, it reduces the size of internal buffers at switching nodes and their queuing delays. This enables ATM to provide the small delay and jitter which is needed for real-time services, such as telephony.

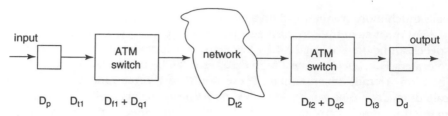

Figure 15.2   Delays in an ATM network

D_p = packetisation delay
D_t = transmission delay
D_f = fixed switching delay
D_q = queueing delay
D_d = depacketisation delay

(*f*) A policing function is included. The first ATM switching node encountered in the network must ensure that the user is not sending and receiving digits, and therefore cells, at a higher rate than has been chosen to pay for on that call.

The delays encountered in an ATM network are shown in Figure 15.2. They are as follows:

(i) Packetisation delay
This is introduced at the sending terminal when samples of a real-time signal (such as speech) are accumulated to form packets. Clearly, the shorter the cells, the smaller is this delay. For cells containing 48 octets sent at 150 Mbit/s, the delay incurred for 64 kbit/s PCM is 6 ms. For a 2 Mbit/s video coder, the delay is only 212 μs. In a pure ATM network, this delay only occurs at the source. If there is a mixture of ATM and circuit switching, this delay occurs at each boundary between them.

(ii) Transmission delay
This depends on the velocity of propagation in the transmission medium and the distance. It is independent of the switching mode employed. For modern transmission systems, it is typically 4–5 μs/km.

(iii) Fixed switching delay
This is the time taken for a cell to traverse a switch, even if only a single packet is being handled. It is of the order of tens of cells per exchange. Since high digit rates are used, it can be small. For cells of 53 bytes and a digit rate of 150 Mbit/s, this delay is of the order of 30 μs per exchange.

(iv) Queuing delay
Queues are necessary in switches to avoid loss of cells. The delay increases with the traffic load on the switch. Proposed switches need maximum queues of 10 cells or less, but several queues may be needed in each exchange. It can be shown that, for switches loaded to 80% of their maximum capacity, a cell-loss probability of 10^{-10} and 50 consecutive queues, the overall delay is 235 cells [12]. For a cell length of 53 bytes and a digit rate of 150 Mbit/s, this gives a delay of 664 μs.

(v) Depacketisation delay
Packets arrive at the receiving terminal with intervals between them. However, for a real-time service, output samples must be generated at a constant rate. A buffer is provided into which cells are written as they arrive and from which samples are read out at a constant rate. This buffer must accommodate delay variations in queuing delay during a call; however, the maximum delay has already been included in (iv).

For a maximum distance of 1000 km, giving a transmission delay of 4 ms, the total of delays (i) to (v) is well within the maximum of 25 ms permitted by CCITT recommendations without the use of echo cancellers [14]. However, transmission over a longer distance or the use of longer cells would necessitate the use of echo cancellers for telephony.

If the header contains n_h bits and the information field contains n_i bits, the proportion of transmitted bits actually carrying information is $\eta = n_i/(n_i + n_h)$. Clearly, the larger n_i, the greater the efficiency η of the system. However, the larger n_i, the greater the delays incurred.

The choice of cell size is a compromise between these conflicting requirements. The ITU-T has recommended a cell containing a 48-octet information field, preceded by a 5-octet header. These are handled at 149.76 Mbit/s, which corresponds to the payload of the synchronous transport module STM-1 in the synchronous digital hierarchy (SDH) of transmission systems. An ATM bit rate for the 622.080 Mbit/s STM-2 transport module has not yet been specified. Details of the ITU-T recommendations are given in References [15–27].

If cells are filled with 48 speech samples, the overall delay would be satisfactory for more than 80% of calls within the UK; for the remainder, echo cancellers would be needed. The need for echo cancellers could be completely avoided by filling fewer than 48 octets in each cell. Thus, telephone calls would transmit cells more frequently and the packetisation delay would be reduced. This leads to the concept of the level of utilisation of the cell being determined by the destination of the call.

15.4.3 *ATM switches*

The basic principles of an ATM switch are shown in Figure 15.3. A space-division switch is required to route cells from an incoming trunk I_j to an outgoing trunk O_k. At the same time, the header value must be changed from an incoming value H_i to a different outgoing value H_o which will be required to operate the next switching node in the connection. The value of each header on an incoming or outgoing trunk is unique. However, the same header value may be present on different trunks.

Two cells arriving simultaneously on different incoming trunks may be intended for the same outgoing trunk. To prevent loss of a cell, one of these two contending cells must be stored in a buffer until the other has been transmitted. In general, several cells for the same destination may arrive

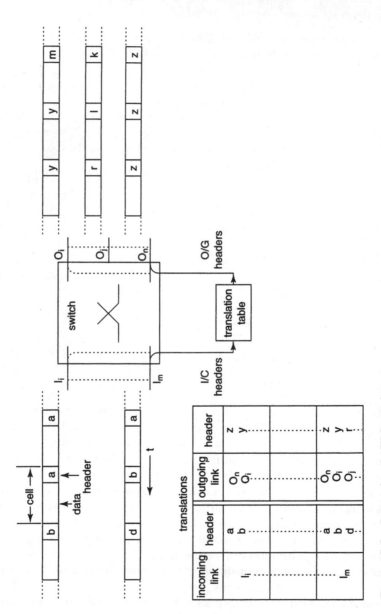

Figure 15.3   Principles of ATM switching

together and buffers must be provided which can store queues of several cells. Thus, the functions to be provided by an ATM switch are:

(*a*) call routing (by space-division switching);
(*b*) header translation; and
(*c*) queuing.

Three different queuing strategies are possible:

(i)   input queuing;
(ii) output queuing; and
(iii) central queuing.

If input queuing is employed, all calls arriving on an incoming trunk enter a buffer. Each remains there until arbitration logic selects it for transfer to the required outgoing trunk, as shown in Figure 15.4*a*.

If output queuing is used, incoming cells are routed directly to an outgoing trunk. The outgoing trunk can only send one cell at a time. Others arriving simultaneously must queue in its buffer, as shown in Figure 15.4*b*. It is possible for all *N* incoming trunks to originate cells for the same outgoing trunk simultaneously. The switch must therefore operate at a bit rate which is *N* times that of the trunks and each output buffer must accommodate *N* cells.

If central queuing is used, the queuing buffers are shared between all incoming and outgoing trunks, as shown in Figure 15.4*c*. Each outgoing trunk must select its cells from this queue. Consequently, this method requires a more complex memory-management system than input queuing or output queuing.

It has been shown that central queuing results in the smallest queue-capacity requirement and that input queuing has the worst performance [12]. These results may be explained as follows. With input queuing and a first-in-first-out (FIFO) queue discipline, 'head-of-the-line' blocking occurs. Cells must wait to be served, even if their outgoing trunks are free, because other cells nearer to the head of the queue are waiting for outgoing trunks to become free. With output queuing, each outgoing trunk dispatches its waiting cells as quickly as possible; thus, queues are shorter and delays are less than with input queuing. Central queuing does not use a FIFO discipline. Calls near the back of the queue may be served, if their outgoing trunks are free, before those near the head of the queue. Thus, the number of queuing places required (although not the mean delay per cell) is less than for output queuing. For a cell-loss rate of 10^{-8}, a central-queuing system needs a capacity of only 10 cells per outgoing trunk, whereas an output-queuing system needs 40 cells per outgoing trunk [12].

Several different switching-network architectures have been proposed, based on the different queuing methods and different basic switching elements [12,28]. Some demonstrator models have been built and tested [1,12].

The multistage switching network shown in Figure 15.5 has a self-routing property. It is known as a delta network [28] and uses switching elements

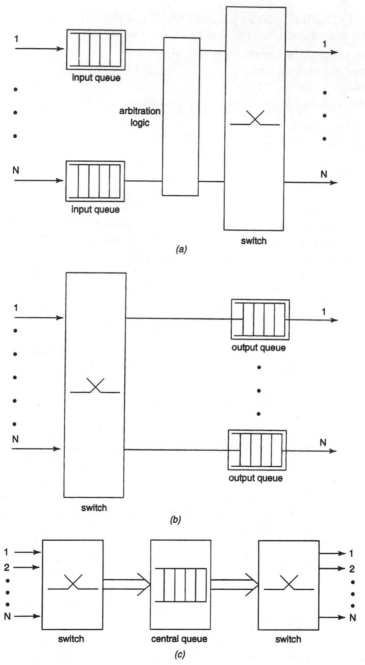

Figure 15.4   Queuing strategies for ATM switches

    a Input queuing
    b Output queuing
    c Central queuing

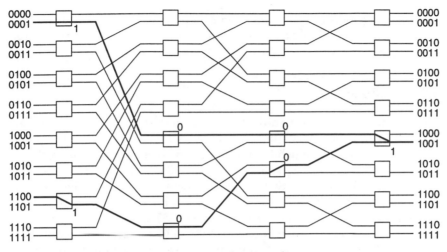

Figure 15.5   Self-routing ATM switch (Delta network)

which have only two inlets and two outlets. The outlet chosen at each stage is determined by a digit in the header of a cell. If this digit is '0', the upper outlet is chosen and, if it is '1', the lower outlet is chosen. Thus, successive bits in the header determine the route for the cell until the required outgoing trunk is reached. Figure 15.5 shows a four-stage network connecting 16 incoming trunks to 16 outgoing trunks (numbered 0–15). It shows two possible connections made to outgoing trunk 9 (i.e. 1001) as a result of reading the digits '1001' in the headers of cells originating from incoming trunks 1 and 12, respectively.

This network has internal blocking, since two cells requiring the same outlet of a 2×2 switching element may arrive together at its two inlets. Figure 15.5 shows this at the final stage. However, it could occur at any switching element in the network. This problem can be solved by:

(a) providing a buffer in each switching element to store one incoming cell while the other is transmitted;
(b) increasing the bit rate on the internal links relative to that on the external trunks; there will be no blocking if the bit rate is speeded up N times (where N is the number of incoming and outgoing trunks);
(c) by providing multiple links between switching elements; or
(d) by providing multiple networks in parallel between the incoming and outgoing trunks.

The network in Figure 15.5, with switching elements of size 2×2, requires $\log_2 N$ stages to connect N incoming trunks to N outgoing trunks and it uses $\log_2 N$ header digits to route the cells. Larger networks can use larger switching elements. For example, if elements of size 16×16 are used, each element will use four header digits. It would replace the whole of the network in Figure 15.5 and three stages would serve $16^3 = 4096$ incoming and outgoing trunks.

Since these networks use switching elements which are all identical and have regular interconnection patterns, they are suitable for manufacture as large-scale integrated circuits.

15.5 Signalling

Signalling over broadband networks can be expected to be based on the now well established common-channel-signalling techniques, although the potential exists for running at much higher bit rates. Early broadband networks will use the digital subscriber signalling system 1 (DSS1) and the ISDN user part (ISUP) of CCITT Signalling System 7 (SS7). Only point-to-point connections will be supported, with few supplementary services. Later, a broadband application part (BAP) will be designed for the SS7 stack.

An alternative under study is to use ATM cells to carry signalling. Even though, by definition, each cell contains routing information, there is still a requirement to carry network signalling information for the establishment of virtual calls and for network-management data.

15.6 Local access

One of the most likely technologies for providing the user with a 600/155 Mbit/s connection to the local exchange is optical fibre, so a considerable amount of work is being expended on resolving the resulting technical problems in as economical a way as possible [1].

The most obvious approach to the customer-access connection (CAC) is simply to replace the telephony copper pair with a single optical fibre, star connected from the exchange to each user. Each fibre carries simultaneous bidirectional signals based on wavelength-division multiplexing: a wavelength of 1300 nm is used in one direction and 1530 nm in the other and filters ensure the directionality of each signal.

This approach to 'fibre to the home' is a *fibre-rich* one. Other methods being considered are based on a *fibre-lean* approach, and one such method is the basis for TPON, i.e. Telephony over a Passive Optical Network [29]. A single fibre leaving the exchange passes through a number of passive optical splitters, each of which has very low loss, until the one source is physically connected to many destinations. The electronics at the source and at each destination share the fibre in a TDM mode, with each destination having a unique address. When a destination unit recognises its own address transmitted on the fibre, it activates its receive circuit to extract information, then transmits on the fibre back to the exchange. The number of end users is typically 32 or 64, although in theory 128 could be accommodated. The TPON system only requires one wavelength for its realisation. Thus, providing the splitters are not wavelength-dependent, other wavelengths can be used over the same fibre for other services, e.g. television distribution. The extension of TPON into broadband is known as BPON (Broadband Passive Optical Network) [30].

The number of different wavelengths which can be used over a fibre is limited by a combination of laser design and frequency-dependent fibre losses. Currently only about four can be achieved, although each one offers up to 600 Mbit/s. However, the use that is made of each wavelength is relatively basic; the light beam is simply turned on and off. As with radio waves, much more information can be carried if the beam is modulated, for example using different modulation frequencies. This technique has been employed for many years in FDM transmission systems, and is now being investigated in relation to light, where it is known as coherent multichannel communication [31] (CMC). Another term for this is optical-frequency multiplexing with coherent detection. Current work is aimed at the realisation of a demonstrator providing, on a single fibre, 10 channels of up to 155 Mbit/s each, and a relatively simple extension of these techniques will yield 100 channels of 155 Mbit/s. It is envisaged that, in the not-too-distant future, CMC techniques will offer 100 channels of 600 Mbit/s each, and this level of capability should satisfy all broadband-network requirements for the foreseeable future.

15.7 References

1. International Conference on *Integrated broadband services and networks*, London, 1990, *IEE Conf. Publ. 329*
2. MADDERN, T.S., and OZDAMAR, M.P: 'Advances in digital switching architecture'. IEE 2nd National Conference on *Telecoms*, York, 1989, *IEE Conf. Publ. 300*
3. CCITT Recommendation G.722: '7 kHz audio coding within 64 kbit/s'
4. WALLACE, A.D., HUMPHREY, L.D., and SEXTON, M.J.: 'Analogue to digital conversion' *in* FLOOD, J.E., and COCHRANE, P. (Eds.): 'Transmission systems' (Peter Peregrinus, 1991), chap. 6
5. CCITT Recommendation H.120: 'Codecs for videoconferencing using primary digital group transmission'
6. CCITT Recommendation H.261: 'Codecs for audiovisual services at n × 384 k bit/s'
7. CCIR Report 629-2: 'Digital coding of colour television signals'
8. CCIR Report 962: 'The filtering, sampling and multiplexing for digital encoding of colour television signals'
9. FLOOD, J.E., and TUCKER, R.C.F.: 'Packet speech multiplexer', *IEE Proc. F*, 1987, **134**, pp. 652–658
10. TURNER, J.S., and WYATT, L.F.: 'A packet network architecture for integrated services'. *Globecom, 83* conference, San Diego, USA, November 1983
11. COUDREUSE, J.P., and BOYER, P.: 'Asynchronous time-division techniques for real-time ISDNs'. *Seminaire international sur les reseaux temps reel*, Bandol, April 1986
12. de PRYCKER, M.: 'Asynchronous transfer mode: solution for broadband ATM' (Ellis Horwood, 1991)
13. CUTHBERT, L.G., and SAPANEL, J-C.: 'ATM: the broadband telecommunications solution' (Institution of Electrical Engineers, 1993)
14. CCITT Recommendation G.131: 'Stability and echo'
15. CCITT Recommendation I.113: 'Vocabulary of terms for broadband aspects of ISDN'

16. CCITT Recommendation I.121: 'Broadband aspects of ISDN'
17. CCITT Recommendation I.150: 'BISDN ATM functional characteristics'
18. CCITT Recommendation I.211: 'BISDN service aspects'
19. CCITT Recommendation I.311: 'BISDN general network aspects'
20. CCITT Recommendation I.321: 'BISDN protocol reference model and its application'
21. CCITT Recommendation I.327: 'BISDN network functional architecture'
22. CCITT Recommendation I.361: 'BISDN layer specification'
23. CCITT Recommendation I.362: 'BISDN ATM adaptation layer (AAL) functional description'
24. CCITT Recommendation I.363: 'BISDN ATM adaptation layer (AAL) specification'
25. CCITT Recommendation I.413: 'BISDN user-network interface'
26. CCITT Recommendation I.432: 'BISDN user-network physical layer specification'
27. CCITT Recommendation I.610: 'OAM principles of BISDN access'
28. AHMADI, H., and DENZEL, W.: 'A survey of modern high-performance switching techniques', *IEEE J. Select. Areas Commun.*, 1989, **SAC-7**, pp. 1091–1103
29. ADAMS, P.F., ROSHER, P.A., and COCHRANE, P.: 'Customer access' *in* FLOOD, J.E., and COCHRANE, P., (Eds.): 'Transmission systems' (Peter Peregrinus, 1991), chap. 15
30. KILLAT, U. (Ed.): 'Access to B-ISDN via PONs' (Wiley, 1996)
31. SHIMADA, S. (Ed.): 'Coherent lightwave technology' (Chapman and Hall, 1994)

Chapter 16
Intelligent networks
M. Eburne

16.1 Introduction

There have been many changes to public switched telephone networks (PSTNs) since the late 1960s, with the introduction of new technologies such as stored-program control (SPC), digital switching, common-channel signalling and, more recently, the integrated-services digital network (ISDN). These changes have brought the benefits of cheaper networks, higher-quality service, better reliability and new customer features.

With the introduction of SPC switching systems into major networks across the world during the 1970s, many people in the industry believed that the floodgates on an ever-expanding range of new customer services were about to open. New services would just require small additions to the switch-control programs, and it was well recognised that software changes were far easier and cheaper to effect than hardware modifications. Reality soon dawned and it became apparent that consistent software upgrades to the many interconnected switching systems which make up a network were every bit as costly, time consuming and potentially disruptive to network operations as hardware changes. Although new technology has brought many benefits to the customer, the envisaged explosion in new services has so far failed to materialise.

Consequently, the operators and suppliers of the more-advanced networks are now turning their attention to real-time control aspects of their PSTNs, or network 'intelligence', in their search for quicker and cheaper ways of introducing new features into the network. These studies aimed at providing more flexible network architectures go under the title 'intelligent network'.

The term 'intelligent network' (IN) originated in work carried out by Bell Communications Research (Bellcore) on behalf of the Bell operating companies in the USA [1,2]. The Bellcore terminology and network models are widely used to describe the concepts involved and they have been adopted for this Chapter.

16.2 Benefits of an IN architecture

Real-time network control or 'intelligence' in current digital switch networks is vested in the call-processing software resident on each switch. Introduction of new call-handling features requires enhancement of the call-control software (and any associated hardware modifications) to every switch within the network. The difficulties of providing new customer services can be summarised as follows:

(*a*) It takes several years to specify, develop, test, and deploy new switch-based services, currently between two and five years. New features have to wait until the necessary timeslot can be made available within the switch suppliers' development cycle and, for most suppliers, these development programmes are fully booked many years in advance.

(*b*) The time taken to upgrade a large network with new software and (possibly) hardware can be in excess of 12 months. The network is highly vulnerable during this period and performance often suffers. Hence, there is a need to limit the number and frequency of major upgrades in the interests of network stability.

(*c*) Because of the time taken to implement network upgrades, and particularly the time it takes to test and trial new switch builds before they can be certified for general release, the introduction of any new service is very costly. It therefore requires a sound business case before development can commence.

(*d*) Although the current structure, where control is distributed across individual switches within the network, is adequate for self-contained single-node features such as short-code dialling, it is less suited to more complex services which require co-operation across a number of nodes. These services often introduce an added complication, with the need to maintain and operate consistent databases across a large number of switching nodes.

These problems make it difficult for telecommunication network operators to react to their customers' needs within acceptable timescales. An intelligent network is intended to overcome these problems.

If an intelligent network architecture is to provide the means to deliver new services and help network operators to become more responsive to their customers' needs, it must address the issues listed above. IN must therefore provide a quicker and cheaper mechanism for new service delivery which does not threaten the underlying stability of the network. In addition, IN must also provide the ability to tailor services to individual customer requirements, and the means to provide real-time service management direct from the customer. Some of the advanced services that can be provided by means of an IN are described in the next Section.

16.3 Intelligent network services

The widespread deployment of a flexible intelligent network structure at all levels of a telecommunication operator's network will assist in the rapid development and deployment of future services. IN architectures are particularly suited to the implementation of value-added services with one or more of the following characteristics:

(*a*) networked services, i.e. services which require co-ordinated action across a number of nodes: examples are wide-area centrex and virtual private networks;

(*b*) services which require large databases and real-time data updates (especially by the customer);

(*c*) services which may require tailoring to meet individual customer requirements;

(*d*) speculative services, i.e. services for which the market demand is unclear and for which a cheap mechanism to develop and trial the service is initially more important than implementation efficiency.

Some of the services that world telecommunications operators have already introduced, or are planning to introduce, using IN architectures include the following:

(i) Freephone

Charges are billed to the called customer, instead of to the calling customer. An IN enables a single number to be used for service centres at different locations. The caller is usually routed to the nearest.

(ii) Virtual private networking (VPN)

Private networks with circuits established on demand using PSTN switching and transmission plant, but offering the customer the same security, availability and other features as conventional private circuits. VPN is closely linked with, and forms an essential part of, wide-area centrex.

(iii) Charge card

This service enables originating calls to be charged to the customer's home account.

(iv) Wide-area centrex

Centrex is the provision of modern PABX service to customers using public switching equipment. Wide-area centrex serves individual customers with sites hosted on two or more network nodes, providing a unified numbering plan and operation of facilities, such as call completion on busy, across all customer sites.

(v) Automatic call distribution

Distribution of incoming calls to a number of geographically-separate answering points, either on a fixed proportional basis or dependent on availability of free circuits.

(vi) Personal numbering

Allocation of a single fixed number which enables calls to be completed to a customer no matter where that customer is located in the network. Call completion is controlled by an updatable personal profile which contains the customer's actual or likely network location. A voice mailbox or radiopaging message can be used as a default option.

(vii) Call-completion services

Simple diversion on busy and diversion on no reply are already available on most modern digital switching systems. IN can offer enhanced features to these existing services, plus network-wide call completion on busy.

(viii) Universal access number

Allocation of a single fixed number to a customer with a number of geographically-separate locations (e.g. a public utility). Incoming calls would be routed according to a predetermined program, e.g. to the nearest office to the caller.

16.4 Intelligent network architecture

16.4.1 General

The key principle of IN is the separation of software which controls basic switch functions (i.e. collecting dialled digits, basic call setup and basic call control) from the software which controls call progression for more advanced features, such as those discussed in Section 16.3. The advanced call control, or 'intelligence', is used to co-ordinate and control the elementary switch functions via a well-defined and rigidly-controlled interface. This enables new services and features to be developed and introduced with minimal impact on the basic switch functions. The intelligence can be located in a centralised processor, called a *service-control point* (SCP), which controls the whole network. Alternatively, it can be distributed over a number of switching systems, either sharing the switch processor or located on a separate processor.

The components which make up an intelligent network are shown in Figure 16.1. They consist of:

(*a*) service switching points (SSP)
(*b*) service control points (SCP)
(*c*) signalling network
(*d*) intelligent peripherals (IP)
(*e*) service-creation environment
(*f*) management systems.

These are discussed in the following Sections.

16.4.2 Service switching point

Major restructuring of the digital-switching-system software is required in order to provide the *service switching point* (SSP) functionality necessary for IN. The SSP software resides on the digital switching system, and can be located at any level of the network (local, trunk or special overlay).

The SSP performs the function known as *triggering*. This is the ability within basic call control to recognise the need for more advanced processing, to suspend normal call processing and refer the call to the *service control point* (SCP). A range of possible triggering events includes: digits dialled,

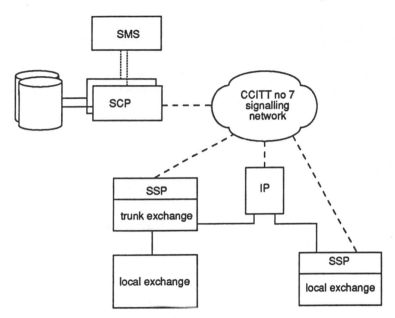

Figure 16.1   *Intelligent-network architecture*

IP = intelligent peripheral
SCP = service control point
SMS = service-management system
SSP = service switching point
. . . . .   X.25 link
——— speech path
------- signalling path

customer's service mark and call events (e.g. ring tone no reply).

Also implemented at the SSP are the basic building blocks from which advanced services are constructed. These are called *functional components* (FC). The activation of various FCs at the SSP during the call is controlled by the *service logic program* at the service control point (SCP). Examples of typical functional components are: connect switch path, break switch path, terminate call, connect call (to specified destination), play announcement, collect digits (from the customer), and report call event (e.g. called-customer answer).

16.4.3 Service control point

The service control point (SCP) is a centralised transaction processor which hosts the control software for the advanced customer services. These control programs are called *service logic programs* (SLP). The SCP must provide high throughput (measured in transactions per second), high reliability (equivalent to that given by modern switching systems), fast response times (transaction responses of less than 500 ms) and high-capacity fast-access database systems. The structure of a typical SCP is shown in Figure 16.2. The *service logic interpreter* (SLI) is the program-execution environment which hosts the SLPs; it is built onto a standard computer and operating-system base which hosts the SLPs. The SLI contains the embedded functional-component control logic that can be called by individual SLPs. To enable rapid introduction of new services, it is essential that the SLI design limits the ability of new SLPs to impact on existing SLPs or the SLI itself.

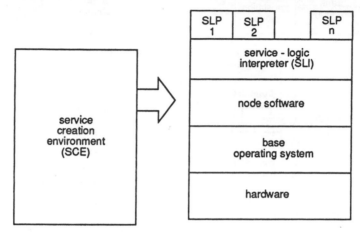

Figure 16.2   Service control point

SLP = service logic program

16.4.4 Signalling network

The CCITT no. 7 common-channel signalling network is the means used to pass control messages between the SCP and the SSPs, and between the SCP and the intelligent peripherals. Current standards for CCITT no. 7 are optimised for circuit switching; for IN, some changes are required to equip it for use in a transaction-processing environment. Two changes are required to the CCITT no. 7 message formats: the addition of the *signalling-connection control part* (SCCP); and the introduction of *transaction capabilities* (TC). The SCCP enables the signalling messages to be routed in a non-circuit-related mode, i.e. the signalling messages no longer need a speech-circuit number as the message identifier. The TC is a message structure optimised for carrying the query/response messages used for transaction working. To pass messages between an SSP sited at a local exchange and a centralised SCP requires a more complex signalling network using signalling transfer points (STP) and SCCP translation nodes, which determine the destination identity from information within the message's SCCP fields. A possible implementation of such a network is shown in Figure 16.3, where the STP and SCCP translation capabilities are associated with the trunk exchanges.

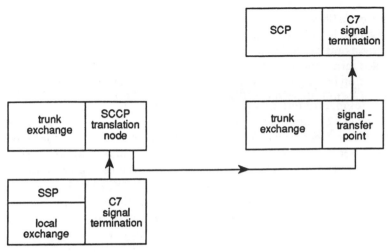

Figure 16.3   Signalling-network structure

SCCP = signalling-connection control part
SCP   = service control point

16.4.5 Intelligent peripheral

The *intelligent peripherals* (IPs) provide any additional specialised functions required for IN services. The most notable examples of IPs used in current IN networks are voice-guidance and digit-collection units, with some suppliers planning to introduce voice-recognition technology in the near future. The original Bellcore IN architecture proposed that the IPs be controlled directly from the SCP via CCITT no. 7 signalling links, with speech circuits connecting the IP to the SSP. However, many current implementations, including the AT&T system discussed in Section 16.7, effect control of the IP via the SSP. The control instructions are received by the SSP from the SCP and the SSP then relays these control messages to the IP.

16.4.6 Service-creation environment

If the intelligent network and its associated management systems are to offer a means of rapid deployment and operation of new services, consideration must also be given to a set of tools to speed the service-development process itself. Systems which assist the development process are collectively referred to as the *service-creation environment* (SCE). The full range of facilities encompassed by the SCE can be considered as a set of tools to manage the whole lifecycle of the service software, including:

(*a*) requirements capture
(*b*) service specification
(*c*) prototyping and demonstration
(*d*) design, coding and test

(*e*) service trialling
(*f*) software release and download
(*g*) upgrade and maintenance
(*h*) service termination.

However, in the near term, most SCE systems will probably be limited to software design and test, and possibly service-prototyping, facilities.

Another key element closely linked to the service-creation environment is the *application-programming interface* (API). The API is the definition of facilities offered by the service-logic interpreter and how to access those facilities [3]. It enables the telecommunication network operator or the supplier to program new services (SLPs) using an appropriate SCE. In the longer term, it may be possible to produce a standard API definition to which all IN-equipment suppliers conform. However, in the near term, API definitions are likely to be specific to each SCP supplier.

16.4.7 Management systems

The Bellcore IN model includes a *service-management system* (SMS). This system interfaces to all the SCPs using X.25 data links, and it provides service-management support for the IN services. The main function performed by the SMS is management of the service data deployed on the SCPs, including: data updates (from the administration or the customer); adding new customers; data audits (to check data consistency across all SCPs); synchronised data updates across all SCPs; and data reload (after an SCP-system crash).

If the SCP and SSP offer the means to deploy new services rapidly and the SCE offers the means to develop new services rapidly, it is important that implementation of the management needs of any new service does not become the bottleneck in the process. Data management, provided by the SMS, is an important component of IN, but consideration must be given to the other management requirements. These include: charge recording and billing; alarms; event management and maintenance; traffic management; capacity planning (using usage statistics); customer enquiries; and processing of new orders. All these systems must provide simple mechanisms for deploying new services rapidly without the need for major new development.

16.5 Example of an IN call

Let us examine a typical example of an IN service and the role of the various network components introduced in Section 16.4. The sequence of diagrams in Figure 16.4 describes the progression of a Freephone call. Note that in this example the SSP function is located at the trunk tier in the network.

The calling customer dials the Freephone number (0800 XXXXXX) and the call is routed from the digital local exchange to the digital trunk

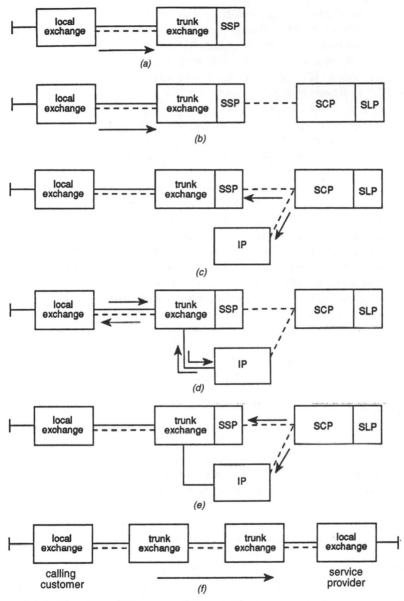

Figure 16.4   Freephone call: sequence of operations

IP=intelligent peripheral
SCP=service control point
SLP=service logic program
a Initial connection to trunk exchange
b Reference to SCP
c Connection to IP
d Interaction between IP and caller
e Release of announcement path
f Final connection to service provider

exchange, as shown in Figure 16.4a. The triggering function in the SSP associated with the trunk exchange detects that this call requires reference to an SCP on the basis of the dialled digits 0800.

Normal call handling is suspended and a query message containing the dialled digits and calling-line identity (CLI) is sent via CCITT no. 7 signalling to the SCP as shown in Figure 16.4b. The SCP determines that the call is for the Freephone service and refers the message to the Freephone SLP. The SLP determines the appropriate call-handling procedures from information in the query message. These will depend on the service provider for whom the call is destined and the geographical location and/or the identity of the calling customer.

In this example, the handling instructions for the requested service provider indicate that further information is required from the calling customer to determine which department is required (e.g. accounts, customer orders, general enquiries). A message is sent from the SCP to the IP, as shown in Figure 16.4c. This requests connection of a guidance announce-ment, followed by collection of further dialled digits. A further message is sent from the SCP to the SSP requesting connection of a speech path from the calling customer to the selected announcement channel on the IP. This path is established as shown in Figure 16.4d.

The guidance announcement is played to the calling customer, requesting further information on the required destination. The customer responds by inputting the appropriate digit by MF keyphone. This information is collected by the IP and returned by a CCITT no. 7 message to the SCP, where the Freephone SLP determines the required destination from this additional information.

The SCP requests that the IP and the SSP disconnect the announcement path, as shown in Figure 16.4e. The SCP passes routing data and a request for call completion to the SSP and, at this point, the SCP relinquishes control. The call is then completed to the required service provider using standard call-setup mechanisms, as shown in Figure 16.4f.

16.6 Distributed architecture

This Chapter has so far concentrated on introducing IN principles based upon a centralised network-control architecture, which uses a pair of centrally-sited SCPs to control service-execution processes across the whole network. IN can also be achieved using distributed control and Figure 16.5 shows such an architecture. The service-control environment, represented by the service-logic interpreter (SLI), the service-logic programs (SLP) and the database, is cosited with the local or trunk switching system rather than at a separate SCP. The SLI can be implemented either on the switch processor or on a separate close-coupled processor.

The main advantages of the distributed architecture are:

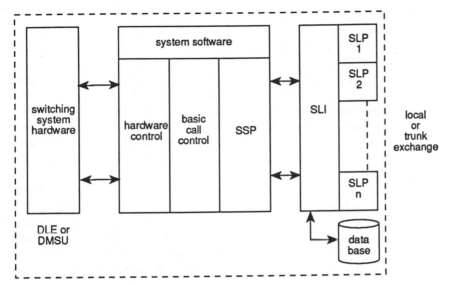

Figure 16.5   Distributed IN architecture

SLI = service logic interpreter
SLP = service logic program
SSP = service switching point

(*a*) faster response times;
(*b*) savings on signalling-network infrastructure; and
(*c*) cheaper implementation for high-volume services.

The advantages of a centralised architecture are:

(i) easier data management;
(ii) faster rollout of new services; and
(iii) cheaper implementation for a large number of low- and medium-throughput services

A possible implementation strategy would be to use centralised SCP control for the rapid rollout of new services, with a gradual migration of proven and more successful (i.e. high-volume) services to a distributed platform.

16.7 Examples of intelligent network implementation

IN principles have now been implemented or are being planned in several countries [3–8]. In the USA, the long-distance network of AT&T has common-channel signalling, and network control points (NCPs) were connected to its toll offices to act as SCPs and provide advanced services, such as freephone and calling-card services [4]. Now, a centralised number administration and service centre (NASC) manages an advanced-intelligent-network (AIN) capability. Its independent database is interrogated to find out the appropriate long-distance carrier for each '800' call. An IN structure is now being introduced in local-exchange areas to extend advanced features to end offices.

One example of an early IN implementation [9] is British Telecom's *Digital Derived Services Network* (DDSN), installed in 1989 to provide advanced telemarketing services (0800 freephone). As shown in Figure 16.6, the DDSN consists of an overlay network of 10 *digital derived-service switching centres* (DDSSC) interconnected to the trunk tier of the PSTN. Two network control points (NCP), located with the London and Manchester DDSSCs, provide the service-control-point function. The service-switching-point function, called action control point (ACP), is present on all 10 DDSSCs. The voice-guidance and digit-collection function (part of the intelligent-peripheral role in the Bellcore model) is performed by two *network service complexes* (NSCs) controlled via their host ACPs. Data-update and management functions are provided by the NETSTAR system (Network Subscriber Transaction Administration and Recording) which broadly equates to the service-management system in the Bellcore model.

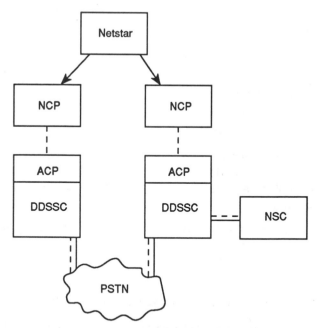

Figure 16.6   BT digital derived-services network

ACP=action control point
DDSSC=digital derived-services switching centre
NCP=network control point
NSC=network service complex
PSTN=public switched telephone network

This intelligent-network implementation has enabled BT to introduce a range of advanced features to supplement the basic 800 linkline service:

(*a*) Time-of-day/day-of-week routing
Answering programmes can vary according to the time of day and day of the week.

(*b*) Geographical routing
Processing of received calls can be varied according to the geographical area in which the call originates.

(*c*) Proportional distribution
Incoming calls can be directed to two or more answering points on a predetermined proportional basis.

(*d*) Call queueing/completion of busy calls
Queueing of calls meeting busy or completion to an alternative destination.

(*e*) Call prompting
Callers accessing a single linkline number can be offered a choice of answering points (e.g. accounts department, sales or customer service) using the voice-guidance and digit-collection facilities of the NSC.

British Telecom has subsequently enhanced all its trunk exchanges [digital main switching units (DMSU)] to provide an SSP capability. The DMSUs are linked to two SCPs by CCITT no. 7 signalling and each SCP is linked to an SMS by an X.25 network. Phase 1 of this project [10] provides two types of number translation:

(*a*) fixed translation, which provides a direct translation from one number to another;
(*b*) geographic-based translation, where a translated number is based on the geography or origin of the call.

This scheme is intended to be an interim step towards providing a full set of IN capabilities [11].

16.8 Conclusions

The intelligent-network concept promises to be one of the most significant influences on switching-system architectures and customer services since the advent of stored-program control. The operators of most of the world's more advanced networks have already introduced, or are planning to introduce, some form of IN architecture.

In the short and medium term, most of these implementations are likely to be limited in their scale of deployment and capability, with centralised SCP control used to introduce a limited range of key services, such as freephone, calling card and VPN. The progress of IN development will be determined by commercial considerations rather than technical limitations. The deployment of IN systems by telecommunications operators must be backed at each stage by sound revenue opportunities generated by the new customer services. The benefits of a fully flexible IN, deployed at all levels of the network and using a mixture of centralised and distributed architectures, is therefore unlikely to be realised until the next generation of switching systems begins to replace current network technology.

16.9 References

1. 'IN1+ network plan'. Bellcore special report SR-NPL-001034, issue 1, 1988
2. 'Advanced intelligent network release 1 proposal'. Bellcore special report SR-NPL-001509, 1989
3. 'Application programming interface'. Joint study by British Telecom and IBM, 1989
4. 'Stored program controlled network', *Bell Syst. Tech. J.*, 1982, **61**, (3), special issue, pp. 1573–1815
5. DOYLE, J.S., and McMAHON, C.S.: 'The intelligent network concept', *IEEE Trans.*, 1988, **TC-36**, pp. 1296–1301
6. 'Intelligent network architecture'. BT, GPT and Ericsson joint study, 1989
7. 'Intelligent networks: dedicated to services', *Electr. Commun.*, 1990, **65**, (1), special issue
8. AMBROSCH, W.D., MAHER, A., and SASSCER, B.: 'The intelligent network'. Bell Atlantic, IBM and Seimens joint study (Springer-Verlag, 1989)
9. WEBSTER, S.: 'The digital derived services intelligent network', *Br. Telecom Eng. J.*, 1989, **8**, pp. 144–149
10. WYATT, T., GOZAL, D., and HANCOCK, J.: 'Intelligent network phase 1: intelligence in the core network', *Br. Telecom Eng.*, 1995, **114**, pp. 202–208
11. ITU-T Recommendation Q.121x: 'IN capability set 1'

Chapter 17

Network performance

G.J. Cook

17.1 Introduction

Network performance can be broadly considered as including all those aspects of a telecommunication network which relate to communication between users. It is part of the overall provision of a telecommunication service, which includes such areas as: installation, network performance, repair service, billing and customer handling. This Chapter concentrates on the network-performance aspects of the overall service, with emphasis on voice telephony. It introduces the general approach being taken by the ITU to quality of service and network performance.

The Chapter considers the principal network-performance parameters covering the various phases of a call, as follows:

(*a*) the access phase (i.e. connection set-up);
(*b*) the information-transfer phase (e.g. conversation); and
(*c*) the disengagement phase (i.e. connection clear-down).

The performance of a switched network for the access and disengagement phases of calls is mainly determined by the performances of its telephone exchanges and signalling systems. Apart from complete failures, transmission systems have negligible impact. On the other hand, except for switching noise in old electromechanical exchanges and delays introduced by digital exchanges, the information-transfer performance is largely determined by the performance of transmission systems.

17.2 General approach to quality of service and network performance

Services can be divided into two categories:

(*a*) teleservices, in which the user experiences the service via a terminal such as a telephone; and
(*b*) bearer services, where the user is presented with transmission capacity which he or she then adapts to support the desired communication function.

The ITU has determined that the quality of the services should be described in terms of parameters which are user oriented and which make no assumption about the network over which the services are supported [1]. On the other hand, network providers need to assure themselves that their networks are capable of offering the desired quality of the services; network performance is therefore described in terms of parameters specific to network elements or connection types. The ITU illustrates these concepts as shown in Figure 17.1.

The ITU makes use of the 3×3 matrix shown in Figure 17.2, which illustrates the outcome of the three phases of a communication function. This matrix can be used for each service type or connection type as a check list to be assured that all relevant aspects of the communication function are being addressed. For some services or connections, certain cells should be null. An important point to note is that the parameters which are to be listed in the various cells describe the quality of service or network performance during the available state. These parameters are known as *primary* or *directly observable parameters*. If any of these parameters, either singly or in combination, exceeds its threshold value for a defined period of time, the service or connection is deemed to be unavailable. The parameters which describe availability performance, e.g. frequency and duration of outages, are known as *secondary* or *derived parameters*. They are not directly measured, but are determined from observation of the primary parameters.

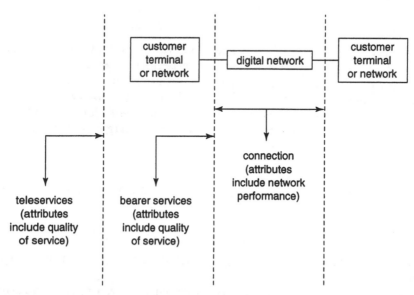

Figure 17.1   Network performance and quality of service
From CCITT 1.350

CCITT Recommendation I.350 [1] defines network performance (NP) as 'the ability of a network (or network portion) to provide functions related to communications between users'. It additionally points out that NP should be defined and measured in terms of parameters meaningful to the network provider and used for the design, configuration, operation etc. of that network. NP should also be defined independently of terminal performance and user action.

Quality of service (QOS), on the other hand, should be expressed in terms which focus on user-perceivable effects which do not depend on assumptions about network design. QOS can be considered as a statement of the performance of a given service as offered to the customer. As indicated in Figure 17.1, this statement must be service specific, with all service aspects under the control of the service provider. For example, it is clearly difficult to give statements concerning the performance of a teleservice if the terminal equipment is not part of that service.

The final aspect, but the most important, is the user's opinion of the service. This will be influenced by the QOS, but it is also affected by:

(*a*) customers' terminal equipment (where it is not part of the service);
(*b*) customer actions; and

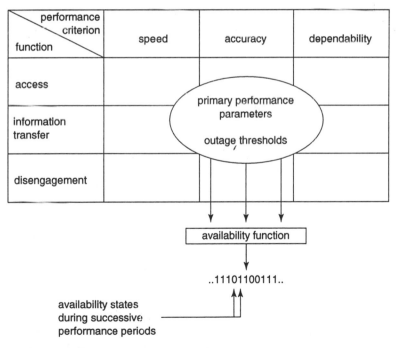

Figure 17.2　*Performance-parameter matrix*

From CCITT I.350

(*c*) external factors, such as the image of the telephone company, any billing disputes etc.

User opinion can be measured by customer surveys or by more direct feedback such as complaints.

It can be seen that QOS and user opinion are different and it is important that the two are not confused. For example, CCITT Recommendation I.350 [1] only considers aspects of QOS which can be directly observed (and measured) at the service-access point. However, QOS is a term which is widely used and there are many examples where it has been used to cover aspects which are subjective and outside of the service provider's control.

17.3 Principal network-performance parameters

The 3×3 matrix shown in Figure 17.2 can be used to derive the network-performance parameters and to relate them to the customer perception of the service. As an example, the telephony service can be used to show the principles. Table 17.1 shows the functions from the 3×3 matrix (access, information transfer and disengagement) and lists the customer expectations

Table 17.1 Principal network-performance parameters

Function	Customer expectations	Network-performance parameters
Access	No loss of service	Network availability
	Fast call setup	Delay to dial tone
		Connection-establishment delay
	No congestion/ call failures	Connections encountering congestion
		Failed connections (no tone)
		Misrouting
Information transfer	Good transmission quality	Transmission loss
		Noise
		Loss/frequency distortion
		Nonlinear distortion
		Echo and delay
		Error performance
Disengagement	No cutoffs	Connections prematurely released

based on the criteria of speed, accuracy and dependability. For example, fast call setup covers the speed criterion for access and good transmission performance (clarity and volume) covers the accuracy criterion for information transfer. A similar exercise can be carried out for the NP parameters as shown in Table 17.1. For example, the principal transmission parameters determining the information-transfer-function performance can be identified and also linked to customer expectation.

In this way, the important end-to-end NP parameters can be identified and linked to the customer's expectation in general terms. The next stage is to consider appropriate end-to-end values for the network-performance parameters.

17.4 Objectives for overall values

17.4.1 General

Objectives for end-to-end values of the identified NP parameters can be found in relevant CCITT Recommendations. References 2–5 give the appropriate recommendations for four of the parameters, i.e. call-setup delay, error performance, propagation delay and talker echo, respectively.

This Section gives some additional background material on the ITU approach to specifying end-to-end limits for key transmission and call-processing parameters.

17.4.2 Analogue transmission

In a telephone connection, the complete path includes the air path from the talker's mouth to a telephone transmitter and from a telephone receiver to the listener's ear, in addition to the telephones and the switched connection between them. The overall attenuation of such a path is expressed in terms of its *overall loudness rating* (OLR) in decibels [6,7]. This is measured by comparing the perceived loudness of the received sound with that from a standard speech path, called the Intermediate Reference System, which was defined by the CCITT [8].

The same OLR can be obtained for a connection having a more sensitive transmitter or a shorter customer's line at the sending end, together with a less sensitive receiver or a longer customer's line at the receiving end. It is therefore necessary to measure separately the *send loudness rating* (SLR) and the *receive loudness rating* (RLR). Then

$$OLR = SLR + RLR + JLR$$

where JLR is the *junction loudness rating*, which represents the attenuation of the connection between the two local exchanges concerned.

The CCITT recommended preferred values of SLR and RLR up to the boundary between a national network and the international network

Table 17.2 CCITT preferred values of SLR and RLR up to boundary between national and international networks

	Traffic-weighted mean	Maximum value (average-sized country)
SLR	7–9 dB	16.5 dB
RLR	1–3 dB	13 dB

(referred to a 0 dBr international switching point) [9], as shown in Table 17.2.

Subjective experiments have been carried out to determine users' opinions on a variety of typical telephone connections. In this way, it has been possible to plot curves of *percentage difficulty* (i.e. the percentage of connections considered unsatisfactory) against the overall loudness ratings of the connections. Some results [10] are shown in Figure 17.3. It shows that there is a preferred range of loudness ratings from about +5 dB to +15 dB.

In an analogue network, four-wire connections must have an overall nominal loss to ensure stability when component links have losses less than their nominal values, as described in Chapter 3. When these component losses are greater than the nominal values, the overall loss of the connection is increased. For a large number of analogue links in tandem, it can be very large and the worst-case connection can have an extremely high OLR. For example, the limiting connection in the former UK analogue network had an OLR of 29 dB. This did not give acceptable performance for some connections. Fortunately, it was encountered by only a small proportion of calls, because it required a combination of links all of which were adverse.

In an integrated digital network (IDN), all the interexchange links have

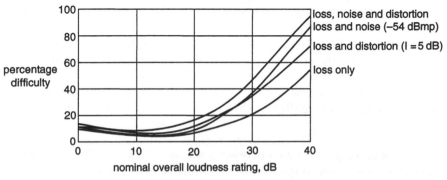

Figure 17.3 Variation of percentage difficulty with nominal OLR for noise and attenuation distortion

Table 17.3 CCITT circuit-noise objectives

Line	normal circuit	3 pWOp/km
	very-long-distance circuit	2 pWOp/km
Modulators	channel	200 pWOp
(per pair)	group	80 pWOp
	supergroup	60 pWOp

zero attenuation with zero tolerance. Consequently, the variation in OLR is due only to customers' lines and telephones and analogue-to-digital conversion in local exchanges. Thus, this variation is much less than in an analogue network and it is possible to ensure that all connections have an OLR within the preferred range. In an ISDN, variation in OLR due to different lengths of customers' lines is also removed.

In analogue transmission, noise tends to increase with the length of the transmission circuits and with the number of multiplexing stages. The CCITT specified noise objectives for multichannel FDM equipment and line systems, respectively [11], as shown in Table 17.3.

In digital telephone transmission, the use of regenerators ensures that noise introduced into a connection is negligible, regardless of the length of the circuit. Thus, the noise is substantially only that due to quantising distortion introduced by coding. The CCITT has defined a *quantisation-distortion unit* (QDU), where 1 QDU represents the quantisation distortion introduced by one PCM-coding stage. For an international connection, the maximum QDU recommended is 14, consisting of 4 for the international link and 5 for each of the originating and terminating national networks [12].

In an integrated digital network, coding and decoding take place only at the originating and terminating exchanges. Consequently, all connections have only 1 QDU and noise is therefore negligible.

In four-wire circuits, propagation delay results in echo, as explained in Chapter 3. As the delay of the echo path increases, the loss of the circuit should increase to avoid the troublesome subjective effects of echo. The ITU-T recommends [5] that echo suppressors or echo cancellers should be fitted if the one-way delay of the echo path exceeds 25 ms. This delay can be exceeded on international circuits. In particular, transmission to and from a geostationary satellite results in a one-way delay of 260 ms. Echo control is thus essential. The ITU-T recommends [4] that the one-way delay for an international connection should not exceed 400 ms. Thus, an international connection should not contain more than one satellite link. Some typical values for delays associated with different equipments are shown in Table 17.4. The values given for digital exchanges are those recommended by the ITU-T [13].

Table 17.4 Typical values for propagation delay

System	Delay
2 Mbit/s; pair cable, μs/km	4.3
140 Mbit/s; coax cable, μs/km	3.6
140 Mbit/s; fibre optic, μs/km	4.9
140 Mbit/s; radio, μs/km	3.3
Digital mux/demux pair, μs	1–12 (depending on hierarchical level)
Digital transit exchange (sum	<900 (mean value)
for both directions; digital-to-	<1500 (95% percentile)
digital), μs	
Digital local exchange (sum for	<3000 (mean value)
both directions; analogue-to-	<3900 (95% percentile)
analogue), μs	
Satellite; geostationary orbit, ms	260

17.4.3 Digital transmission

Digital transmission links introduce a number of impairments, including bit errors, controlled and uncontrolled slip, short breaks, jitter and wander. These are usually unnoticeable for speech transmission, but they can cause unacceptable errors in data transmission.

The CCITT specified a hypothetical reference connection (HRX) for the purpose of determining the permissible error performance for ISDN connections [3,14]. This is shown in Figure 17.4. The error parameters specified and the objectives set for them are shown in Table 17.5. Parameters (*a*) and (*b*) are most relevant to telephony and parameter (*c*) is more relevant

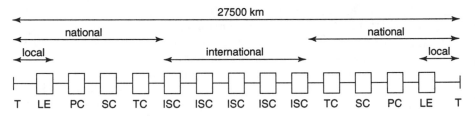

Figure 17.4   Hypothetical reference connection (HRX) used by CCITT

    □ digital exchange
    – transmission element
    ISC= international switching centre
    LE=local exchange
    PC=primary centre
    SC=secondary centre
    T=user/network interface for ISDN
    TC=tertiary centre

Table 17.5 Error-performance objectives for an international HRX at 64 kbit/s

Performance classification	Objective
(a) Degraded minutes	< 10% of 1 min intervals to have a bit error ratio $> 1 \times 10^{-6}$
(b) Severely errored seconds	< 0.2% of 1 s intervals to have a bit error ratio $> 1 \times 10^{-3}$
(c) Errored seconds	< 8% of 1 s intervals to have any errors (i.e. > 92% to be error free)

A total averaging period of any one month is suggested as a reference

to data services.

A *slip* is the deletion or repetition of consecutive bits in a digital signal. In an IDN, all the signals entering and leaving exchanges are normally in synchronism. However, for connections between different clock regimes or in the event of a network losing synchronism, signals arriving at an exchange are no longer at exactly the same rate as the exchange clock. Eventually, the exchange implements a controlled slip and this results in the repetition or deletion of an octet (8-bit byte) in each 64 kbit/s channel. The CCITT established objectives for controlled slips in the hypothetical reference connection [15]. These are shown in Table 17.6.

Uncontrolled slips are unplanned and unforeseen. They can arise from faults in the network causing protection switching or from excessive jitter or wander. Unlike a controlled slip, an uncontrolled slip is almost certain to cause loss of alignment and results in a long recovery time while multiplexers realign. ITU-T recommendations do not yet embrace uncontrolled slips.

If a short break occurs somewhere in a network, it usually leads to a loss of signal being detected at the subsequent equipment, which initiates an automatic fault-reporting procedure. However, a break of a few microseconds in a high-bit-rate system (e.g. at 140 Mbit/s) can result in a break of several tens of milliseconds at the end of the connection. This extension arises

Table 17.6 Slip-rate objectives for an international HRX at 64 kbit/s

Performance category	Mean slip rate	Proportion of time
(a)	< 5 slips in 24 h	> 98.9%
(b)	> 5 slips in 24 h and	< 1.0%
(c)	> 30 slips in 1 h	< 0.1%

Total averaging time is at least one year

because of the time taken by the chain of subsequent demultiplexers to recover frame alignment.

Jitter is produced by imperfect timing extraction in regenerators [16] and by justification in higher-order multiplexers [17]. It accumulates as a signal passes along a chain of systems. However, it is not damaging as long as the input tolerance to jitter of any system exceeds the jitter produced by previous systems. The CCITT published a recommendation for the control of jitter in networks based on the 2 Mbit/s hierarchy of systems [18].

Wander is a long-term variation of timing produced, for example, by changes in propagation time caused by temperature variation in cables. CCITT Recommendation G.823 contains a requirement on input tolerance to wander in addition to jitter [18]. Table 17.7 gives estimated values for wander for some typical systems [7].

Table 17.7 Estimated values for wander

System	Typical wander	
	Monthly variation	Annual variation
2 Mbit/s; pair cable	30–42 ns/km	62–77 ns/km
140 Mbit/s; coax cable	0.5–0.7 ns/km	1.2–1.3 ns/km
140 Mbit/s; fibre optic	0.3–0.4 ns/km	0.6–0.8 ns/km
140 Mbit/s; radio	very small	very small
Digital mux/demux pair	negligible	negligible
Digital transit exchange	negligible	negligible
Satellite; geostationary orbit	daily variation 0.15–1.5 ms	

17.4.4 Call processing

From Table 17.1 two of the important call processing parameters are:

(*a*) call-setup delay and
(*b*) probability of encountering congestion.

Target values for these parameters can be determined from CCITT Recommendation E.721, which covers grade-of-service parameters for circuit-switched services in the evolving ISDN, i.e. networks which may include both PSTN and ISDN components [2]. In this case, CCITT terminology is not completely consistent and E.721 uses the terms 'post-selection delay' and 'end-to-end blocking' which are broadly equivalent to 'call-setup delay' and 'congestion', respectively.

The target values for call-setup delay in E.721 range from 3 s up to 8 s, depending on the type of connection. National connections would expect to achieve values at the lower end of this range. However, adverse international connections, with perhaps 10 or 12 switching nodes, will result in longer delays. Ideally, customers would prefer uniform delays irrespective of the destination or routing, but this is unlikely to be achieved over the whole international network with existing circuit-switched technology. Current best practice indicates that connection-establishment delays of less than 1 s can be achieved on connections which do not include any analogue switching stages.

The E.721 target values for congestion indicate that the probability of encountering congestion in the busy hour should range from 2% on local connections to no more than 5% on an adverse international connection. In practice, these target values will be used to determine the dimensioning of the circuits between switches. Current best practice tends to achieve better than these targets, with modern all-digital networks providing probabilities of encountering congestion of less than 1%.

17.5 Network design to meet end-to-end requirements

For a network to deliver the required end-to-end performance, it is important to understand the factors influencing the performance. There are three aspects to consider:

(*a*) network structure
(*b*) interface requirements
(*c*) performance of individual network components.

Essentially, the end-to-end performance is determined by the performance of individual components, the effect of interactions between the components and the number of components in the overall connection as determined by the network structure. The network structure allows a variety of routings and the performances on the different routings will be slightly different. It is important to understand the likely distribution of performance as this will have an impact on customer opinion. It can be seen that end-to-end performance is a function of network structure and network-component performance. The significance of interface requirements is less obvious and best illustrated by examples.

Each digital network component is potentially a source of transmission errors as a result of electrical or environmental disturbances. Errors can therefore be introduced by digital line systems, multiplexers and digital switching systems and there would be an appropriate specification for the contribution from each system.

Another potential source of errors is at the system interfaces. The input port of one system must be able to recognise and act on the digital signal presented to it from the output port of the preceding interface. Key areas of the interface specification include jitter and wander characteristics. CCITT

Recommendation G.823 contains information about the jitter-control philosophy recommended for digital networks based on the 2048 kbit/s hierarchy [18]. Additional contributors to error performance include the network-synchronization arrangements, as timing inaccuracies can lead to bursts of errors.

It can be seen that there are a number of potential contributors to error performance. Successful control of end-to-end performance requires the identification of all these contributory mechanisms and appropriate control and specification.

A similar approach can be used for the call-processing parameters. The overall call-setup delay will be a function of:

(*a*) exchange performance,
(*b*) number of links in the overall connection,
(*c*) interexchange-signalling performance, and
(*d*) local-access signalling performance.

Each of these aspects requires appropriate control and specification, and it is important that the network designer understands the contribution that each can make to the overall performance. CCITT Recommendation I.352 gives additional information on this topic [1].

Availability might appear to be a straightforward parameter to describe and measure. However, little progress has been made by the ITU-T in recommending objectives, even for analogue systems. Using the 3×3 matrix of Figure 17.2, the available state is determined by observing the primary parameters. When a primary parameter exceeds a threshold value for a predetermined period of time, the connection is considered to be unavailable. For example, the primary parameter for a 64 kbit/s connection is the severely errored second (see Table 17.5). If the error rate exceeds 1×10^{-3} in each second for more than ten consecutive seconds, a failure is assumed to have occurred and the connection is deemed to be unavailable.

As an example of network design, the IDN of British Telecom will be considered. The structure of this is shown in Figure 2.10. National calls are completed using the fully-interconnected trunk network of digital main switching units (DMSU). Calls requiring additional network features (e.g. 0800 calls) use the derived-services switching units (DDSU). Calls destined for other countries are routed to the digital international switching units (DISU).

The transmission path for a limiting national connection is shown in Figure 17.5. Since all interexchange links have zero loss, it has been possible to ensure that all connections have an overall loudness rating less than 16 dB and a distribution [9] which meets CCITT Recommendation G.111. Thus, all national calls have a loudness rating which gives minimum dissatisfaction. To obtain this OLR, the two-wire-to-two-wire loss between local exchanges [19] is 2 dB.

Although end-to-end losses have decreased in the digital era, additional

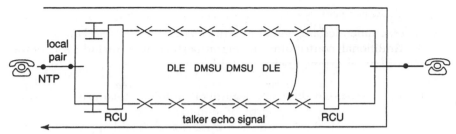

Figure 17.5   *Transmission path for limiting national connection in BT integrated digital network*

NTP   = network termination point
RCU   = remote concentrator unit
DLE   = digital local exchange
DMSU = digital main switching unit

delays have been introduced by digital switches. In the BT network, a digital trunk exchange (DMSU) can introduce a delay of 0.5 ms and a remote concentrator unit (RCU) can introduce a delay of 1.5 ms (including analogue-to-digital conversion). The decrease in loss and increase in delay makes it necessary to ensure that the echo performance is adequate. Total one-way delays exceeding 15 ms are unlikely to be encountered in the UK, so echo-control devices are not needed for national calls. Since four-wire-to-two-wire terminations are located only in customers' line circuits at local exchanges and permanently associated with the lines, it has been possible to improve balance-return losses. Thus, an open-loop loss of 12 dB is provided, instead of the 6 dB loss obtained with the previous analogue network [19].

17.6 Conclusions

In this Chapter, it has only been possible to give a brief introduction to some aspects of network performance and its relation to the quality of service seen by customers, including the role of the ITU in these areas. Key network-performance parameters have been identified, together with recommended objectives for appropriate end-to-end values to meet customer requirements. Finally, the Chapter gives some indication as to how a network may be designed to meet the end-to-end objectives, including the need to identify and specify all the contributing factors.

17.7 References

1. CCITT Recommendations I.310–I.470: 'Integrated services digital network (ISDN): overall network aspects and functions'
2. CCITT Recommendation E.721: 'Network grade of service parameters and target values for circuit switched services in the evolving ISDN'

3. CCITT Recommendation G.821: 'ISDN error performance'
4. CCITT Recommendation G.114: 'Mean one-way propagation delay'
5. CCITT Recommendation G.131: 'Stability and echo'
6. RICHARDS, D.L.: 'Telecommunications by speech: the transmission performance of telephone networks' (Butterworth, 1973)
7. McLINTOCK, R.W.: 'Transmission performance' *in* FLOOD, J.E., and COCHRANE, P. (Eds.): 'Transmission systems' (Peter Peregrinus, 1991), chap. 4
8. CCITT Recommendation P.48: 'Specification for an intermediate reference system'
9. CCITT Recommendation G.111: 'Loudness ratings in an international connection'
10. FRY, R.A., and BANNER, I.H.: 'Transmission performance assessment of telephone networks', *Post Off. Electr. Eng. J.*, 1972, **65**, p. 145
11. CCITT Recommendation G.123: 'Circuit noise in national networks'
12. CCITT Recommendation G.113: 'Transmission impairments'
13. CCITT Recommendation Q.551: 'Transmission characteristics of digital exchanges'
14. McLINTOCK, R.W., and KEARSEY, B.N.: 'Error performance objectives for digital networks', *Br. Telecom Eng. J.*, 1984, **3**, pp. 92–98
15. CCITT Recommendation G.822: 'Controlled slip rate objectives on an international digital connection'
16. DORWARD, R.M.: 'Digital transmission principles' *in* FLOOD, J.E., and COCHRANE, P. (Eds.): 'Transmission systems' (Peter Peregrinus, 1991), chap. 7
17. FERGUSON, S.P.: 'Plesiochronous higher-order digital multiplexing' *in* FLOOD, J.E., and COCHRANE, P. (Eds.): 'Transmission systems' (Peter Peregrinus, 1991), chap. 8
18. CCITT Recommendation G.823: 'The control of jitter and wander within digital networks based on the 2048 kbit/s hierarchy'
19. HARRISON, K.R.: 'Telephony transmission standards in the evolving digital network', *Post Off. Electr. Eng. J.*, 1980, **73**, pp. 74–81

Chapter 18

Network management

J.M. Fairley

18.1 Introduction

Network management (NM) is, on the one hand, the control of the cost of ownership of networks and, on the other, the provision of services to the users of networks. Although network management is of fundamental importance to network-operating authorities and has a direct bearing on costs and income, until recently it has been addressed largely in a piecemeal way with no comprehensive attempt to provide a co-ordinated system. Operators' first priority has been to install their networks and get them running. Now, in the face of competition, attitudes are changing and moves towards automation of network management are taking place [1,2,3].

Network management can comprise up to 30% of the cost of installation of any telecommunication network, while the cost of operation can easily exceed twice the cost of purchase over a ten-year period. Since a large telecommunications company can have a turnover in excess of £1 billion, this represents potential business with just that one operator of some £300 million.

This market has been made possible by the considerable investment which has been made in the definition of international standards over the last ten years and it received a major boost in 1988 with the formation of the Network Management Forum (NMF). Since then, membership has risen from the handful of founding companies to include most of the major suppliers and network operators in five continents.

Interest in international NM standards started in the mid 1980s, with both CCITT (now the ITU-T) and ISO starting to write drafts. In 1987 Bell Communications Research (Bellcore) issued its 'Specification of system maintenance messages at the OS/NE interface' which aimed to give Bell operating companies considerable flexibility in their choice of supplier. This was part of a major long-term strategy in the Bell operating companies to standardise all major interfaces within their networks [4,5,6,7].

These first specifications are primarily concerned with maintenance, but there is recognition of wider issues. These were first outlined by ISO, which prepared draft OSI standards that include the following aspects of managing the cost of ownership of networks and in providing timely service to users

[10–15]:

(*a*) fault management
(*b*) configuration management
(*c*) accounting management
(*d*) performance management
(*e*) security management.

These aspects, which are frequently referred to as 'FCAPS', are discussed later in this Chapter. Research into these has been undertaken in several RACE projects, namely ADVANCE, AIM, NEMESYS, NETMAN and TERRACE [1,16].

In March 1990, British Telecom announced its 'Concert' programme (now called 'Syncordia') which could eventually provide corporate network users with the ability to manage resources within the public network as if they were their own. This is based on NMF interfaces and will have a significant impact on private networks as corporate network managers realise the benefits to be gained from being able to manage both their own equipment and the resources they lease from the public network in a seamless way [1,3].

18.2 Network operators and users

18.2.1 General

Three viewpoints need to be considered in the business of operating telecommunication networks. These are:

(*a*) the networks themselves and all the equipment in them;
(*b*) the network operating authorities and organizations; and
(*c*) the users of the networks.

Relationships between them are shown in Figure 18.1.

18.2.2 Networks

Telecommunication networks are basically large programmable machines. Their operation can be altered, i.e. programmed, in many ways and by many groups of people. In a sense, customers dialling numbers to obtain connection to other customers or to services, are programming the network. So are system-operations engineers when altering its overall configuration by changing routing tables at switching centres, by bulk loading of software algorithms and data, by incremental changes via control desks, or by physical alteration of equipment.

'Programming' of the local distribution network is largely achieved by means of jumpers in main distribution frames in exchange buildings, crossconnect points in street cabinets and manholes, and house distribution points. With the provision of suitable crossconnect equipment, all of this

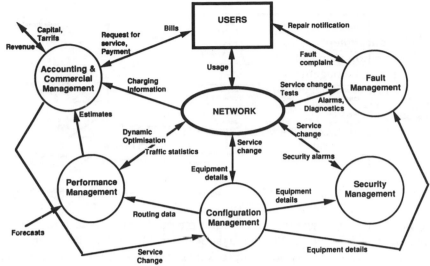

Figure 18.1   Network-management functions

could be done remotely, with a consequent improvement in the speed of response to new customer requests, which is always an important consideration in a competitive environment.

18.2.3 Operators

Network-operating authorities and organisations control the costs of operation of their network to maximise profits. This is achieved in part by managing the network; in part, by managing all operations which affect the maintenance of service and the economic operation of the network. As shown in Figure 18.2, this takes place at four levels as follows:

(*a*) the business level:
This is the management of a network-operating business itself, i.e. sales, customer administration and billing, profit-and-loss accounting, inventory control, investment planning etc.;

(*b*) the service level
This is the management of services offered to users of a network. Many services are provided over networks, apart from the basic telephone or data transport switching and bearer service. Service provision is separate from network operation and has to be separately managed. Proposed legislation will require a clear separation of basic service management and value-added service management;

(*c*) the network level:
This requires maintaining a database which describes the installed system, route optimisation and flow control, planning changes to extend or enhance the system, contingency planning to cope with emergencies and financial management of the installed equipment base; and

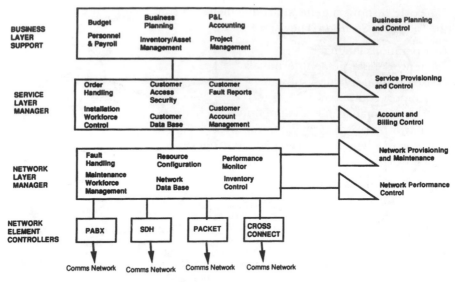

Figure 18.2   Levels of management

(*d*) the network-element level:
This includes the installation of equipment, the prediction, detection and repair of faults, and the management of maintenance, repairs and alterations.

18.2.4 Users

Management of networks is necessary to ensure that an adequate quality of service (QOS) is provided to network users [2,3]. Quality of service could be defined in terms of value for money and as such is an ever-varying quantity. Corporate users will see access to NM functions as part of this quality of service.

Class of access to networks and services is a matter of business policy and is not fixed for all time. Changing market demand and competition bring about changes in who is offered access to what. Thus, some users may be allowed to alter the configuration of their (virtual) networks; others may be able to alter their access to bandwidth on demand. These facilities were regarded until recently as being firmly within the control of the operating authority.

18.3 Network management architecture

18.3.1 The telecommunication-management network

The components and interfaces involved in a telecommunication management network (TMN) are described in CCITT Recommendation M.30 [17].

Figure 18.3 shows a general arrangement and the features are described in the following Sections. They are:

(*a*) operations language
(*b*) network-management terminals
(*c*) application-function software
(*d*) support computers
(*e*) network-management communication network
(*f*) network-management communication language
(*g*) network-element-management controllers
(*h*) network elements.

Control of telecommunication networks involves reading the values of data held in network elements, accepting unsolicited alerts or event reports, interpreting the information thus gained, and altering the status of these or other network elements by altering their control data to achieve a desired

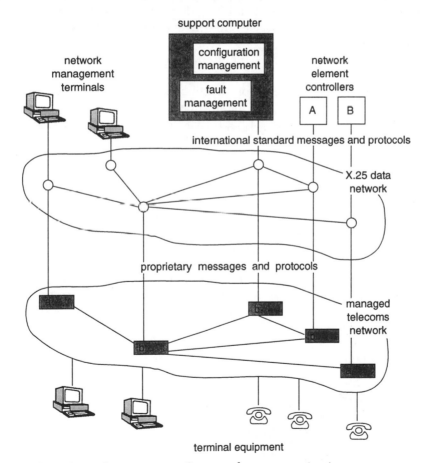

Figure 18.3   General arrangement of a network-management system

result (for instance, to change alarm thresholds or to switch elements in or out of service). In the past, this was achieved by manual interaction through control ports, element by element and data item by data item. All too frequently this is still the case today. For many network-wide changes, long sequences of many changes need to be made and it is essential to carry out such changes automatically under software control.

Telecommunication networks are distributed over large areas; thus, to effect control over them, the network management system itself must use a telecommunication network to reach the elements it controls. This tele-communication management network may be a separate overlay network. Alternatively, it may make use of the network it controls; in this case, the protocols used must ensure that the management functions can never disconnect themselves irrevocably from the network under control.

Since the network-management system makes use of a telecommunication-management network, its support computers and their terminals may be distributed throughout the country and need not be located centrally. Terminals need not be colocated with the computers and may be located at the most convenient sites.

As shown in Figure 18.4, the network-management systems for the telecommunication network in any area may need to co-operate with the network-management systems of neighbouring areas and, perhaps with national, or even international, master network-management systems. Many networks will be so large that it is impracticable to control them from only one central point and the network-management system will have to be arranged hierarchically or as a distributed management system. As much functionality as possible then is distributed to lower levels, with only summarised information and urgent alarms with overall network significance ever reaching the centre.

Private networks which use the public systems as a transmission medium may cross national boundaries and may have to interwork with several differing national network-management systems, as shown in Figure 18.5.

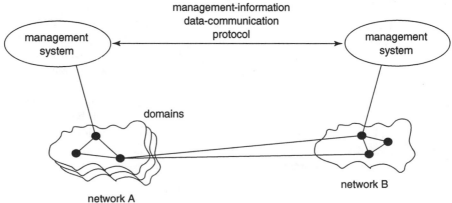

Figure 18.4   Network-management domains

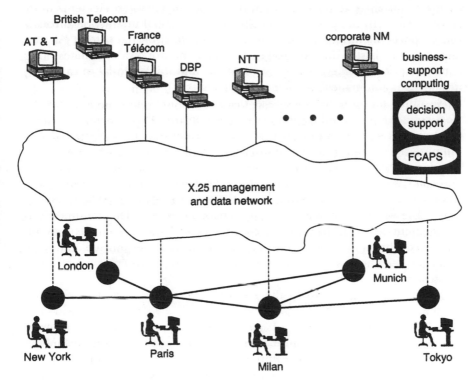

Figure 18.5   Cross-border private network

Thus, different domains of interest can be perceived in which there will be differing requirements and differing business opportunities.

18.3.2 Operations language

The operations language is the only means whereby network-management users, be they operators, maintainers, constructors or network planners, may communicate with the network-management system and the network. Its design is of paramount importance in ensuring that the system is logical and easy to use, so that personnel with widely differing levels of skill can use it without difficulty. Special vocabulary will be necessary to cater for differing functional requirements (e.g. accounts, engineering).

Human factors are very important in the design of the operations language in order to avoid a muddled and confusing user interface. Programmers are generally the poorest designers of user interfaces. Because of their intimate knowledge of the structure of the underlying software, they tolerate far greater inconsistency than will operations staff. The language must have a consistent grammar throughout all applications. It should preferably make use of simple window, icon, menu and pointer (WIMP) techniques, together with fast keyboard-entry shortcuts which can be learned at leisure. 'Show-and-point' rather than 'remember-and-type' interfaces are being looked for more and more by managers of networks.

Pictorial representation of connectivity represents the real world by displaying network resources against an appropriate map or equipment-plan background. Elements in the pictures can be colour coded and annotated to show status, statistics and facilities available or in use. This information can be represented and manipulated in either graphical or tabular ways as appropriate. Modifications to resource status or even to network arrangement may be made by alteration of pictures. New information may be added by completing forms. Context-sensitive pull-down menus provide choices of functions as appropriate, with key-stroke shortcuts also available. Users can create and edit sequences of commands recallable by invoking single icons or commands. Report generation will be based on the various data gathered by the system, with support from proprietary spreadsheets, graphics routines and other support software.

18.3.3 Application software and support computers

Network management comprises several functions, all of which are necessary in controlling the cost of operation of networks. Their relationships are shown in Figure 18.1 which shows the main flows of data between them.

The applications functions used for any network will be implemented in software. They depend on the requirements of the network owners and, for any installation, they may be only a subset of the total. The functions and the algorithms they use also depend on the topology of the network and require a database which describes the network to be managed.

Modern systems provide a set of core functions which will satisfy the needs of the majority of systems. These are customised by means of macro scripts or the graphical interfaces of a fourth-generation language. Database dimensioning must be adaptable to suit different sizes of telecommunication system. Its performance must take into account the enormous volume and rate of use of data required by systems and generated by them. Hardware performance is critical, since data must be handled in real time without the aid of flow control. In many cases, remote data-concentration equipment will be necessary.

18.3.4 Fault management

Fault management performs the detection, diagnosis, isolation and correction of abnormal operation of the network and of network elements. Such abnormal operation may be transient or persistent. Tracking of maintenance work is an integral part of this.

Fault management might make use of artificial-intelligence techniques and automated interactive maintenance manuals to assist staff in locating and curing faults speedily and in interpreting multiple alarms [1,16]. The identities and positions of faulty parts will be identified automatically, as will the identities of spares to be drawn from stores. The system will issue work tickets and will monitor the execution of repair work on-site and on parts which have been returned for repair.

Alarms may be generated automatically as equipment alarms, as customer complaints or as environmental alarms (fire, intruder etc.). Audible and visual indication may be given. The network must be monitored to detect and analyse alarm conditions, as follows:

(*a*) collect alarms/exception reports;
(*b*) filter alarms;
(*c*) alter alarm-filter thresholds;
(*d*) 'prioritise' alarms;
(*e*) report and route alarms;
(*f*) accept/acknowledge alarms;
(*g*) reconfigure round faults (either manually or automatically);
(*h*) clear alarm reports;
(*i*) identify inventory identity of faulty parts;
(*j*) display network status (nodes and links);
(*k*) analyse alarms; and
(*l*) analyse trends.

Facilities are also needed to perform the appropriate actions, as follows:

(i)   maintain error logs;
(ii)   analyse error logs and predict faults;
(iii)   recognise and accept fault conditions and alarms;
(iv)   schedule immediate and deferred repair work;
(v)   trace faults and isolate them;
(vi)   carry out diagnostic tests and continuity tests;
(vii)   correct faults;
(viii)   avoid faults by reconfiguration;
(ix)   accept alarms/register faults;
(x)   schedule trouble tickets;
(xi)   diagnose faults/carry out remote tests;
(xii)   track fault repair (both for equipment replacement and for equipment repair);
(xiii)   patch/debug code bugs (supplier);
(xiv)   track code bugs (supplier);
(xv)   replace data errors with archive data;
(xvi)   track data errors (operator);
(xvii)   carry out network-wide testing; and
(xviii) create tracking and management summaries.

18.3.5 *Configuration and name management*

Precise information about the identity and status of every network resource is essential throughout the life of every network. Administering this information is the function of Configuration and Name Management, which maintains a running record of all relevant data including moves and changes, equipment alteration and inventory control.

A master database is maintained which is used to control the network. This database must be kept in step with changes to the network, whether due to maintenance action, automatic protection switching, provisioning or faults. The amounts of data can be very large, as can the amount of change traffic due to enhancements, extensions, rearrangements or day-to-day routine customer changes. For instance, circa 20M bytes are necessary to control each System X exchange. In the first extensions to some System X exchanges more than 20 000 lines of code were added.

The functions involved include:

(*a*) equipment number/directory number mappings;
(*b*) identity, status and position of every network element;
(*c*) elements to which it is connected;
(*d*) routing-table contents etc.; and
(*e*) writing of mapping rules.

Data must be collected relating to:

(i)   environment
(ii)   routing
(iii) configuration.

Its compilation involves the following:

(*a*) data loading to network;
(*b*) scheduling of work;
(*c*) tracking/reporting/audit trail;
(*d*) verification of data before loading;
(*e*) decompiling; and
(*f*) data archiving.

18.3.6 *Accounting and commercial management*

Accounting management includes the means to support sales of network services, to manage accounts and billing, to maintain customer records, and to permit financial control of the installed system. It makes use of other network-management facilities to control access to the network, e.g. barring incoming or outgoing calls etc.

The facilities required include:

(*a*) collection of call data (call records);
(*b*) combination of call data for a single call;

(*c*) costing a call, either direct (meters/pulses) or derived (charging algorithm);

(*d*) customer accumulating meters;

(*e*) equipment rentals (lines, terminals etc.);

(*f*) charging (to people or project);

(*g*) administering customer records;

(*h*) managing accounts (charging and credit control);

(*i*) managing administration records;

(*j*) collecting payment;

(*k*) archiving raw data;

(*l*) managing tariffs;

(*m*)credit control;

(*n*) billing;

(*o*) receipt collection;

(*p*) control value of installed equipment; and

(*q*) monitoring operating costs and installed capital value.

18.3.7 Performance management

Performance management provides the means to model, evaluate and optimise the performance of the network. It gathers statistical data for the purpose of long-term planning and for short-term intervention in the event of unexpected surges in traffic. These may be due to, say, a 'disaster number' being broadcast on television news or, for cellular radio, the effect of traffic jams on a motorway. The following steps are needed to determine the appropriate actions:

(*a*) collect traffic statistics;

(*b*) determine correlation between nodes; and

(*c*) analyse traffic flow.

Short-term intervention applies controls which may be manually or automatically applied. Extra trunk capacity may be switched into service, hard-to-reach-destination markings may be broadcast, or call gapping may be applied. In the event of a national disaster, nonpriority calls in progress might be disconnected.

Medium-term intervention can consist of load balancing, lost-call analysis and exchange reorganisation.

Long-term action implies physical alteration of the network to install extra capacity, to remove redundant equipment or to re-arrange the network to redistribute load. Long-term alterations such as these involve alteration of network data under the control of configuration management. The data should include:

(*a*) graphical summaries;

(*b*) network-traffic display;

(*c*) analysis of call records for ineffective calls and unbillable calls;

(*d*) remote configuration and provisioning;
(*e*) availability;
(*f*) absolute limits of acceptable operation; and
(*g*) traffic versus time-of-day profiles.

These enable capacity planning and performance optimisation to be carried out.

18.3.8 Security management

Security of management is essential in open distributed systems. There may be security problems in protecting against malicious attacks (hacking) unless a ring-back, or other security protocol, is used. Hacking is avoided if a separate private overlay network with no external access is used. However, security protection for the management system must still be provided against internal threats.

The extent of the effects of changes must largely determine who may do what, from where, how and when. Types of access might be:

(*a*) passive (read only);
(*b*) active (read/write); or
(*c*) command (execute operation).

Security management defines who may do what in controlling a network. Thus, it defines the types of access allowed to objects in the network and the rules of access to them. Passive access is usually allowed to more resources than is active access. Command access could be expected to be even more restricted. Authentication is necessary in all components and in the network-management applications, to protect against accidental or malicious attacks on the network. There may be different levels of security. Each must be suspicious of higher and lower levels, applying verification to all transactions between them to mitigate the effect of attacks on the system.

Since each network-management terminal can gain access to any support computer and, via it, to any network element using the telecommunication-management network, access will have to be by means of security protocols, authorisation keys, territory, allowed functions and command subset.

Security management includes the following facilities:

(*a*) control of access to the management system, by log-in password and allocation of service access level (personal, terminal, territorial, functions and commands);
(*b*) control of access to premises;
(*c*) control of access to network objects;
(*d*) control of encryption keys;
(*e*) authorisation;
(*f*) encryption and key management;
(*g*) authentication;

(*h*) maintenance and analysis of security logs;

(*i*) logging of user-interface commands for detection of attacks on system (call-record analysis of ineffective calls, unbillable calls etc.); and

(*j*) authentication of data to and from system.

18.3.9 Network-management-communication language

The machine-to-machine language of network-management systems is where the centre of gravity of network-management standardisation lies. At present there is considerable variation between the languages used to communicate with the individual components of networks (if indeed they can be called languages at all). In the past, manufacturers gave little thought to language design and many languages are little better than primitive sequences of codes and parameters. However, the situation is changing and considerable effort is being invested internationally to specifying the necessary language in terms of architecture, protocols, messages and managed objects. Other systems of note are those based on SNMP (Simple Network Management Protocol) for TCP/IP networks [18].

The Network Management Forum adopted an entirely pragmatic approach and chose to concentrate its efforts first on solving the problems of managing not only new equipment but also the currently installed base of networks [9]. Recognising the need to enable the many network-management 'islands' to communicate with each other, the Forum adopted a simple architecture as shown in Figure 18.6.

This provided a means for network-management equipment from different vendors, called *conformant management entities* (CME), to be interconnected, thus providing a 'causeway' to link the 'islands' such that integrated systems can be achieved. This 'causeway' consists of a profiled seven-layer OSI stack plus a specially defined set of management messages. Meanwhile, as shown in Figure 18.7, the CMEs continue to communicate with their respective network elements using their previous proprietary protocols and messages. As shown in Figure 18.8, the upper layers support transaction services and file-transfer services (CMIP, ROSE, ACSE and FTAM). The lower three layers are connection orientated according to CCITT Recommendations X.25 (1984) and X.21bis. Optionally, CMEs can communicate via connectionless networks according to IEEE 802.2/802.3 LAN link-layer services with the network-layer

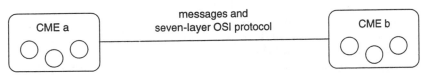

Figure 18.6 NMF architecture showing interoperating conformant management entities (CMEs) containing managed objects

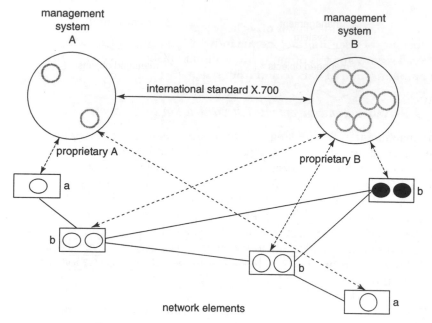

*Figure 18.7   Interfaces between network-element controllers (A and B) and correspond-
ing network elements (a and b), showing the managed objects in the
network elements and their images in the controllers*

services defined in ISO 8348/AD.

The CMEs contain standard managed objects. These may be software invocations, such as data sieves running in the CMEs themselves, or as in Figure 18.7, might be images of real objects out in the network elements themselves. The managed objects are defined according to object-oriented design principles [15]. Inherent in object-oriented design is the principle of inheritance. Thus, it is possible for any vendor to design a CME which can manage at least the basic functionality of objects provided in another vendor's CME because of inheritance from the objects' super-classes, even though additional functionality might exist which can only be managed privately. This approach allows innovation whilst preserving openness. However, it bears with it the risk of renewed divergences from standards.

In Figure 18.7, the network-element controllers (i.e. the CMEs) monitor and control so-called managed objects within the network equipment, providing images of these objects to each other and to the network-management software across the NM interface. Network management involves interrogating managed objects, altering their status as appropriate (for example, to change thresholds or to switch elements in or out of service), and accepting alarm messages, all using standardised messages and protocol profiles.

The managed objects contain data which describe the identity and

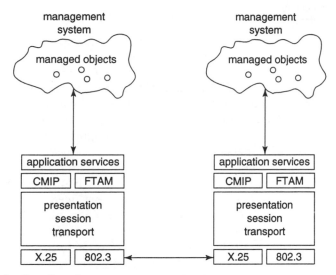

Figure 18.8   Interfaces between management systems

attributes of each. Each type of object can perform certain actions to change its in/out-of-service state, set error-filter thresholds, change routing tables etc. Objects report their status on command in monitoring usage of the network (test, measurements, traffic statistics, metering) and in confirming that settings are as requested ('sanity' checking). Finally, objects report alarms and out-of-bounds conditions such as excessive error rates [7,16].

For the installed network base, CMEs as network-element controllers communicate with their respective network elements using their existing proprietary management interfaces. It will be appreciated that, in most existing networks, it will be economically and physically impossible to retrofit these protocols into all the network elements. At first, achieving the ability to control all network elements from one terminal might be a sufficient advance in the state of the art to satisfy everyone concerned. However, in due course additional needs will arise. Here, the inherent flexibility of the open approach becomes evident. It is possible now to add additional support computing which, provided that it conforms to international open interfaces, can communicate directly with the CMEs already installed.

In Figure 18.3, a support computer is shown running various types of business-decision-support software as well as network-administration software including fault management and configuration management. Later, accounting management, performance management and security management might be provided, either on this computer or on others connected to the network. The role of the plant-level controllers changes to providing data-concentration and logging functions, monitoring and controlling their corresponding network elements by means of the current proprietary protocols as before.

18.3.10 Network-management communication network

Communication is needed between the network-management systems and the objects to be managed. This may be achieved either by using a separate overlay network or by using the network being managed. Whether the physical interconnection is best provided by a separate overlay network (which itself must be managed), or whether use can be made of the network being managed depends on several considerations, among them management traffic, cost and security. Route diversity is important to protect against physical damage to the management network.

In Figure 18.3, the whole of this management system is interconnected by an X.25 packet-switched wide-area network (WAN). If the network-management terminals are also connected via the WAN, they need not be co-located with the support computers. This gives more flexibility over the position of the operations room and even allows control to be handed on from time zone to time zone around the world as the earth turns on its axis.

In large networks, the traffic generated by the management system itself could be quite significant and it may be preferable to use a separate overlay to avoid degrading the service offered by the network being controlled. Overlay network-management networks will need to be managed as well.

18.3.11 Network-element controller

In any practical installation, there is a need to concentrate data at remote sites, sending only what is significant to central network-management controllers. Control information from centres will be distributed to the network elements at the remote sites. Local emergency access must be provided to equipment on site, as must remote access from site to the network-management support computers.

There will be an interim period, lasting for many years, when network-management systems will have to interwork with existing equipment which does not comply with the standard. During this period, translation equipment (mediation equipment) will be necessary between the network-management system and nonstandardised equipment. The use of mediation equipment provides a means of controlling equipment from other manufacturers.

18.3.12 Network elements

As discussed in Chapter 2, modern telecommunication networks may contain the following components, or combinations thereof:

(*a*) circuit-switching exchanges
(*b*) packet-switching exchanges
(*c*) semipermanent private circuits
(*d*) common-channel signalling
(*e*) flexible-access points
(*f*) digital cross-connect equipment

(*g*) add/drop multiplexers

(*h*) local and backbone transmission bearers

(*i*) network operations and management controllers

(*j*) customers' terminals and customer premises networks

(*k*) metallic and optical-fibre cables

(*l*) satellites

(*m*)software algorithms

(*n*) software data.

Customers' terminals or terminal networks (customer premises networks) may connect to more than one carrier network (e.g. BT, Mercury, satellite), through which information may be transmitted transparently or may be processed. Access to external networks may be controlled by the customer premises network's own network-management system (e.g. to provide least-cost routing).

Access to these networks might be controlled by the operating authority to allow either on-demand customer-programmable connection or preselected, quasipermanent connection to various types of access which have to be ordered in advance as follows:

(*a*) user to user;

(*b*) user to service;

(*c*) service to service;

(*d*) user to network; and

(*e*) network to network.

For the purpose of network management, the network elements must provide a control and monitoring interface for:

(i)   state control (in/out of service, access control);

(ii)   monitoring quality of operation (error rates, faults); and

(iii)   monitoring of usage of the network (traffic statistics, metering).

18.3.13 Operating and maintenance instructions

Operation and maintenance (O&M) manuals are the *only* information available to network-operating authorities on how the various equipment within their networks are to be operated and maintained. They are therefore of great importance to them, especially when new equipment is installed and whenever out-of-the-ordinary events occur. In use, printed text and diagrams require a fair degree of familiarity if efficient use is to be made of them. Familiarity depends on use. When dealing with extremely reliable equipment, there is little opportunity to obtain this familiarity, since the equipment does not go wrong often enough to allow expertise to be built up.

The O&M manual for a large public telecommunication system can consist of several volumes and cover many procedures. Access to these volumes might be eased by grouping them under keywords and structuring their text. However, a more effective method is provided by online systems using multimedia hypertext systems, which automate the presentation of the texts and diagrams so that information appropriate to the network component of interest is selected automatically. This information is relevant to the state of the component and the operations desired by the user of the system and shows the expected results of operations in advance of their being applied. Thus, operators can be confident that they are about to select the correct operation and, during fault finding, technicians can compare actual operation with correct operation.

Such automated manuals might make use of the techniques of artificial intelligence [19]. These can both provide a simulation of operation and make inferences about the likely state of faulty or healthy systems and the allowable or advisable operations which can be carried out. Such a system is under development in the RACE project AIM [16].

18.4 Conclusions

Network management is the cornerstone of not just networks, but economic network business operation [20,21]. Although the functions required for the public and private network-management markets are similar, there is considerable difference in scale and emphasis. Thus, we are most likely to require a family of network-management components which can be mixed and matched in a modular fashion, in order to be able to satisfy all requirements and to be able to provide a simple and logical means of network expansion and enhancement.

18.5 References

1. *BT Technol J.*, 1991. **9**, (3), special issue
2. MEANDZIJA, B., and WESTCOTT, J. (Eds.): 'Integrated network management' (North Holland, 1989)
3. *Br. Telecom. Eng.*, **10** (3), special issue
4. 'Generic requirements for operations interfaces using OSI tools: transport and network element surveillance'. Bellcore Technical Advisory Service, document TA-NWT-001030
5. 'Generic requirements for operations interfaces using OSI tools: synchronous optical network (SONET) transport'. Bellcore Technical Advisory Service, document TA-NWT-001042
6. 'Generic requirements for operations interfaces using OSI tools: ISDN basic rate access testing'. Bellcore Technical Advisory Service, document TA-NWT-001057
7. 'Generic state model for managing network elements'. Bellcore Technical Advisory Service, document TA-NWT-001093
8. EMBRY, J., MANSON, P., and MILHAM, D.: 'An open network management architecture: OSI/NM Forum architecture and concepts', *IEEE Netw. Mag.*, 1990, **4**, (4)
9. Network Management Forum: 'Discovering OMNI *Point*' (Prentice Hall, 1993)
10. ISO/IEC 7498-8; (CCITT Recommendation X.700): 'Information technology. Open systems interconnection. Management framework'
11. ISO/IEC 10040: 1992 (CCITT Recommendation X.701): 'Information technology. Open systems interconnection. System management overview'

12. ISO/IEC 10164 Parts 1–7 (CCITT Recommendations X.730 - X.736): 'Information technology. Open systems interconnection. System management functions'
13. ISO/IEC 9595: 1991 (CCITT Recommendation X.710): 'Information technology. Open systems interconnection. Common management information service definition (CMIS)'
14. ISO/IEC 9596 (CCITT Recommendation X.711): 'Information technology. Open systems interconnection. Common management information protocol (CMIP)'
15. ISO/IEC 10165-4: 1992 (CCITT Recommendation X.722): 'Information technology. Open systems interconnection. Structure of management information. Guidelines for the definition of managed objects (GDMO)'
16. Proceedings of 4th RACE TMN Conference (Unicom Seminars, 1990)
17. CCITT Recommendation M.30: 'Principles for a telecommunications management network (TMN)'
18. CASE, J., FEDOR, M., and DAVIN, C.: 'A simple network management protocol (SNMP)'. Internet Activities Board RCF 1067
19. TAYLOR, A.: 'How expert systems can support network diagnostics, management and control'. Proceedings of international conference on *Network Management*, London (Blenheim Online Publications, 1988) pp. 123–135
20. KAUFFELS, F.-J.: 'Network management: problems, standards and solutions' (Addison-Wesley, 1992)
21. AIDAROUS, S., and PLEVYAK, T. (Eds.): 'Telecommunications management into the 21st century' (IEE and IEEE Press, 1994)

Chapter 19

Network economics

B.R.S. Panesar

19.1 Introduction

A public telecommunications operator (PTO), whether government owned or privately owned, has to make a substantial capital commitment in establishing a network and subsequent system expansion. Also, de-installation and rearrangement of plant and equipment can be disruptive and costly. As the equipment normally has a long economic life, it is critical that careful planning and evaluation is undertaken prior to deciding upon the size and configuration of the system [1,2], in terms not only of the technology available but also market expectations and future trends.

This Chapter describes some methods of investment appraisal in order to evaluate financially the alternative system solutions that may be designed [3–6]. Clearly, there will also be unquantifiable factors which may influence the decision, such as government or regulatory issues. Moreover, long-term strategic considerations may bias the decision in favour of one particular solution.

The problems to which project or investment appraisal can be applied are numerous. The monetary value of the initial investment will dictate the degree of sophistication and resources to be applied in deciding on the best solution. For example, the system planner may be confronted with issues such as:

(*a*) digital switches to replace existing analogue switches;
(*b*) introducing further local switches;
(*c*) expanding the network geographically to exploit opportunities in growing new towns; and
(*d*) provision of a mobile telephone service.

All such projects can be dealt with by the same sort of analysis. However, in each case, the traffic planners, system designers and operations staff will have to work together to overcome the conflicting requirements of the network configuration, trunk and local switching capacity, equipment matching and grade of service. For each of the above types of problem there will be a number of solutions. Some of these solutions are easily dispensed with because of technology or cost constraints, but others will require detailed evaluation.

19.2 Key factors

19.2.1 Demand

The proper assessment rests on the future demand in terms of the level of demand, the rate of growth and the mix of traffic types. A system needs to cope with the maximum customer base at a given grade of service for a significant proportion of its design life. The nature of the demand is now being extended beyond the simple voice and data requirements as customers expect more information and at a higher level of availability and more consistent reliability.

The demand can be forecast using various statistical techniques [7,8]. These will not be discussed here, but the key to all such demand forecasts is not to project for longer periods than there is history available. This does entail some restrictions. Furthermore, care must be taken in the projection of demand for new services. The demand for services has in the past been driven by technology and cost limitations. This is now changing rapidly as the price and performance characteristics improve, as discussed below.

19.2.2 Capital costs

The major item of cost will usually be the initial capital cost of the equipment coupled with the necessary software costs. Certain types of new services are entirely dependent only on software or upgrades of existing software on switches. However, future trends need to be taken into account both for the operating costs and the addition of further modules of equipment.

Historically, the cost of various types of equipment has declined in real terms. Indeed, feature for feature, telecommunication equipment is becoming more sophisticated, smaller, faster and cheaper with the passing of each generation, as evidenced by the explosive growth in digital technology and its application to network operations and especially switching.

For example, the cost of optical fibre has declined, while the cost of copper has continued to rise. The future is expected to continue this pressure downwards (e.g. low-count optical-fibre-cable cost in 1989 was US 25 c/m and is expected to decline to about 15 c by 1999 in 1989 money terms). Similar behaviour is expected to occur for other equipment, e.g. optical-line-terminating equipment and multiplexer line cards. Of course this continues to enhance the price/performance ratios of these types of equipment.

19.2.3 Operating costs

The initial capital or nonrecurring costs are usually the most significant, but the operating and maintenance costs, which are recurring, must also be compared. One needs to take a total life-cycle-cost approach [9]. Even though the capital cost of one option may be less, the related operating costs may be much higher over the time period being evaluated. The operating costs each year will include staff, rent, electricity and other administrative items.

The introduction of digital technology has reduced the need for large amounts of space. The elimination of electromechanical components has helped to reduce the operating costs in terms of manpower, power consumption and accommodation costs. The types of test equipment have increased in complexity and hence cost. The increased skill level of the lower numbers of staff now needed to support a switch and a digital network means that their salary cost is higher.. All these parameters need to be taken into account in the assessment of a project.

19.2.4 *Benefits*

The net benefits of a project might be improved performance in terms of the grade of service, reliability, quality of signal, or enhancement of revenue by the addition of service features. These need to be quantified and stated in cash-flow terms. Where a project results in cost savings, those savings can be quantified and defined as a benefit or positive cash flow.

The general reduction of unit costs in the electronics industry has led to improved price/performance ratios, which means that the operator can provide more capability at the disposal of the customer at only a small extra cost. The enhancement of software means that the operator can provide additional features, such as call waiting, call divert and others, to every customer. The expectations of the general public for more information and processing power have been evident in the computing world for some time. As they overflow into the telecommunications arena, these expectations lead to more revenue-earning potential for the telecommunications operator.

Other service features which provide value added to the operator, such as home shopping via telephone and cable television, have yet to materialise. However, they are just round the corner. As these value-added services tend to require more software than hardware, they can be added later after implementation of the initial hardware. The call-waiting feature effectively provides for two lines at the customer's apparatus without the addition of another line!

19.3 Cash flow

Investment appraisal is primarily concerned with cash rather than profitability and accounting returns. A good cash-flow profile should lead to good business.

Because of the capital-intensive nature of the telecommunications business and the long economic lives of much of the equipment, it is necessary to consider a time horizon of at least 10 years, and sometimes 20 years. For each

The investment will include the capital equipment, software and transport and installation costs. In a modernisation programme, the net cost of de-installing and scrapping existing equipment must be allowed for in the initial investment. Subsequent additions of equipment should take into consideration the expected reductions in real costs described above.

Operating costs each year will include staff, rent, power and administrative costs. Some costs, such as staff salaries, may be expected to rise in the years ahead, so the costs should be adjusted accordingly.

Revenue will include line-rental charges for the growing customer base and be adjusted for price changes. The timing of the receipts should be reflected in the figures by placing the adjusted cash amounts in the appropriate periods.

For the years ahead, pricing decisions on the call charges and cost escalation in terms of the recurring operations and maintenance costs need to be assembled with care. Liberalisation and increased awareness around the world is forcing governments to impose constraints on returns or on prices which PTOs can charge, while staff and related costs continue to rise. It is important not to apply the general retail-price index (RPI) to these parameters.

A typical cash-flow statement for a project is given in Table 19.1. This is also illustrated in Figure 19.1, which shows the net cash flow in each period and the cumulative cash flow.

Table 19.1 Cash-flow statement for a typical project

Period, year	0	1	2	3	4	5	6	7	8	9	10	RV
Investment, £M	50											
Operating costs, £M		1	2	4	8	10	15	17	19	21	23	
Subtotal outflow, £M	50	1	2	4	8	10	15	17	19	21	23	
Revenue, £M	-	2	5	9	14	20	30	35	40	45	50	
Net cash flow, £M	(50)	1	3	5	6	10	15	18	21	24	27	13
Cumulative cash flow, £M	(50)	(49)	(46)	(41)	(35)	(25)	(10)	8	29	53	80	93

19.4 The time value of money

One could lay out the revenues and costs expected for each year of the economic life of the alternative projects and perhaps add them up to get a sum of the flows. However, this comparison would ignore the fact that a pound today is worth more than a pound a year hence.

If, say, £100 is available today and can be invested to provide a 10% return, then at the end of one year it will amount to £110. Furthermore, if all this money is left for another year, it will amount to £121 (i.e. £110 × 1.1). This is summarised in the compound-interest formula:

future value = present value $\times (1+r)^n$

> where r = interest rate
> n = number of years.

Rearranging the above formula provides the relationship:

$$\text{present value} = \frac{\text{future value}}{(1+r)^r}$$

Thus, any future cash flow can be discounted back to today's present value to match the period in which the initial capital investment is made. This is called the *discounted-cash-flow technique*. The technique is made easier by the development of discount-factor tables. An example is shown in Table 19.2. Also, some programmable electronic calculators have suitable built-in functions, as do the spreadsheet packages available for personal computers. In the context of investment appraisal, r is often termed the *discount rate*, or the *opportunity cost of capital*, or the *weighted average cost of capital*.

The discounted cumulative cash flow for the example of Table 19.1, using a 10% discount rate, is shown in Table 19.3. It is also plotted in Figure 19.1. In this example, the result of discounting is to increase the *payback period*, i.e. the time taken before the cumulative net cash flow becomes positive, from only seven years to nearly nine years.

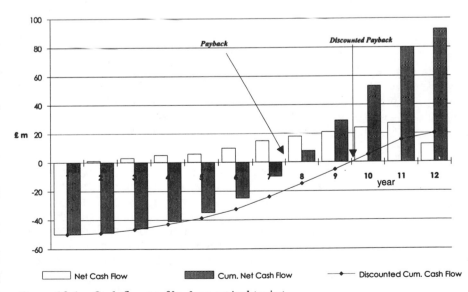

Figure 19.1   Cash-flow profiles for a typical project

Table 19.2 Discount factors
Present value of £1 received after *n* years at annual interest rate *r*

Rate(r)	1%	2%	3%	4%	5%	6%	7%	8%	9%	10%	11%	12%	13%	14%	15%
Years (n)															
1	.990	.980	.971	.962	.952	.943	.935	.926	.917	.909	.901	.893	.885	.877	.870
2	.980	.961	.943	.925	.907	.890	.873	.857	.842	.826	.812	.797	.783	.769	.756
3	.971	.942	.915	.889	.864	.840	.816	.794	.772	.751	.731	.712	.693	.675	.658
4	.961	.924	.888	.855	.823	.792	.763	.735	.708	.683	.659	.636	.613	.592	.572
5	.951	.906	.863	.822	.784	.747	.713	.681	.650	.621	.593	.567	.543	.519	.497
6	.942	.888	.837	.790	.746	.705	.666	.630	.596	.564	.535	.507	.480	.456	.432
7	.933	.871	.813	.760	.711	.665	.623	.583	.547	.513	.482	.452	.425	.400	.376
8	.923	.853	.789	.731	.677	.627	.582	.540	.502	.467	.434	.404	.376	.351	.327
9	.914	.837	.766	.703	.645	.592	.544	.500	.460	.424	.391	.361	.333	.308	.284
10	.905	.820	.744	.676	.614	.558	.508	.463	.422	.386	.352	.322	.295	.270	.247
11	.896	.804	.722	.650	.585	.527	.475	.429	.388	.350	.317	.287	.261	.237	.215
12	.887	.788	.701	.625	.557	.497	.444	.397	.356	.319	.286	.257	.231	.208	.187
13	.879	.773	.681	.601	.530	.469	.415	.368	.326	.290	.258	.229	.204	.182	.163
14	.670	.758	.661	.577	.505	.442	.388	.340	.299	.263	.232	.205	.181	.160	.141
15	.861	.743	.642	.555	.481	.417	.362	.315	.275	.239	.209	.183	.160	.140	.123
16	.853	.728	.623	.534	.458	.394	.339	.292	.252	.218	.188	.163	.141	.123	.107
17	.844	.714	.605	.513	.436	.371	.317	.270	.231	.198	.170	.146	.125	.108	.093
18	.836	.700	.587	.494	.416	.350	.296	.250	.212	.180	.153	.130	.111	.095	.081
19	.828	.686	.570	.475	.396	.331	.277	.232	.194	.164	.138	.116	.098	.083	.070
20	.820	.673	.554	.456	.377	.312	.258	.215	.178	.149	.124	.104	.087	.073	.061

Rate (r)	16%	17%	18%	19%	20%	21%	22%	23%	24%	25%	26%	27%	28%	29%	30%
Years (n)															
1	.862	.855	.847	.840	.833	.826	.820	.813	.806	.800	.794	.787	.781	.775	.769
2	.743	.731	.718	.706	.694	.683	.672	.661	.650	.640	.630	.620	.610	.601	.592
3	.641	.624	.609	.593	.579	.564	.551	.537	.524	.512	.500	.488	.477	.466	.455
4	.552	.534	.516	.499	.482	.467	.451	.437	.423	.410	.397	.384	.373	.361	.350
5	.476	.456	.437	.419	.402	.386	.370	.355	.341	.328	.315	.303	.291	.280	.269
6	.410	.390	.370	.352	.335	.319	.303	.289	.275	.262	.250	.238	.227	.217	.207
7	.354	.333	.314	.296	.279	.263	.249	.235	.222	.210	.198	.188	.178	.168	.159
8	.305	.285	.266	.249	.233	.218	.204	.191	.179	.168	.157	.148	.139	.130	.123
9	.263	.243	.225	.209	.194	.180	.167	.155	.144	.134	.125	.116	.108	.101	.094
10	.227	.208	.191	.176	.162	.149	.137	.126	.116	.107	.099	.092	.085	.078	.073
11	.195	.178	.162	.148	.135	.123	.112	.103	.094	.086	.079	.072	.066	.061	.056
12	.168	.152	.137	.124	.112	.102	.092	.083	.076	.069	.062	.057	.052	.047	.043
13	.145	.130	.116	.104	.093	.084	.075	.068	.061	.055	.050	.045	.040	.037	.033
14	.125	.111	.099	.088	.078	.069	.062	.055	.049	.044	.039	.035	.032	.028	.025
15	.108	.095	.084	.074	.065	.057	.051	.045	.040	.035	.031	.028	.025	.022	.020
16	.093	.081	.071	.062	.054	.047	.042	.036	.032	.028	.025	.022	.019	.017	.015
17	.080	.069	.060	.052	.045	.039	.034	.030	.026	.023	.020	.017	.015	.013	.012
18	.069	.059	.051	.044	.038	.032	.028	.024	.021	.018	.016	.014	.012	.010	.009
19	.060	.051	.043	.037	.031	.027	.023	.020	.017	.014	.012	.011	.009	.008	.007
20	.051	.043	.037	.031	.026	.022	.019	.016	.014	.012	.010	.008	.007	.006	.005

e.g. If the interest rate is 10% per year, the present value of £1.00 received at the end of year 5 is £0.621

Table 19.3 *Discounted cash flow statement*

Period, year	0	1	2	3	4	5	6	7	8	9	10	RV
Investment, £M	50											
Operating costs, £M		1	2	4	8	10	15	17	19	21	23	
Subtotal outflow, £M	50	1	2	4	8	10	15	17	19	21	23	
Revenue, £M	-	2	5	9	14	20	30	35	40	45	50	13
Net cash flow, £M	(50)	1	3	5	6	-0	15	18	21	24	27	13
Cumulative cash flow	(50)	(49)	(46)	(41)	(35)	(25)	(10)	8	29	53	80	93
Discount factor @10%	1.0	0.909	0.826	0.751	0.683	0.521	0.564	0.513	0.467	0.424	0.386	0.350
Discounted net cash, £M	(50)	0.9	2.5	3.8	4.1	6.2	8.5	9.2	9.8	10.2	10.4	4.7
Discounted cumulative, £M	(50)	(49.1)	(46.6)	(42.8)	(38.7)	(32.5)	(24.0)	(14.8)	(5.0)	5.2	15.6	20.3

Net present value, £M 15.6
Net present value with residual value, £M 20.3

Government-owned PTOs are advised of the rate to be used for project appraisal. This is often related to the pretax return on investment being achieved in the private sector and then adjusted upwards to reflect certain unprofitable investments (e.g. obligations to provide universal service and hence locate payphones in remote locations). In the commercial environment, the rate can be related to the cost of borrowing money for that firm and the cost of the dividend it has to provide to the shareholders.

In either case, a set rate is usually defined by the finance or treasury department of the PTO. In certain private-sector organisations different rates may be set, depending on the nature of the investment and the risk associated with it. (For example, adding trunk switching capacity is less risky than going into mobile telephones in a new city.)

In order to borrow money, the company must pay a premium to the bank. This will depend on the bank's view of the company, the project and other factors (e.g. 14% may be paid when the base rate is 10%).

19.5 Net present value, time horizon and residual value

19.5.1 Net present value

The *net present value* (NPV) is the sum of the discounted cash flows described above. This brings the monetary values (whether they be nonrecurring or recurring) of different periods to the value of money today. The project should be approved and implemented if the NPV is positive, since it provides a return in excess of the rate which finance would cost. For a set of alternative projects, clearly, the one which provides the maximum NPV should be selected, assuming all other factors are equal.

19.5.2 Time horizon

The number of periods over which the project is to be evaluated must relate to the economic life of the equipment, the nature of the system and the organisation's policy and objectives.

Most telecommunications investment relates to the laying of ducts and cable and installation of switches. Ducts and cable last over 20 years, maybe even up to 40 years. Analogue switches are now continuously being upgraded to digital technology. It is too early to establish definitive economic lives for these, but 10 years is a reasonable value. New digital exchanges may be expected to last 10–15 years; however, the software may need updating, at significant cost, after only a few years. (This aspect can be handled in the cash-flow analysis by inserting a further investment in the appropriate period.)

The system may be licensed or franchised in a small area and the franchise will be of a limited duration which may not exceed the useful economic life of the equipment. The policy for replacement and depreciation of plant and equipment and subsequent disposal will also affect the time horizon.

Another issue to remember is the rapid decline in the value of the discount factor after year 10, even at modest levels of interest rate. Therefore, it is recommended that 10 years beyond the period of investment is a reasonable time horizon to select.

19.5.3 Residual value

At the end of the 10 years of operation, the system may be still earning revenue, and the equipment is still in use with a significant useful life remaining. It is therefore necessary to take some estimate of the value outstanding for this. This is known as the *residual value* of the system.

The range of possibilities for this are:

(*a*) zero
(or even negative, since it may cost more to close down and de-install than the equipment is worth as scrap).
(*b*) net book value of assets
In our example, if the £50M in investment after 10 years still has another 10 years of useful life, the net book value on the register of plant and equipment would show £25M. Apart from the government or franchisor buying the asset at this value, it may be possible to de-install and sell it (to, say, a developing country).
(*c*) valuation as a going concern
This concept is based on the assumption that the investment is self-contained and can continue to generate the net cash flow at the year-10 level for a further period (say five years). The valuation would then be based on the NPV in year 11 of the future five years of cash flows. This would be more appropriate in the sell off of the investment at year 10 to another PTO or licensee. Of course, the investment would have to be completely self-contained, so this valuation is more likely to be applied to a system. (This approach is often applied in the acquisition of companies.)

Method (*c*) is likely to produce the highest valuation, but even method (*b*) can be significant in the assessment of NPV. The residual value (RV) is regarded as a cash inflow in year 11. Therefore, it increases the NPV by the amount of its discounted value.

19.6 Alternatives to net present value

19.6.1 Internal rate of return

As discussed above, if the project NPV is positive at a selected discount rate over a period of 10 years including a residual value, then one can go ahead. It implies a surplus over and above the cost of funding the project.

If the NPV is zero, a breakeven point or position of indifference is reached. This point of indifference can also be established by iterating on the value of the discount rate for a given cash flow profile. The discount rate which produces a zero NPV is called the *internal rate of return* (IRR) of the project.

Governments and private-sector organisations often establish a value for the IRR as a minimum (hurdle rate) for investments. The IRR indicates the maximum rate of interest at which the organisation could borrow in order to break even on the investment. It also implies that surplus cash flow can be re-invested at the same rate during the life of the project. However, in cases where net cash flows become negative beyond the initial investment, the result is a multiple solution for the IRR.

The IRR does not always provide the correct choice between projects, as illustrated in Figure 19.2. At the selected corporate discount rate the NPV of project B is higher than that of project A. The IRR of project A is, however, greater than for project B. Nevertheless, project B should be chosen.

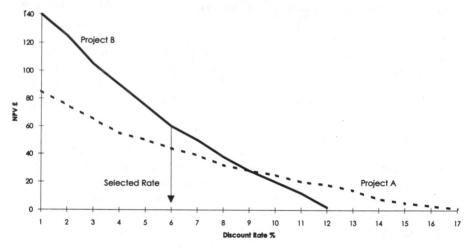

Figure 19.2   *Conflict between NPV and internal rate of return*

19.6.2 Payback period

The payback period relates to the point at which the cumulative cash flow becomes zero. The discounted values for the cash flows should be used in this instance (see Table 19.2). This is a simplistic measure and ignores the varying profile and magnitude of the subsequent cash flows. Figure 19.3 shows three profiles of cumulative cash flow all with the same payback period. Project C represents the highest NPV and should be chosen.

19.6.3 Profitability index

The conflict of having a number of good investment projects which may be of similar NPV is resolved by ranking the projects using an index. This index,

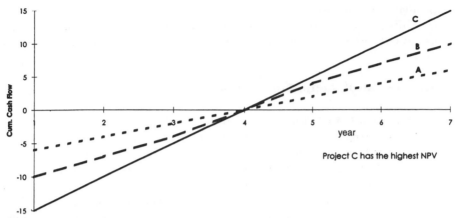

Figure 19.3   Conflict between NPV and payback period

known as the *profitability index* (PI), is calculated as NPV divided by the initial investment. An example is shown below:

Project	NPV	Investment	PI
A	25	5.0	5.0
B	25	4.0	6.3
C	25	3.5	7.1

In this example, project C represents the optimum solution.

19.7 Sensitivity analysis

Because of the substantial commitment to be made in the capital investment and the degree of forecasting involved, it is essential to investigate how sensitive the project's NPV is to changes in key parameters [10], in particular, the cost of the plant and equipment, the volume of traffic (and hence revenue) and the residual value of the equipment at the end of the time horizon being evaluated. There may also be a need to evaluate the project using different discount rates.

A systematic approach to the analysis requires the recalculation of the NPV for changes in each parameter. This can be summarised in graphical or matrix form as in the following example:

Project	NPV (base)	Capital costs +5%	Capital costs −5%	Revenue +5%	Revenue −5%	Discount rate +2%	Discount rate −2%
A	52	50	54	55	49	48	55
B	54	48	60	59	49	48	60
C	60	56	64	66	54	54	67

Sensitivity analysis provides evidence that the preference between projects may shift. In the above example, although C remains the best choice throughout, project B is more sensitive to the capital costs than project A.

19.8 Post-implementation audits

Once a project has been implemented and is operational, it is important to assess any changes in the final parameters from those planned and to establish whether the benefits and costs are being realised. This requires careful monitoring and investigation at a number of points after the project is operational.

The information gleaned will be useful to modify assumptions for other projects which have yet to be planned. Key issues are final cost of installation, the take up of traffic and its mix of calls and, of course, the operating costs actually needed to support the project.

19.9 References

1. MORGAN, T.J.: 'Telecommunications economics' (Technology Press, 1975)
2. AT&T: 'Engineering economy, a manager's guide to economic decision making' (McGraw-Hill, 1977)
3. BREALEY, R.A., and MYERS, B.: 'Principles of corporate finance' (McGraw-Hill, 1981)
4. MERRETT, A.J.: 'The finance and analysis of capital projects' (Longmans, 1973)
5. LEVY, H., and SARNAT, M.: 'Capital investment and financial decisions' (Prentice Hall, 1978)
6. HORNGREN, C.T.: 'Introduction to management accounting' (Prentice Hall, 1981), 5th edn.
7. BOX, G.E.P., and JENKINS, G.M.: 'Time series analysis, forecasting and control' (Holden–Day, 1976)
8. CCITT Recommendation E.506: 'Forecasting international telephone traffic'
9. DELL'ISOLA, A.J., and KIRK, S.J.: 'Life-cycle costing for design professionals' (McGraw-Hill, 1981)
10. DAVIES, N.J., and NAPIER, I.: 'New directions in investment appraisal', *BT Technol. J.*, 1994, **12**, pp. 42–51

Chapter 20
Network planning
K.E. Ward

20.1 Introduction

A telecommunication network can be considered as a layered model, as shown in Figure 20.1. The physical transmission-bearer network supports a number of functional networks which may be switched or nonswitched, e.g. the public switched telephone network (PSTN), the telex network, the switched-data network (circuit-switched or packet-switched), the private circuit network, the switched wideband network (mainly for visual services) and the mobile network. Gateways may be provided between the networks to allow customers parented on one network to gain access to the international network or other functional networks. Also, in a liberalised environment, gateways may have to be provided to give interconnection to networks of other licensed telecommunications operators [1].

An administration network is provided to gain access to embedded network intelligence, to input commands from operations and maintenance centres (OMC) or network management centres (NMC) and to obtain data from the network (e.g. traffic statistics, billing information etc.). Switched digital networks require a synchronisation network to ensure that incoming bit streams and exchange clocks operate at the same average frequency. Common-channel interprocessor signalling (e.g. CCITT no. 7) may also be provided on a separate signalling network. Each of the bearer, functional and ancillary networks requires a separate planning treatment.

All the networks have their own peculiarities and can be considered separately, but their planning must take account of their inter-relationship with the total network. The component networks can also be considered in terms of their spatial disposition, as shown in Figure 20.2. They form the following hierarchy:

(a) Local distribution networks
A local network (commonly known as the access network or 'loop') consists of customers' terminal equipment, lines to the local exchanges and the local exchange itself. Normally, each telephone is connected to the local exchange by an individual pair of wires, but sharing of lines is occasionally employed. Line economies can be achieved by concentrating subscribers' lines onto a

primary multiplexer or putting part of the local exchange (a concentrator) out into the local network connected to the exchange via a fibre-optic transmission system. In future, wideband links for the total communication requirements of major customers will result in the local network penetrating customer premises.

(*b*) *Junction networks*

A junction network consists of the junctions between local exchanges, together with junction tandem exchanges and the trunk junctions between local and trunk exchanges. In the UK, junction exchanges are frequently combined with local or trunk exchanges.

(*c*) *The trunk network*

The trunk network consists of trunk exchanges and interconnecting links.

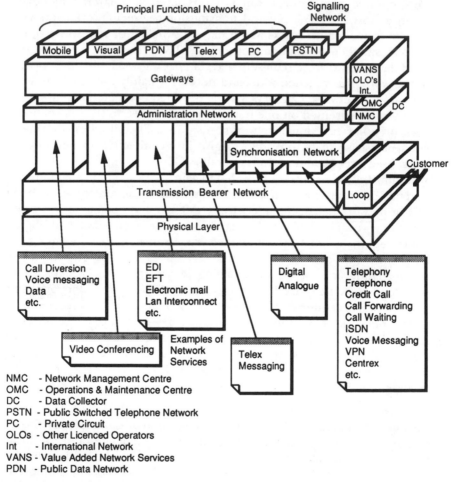

NMC   - Network Management Centre
OMC   - Operations & Maintenance Centre
DC   - Data Collector
PSTN - Public Switched Telephone Network
PC   - Private Circuit
OLOs - Other Licenced Operators
Int   - International Network
VANS - Value Added Network Services
PDN   - Public Data Network

Figure 20.1   Networks and network services

The network is often arranged in a hierarchical manner with trunk exchanges in upper tiers switching traffic from groups of trunk exchanges in lower tiers.

(*d*) *The international network*

The international network [2] consists of international switching centres (ISCs) and transmission links to other countries.

Local and junction networks may also be categorised according to the size of

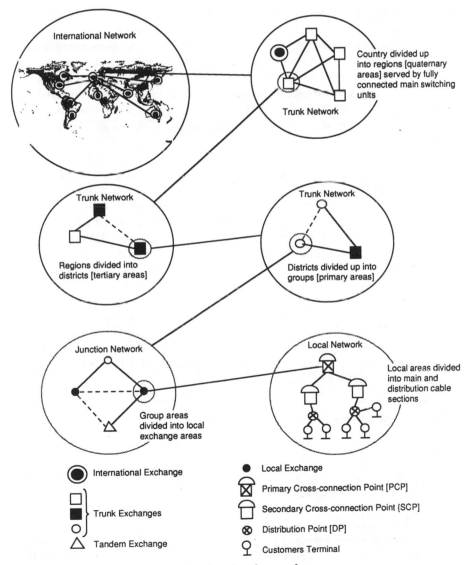

Figure 20.2   Component networks of national network

area and penetration of connections, namely:

(i) rural networks, characterised by a sparse population, wide dispersion of residential customers and low calling rates;

(ii) urban networks, with a high density of connections, evenly distributed over a wide area; and

(iii) metropolitan networks, where the density of connections is extremely high and customers are mainly businesses with high calling rates and special service requirements.

The characteristics of each of these networks are different; each has unique problems and requires a different structure and planning approach.

The component networks arranged in this hierarchy are controlled by plans (transmission, numbering, signalling etc.) which set the parameters within which the network planner must operate. For operational purposes, the division has the advantage of producing manageable units in terms of size and complexity, enabling each level of the hierarchy to be developed in a manner best suited to that level taking account of local and national conditions.

This model does not represent the modern digital network, which provides a large number of services facilitated by the software embedded in the exchanges and intelligent network databases or service control points (SCPs). Such a network can be represented by a layered model, as shown in Figure 20.3. At its lowest level are the cables, ducts buildings and power etc., then the common transmission-bearer network, which supports a multilayer functional network within which the switched layer consists of the exchanges and interconnecting traffic routes. The logical layer represents the embedded intelligence in the switching nodes and the interprocessor signalling. The service-specific layers describe the application software in each switching node and the call-control protocols necessary to support specific services. A common orthogonal network-management layer provides access to the various layers and support software for network- and service-control functions. Various network-planning processes can be applied to this model, as shown in Figure 20.3.

20.2 Planning stages

20.2.1 General

Network planning is a continuous interactive process of monitoring the state of the network, producing plans to meet the requirements for growth and enhancement, implementing those plans and auditing the outcome against expectations [3], as shown in Figure 20.4.

There are obviously interactions between the various hierarchical planning levels for control and flexibility to meet changing circumstances. As the timescales are extended into the future, the data and assumptions become

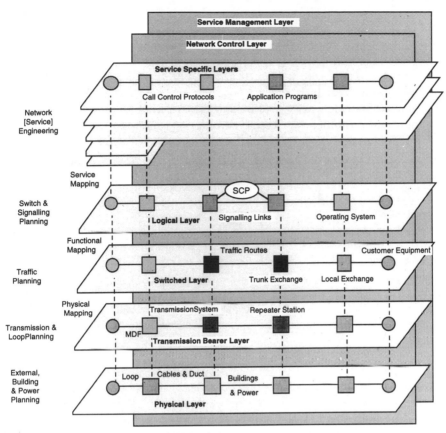

Figure 20.3 Relationship between planning and layered network model

more 'soft' and planning moves from the detailed 'concrete' aspects to the wider implications.

The end product of network planning is heavy capital investment on plants having economic lives, in some cases, of several decades; hence, the development pattern, once fixed, is difficult and expensive to change. The economics of planning are therefore of paramount importance. Telecommunication installations also involve extensions during their life and incur running costs. Hence, the plant with the least initial capital cost could result in heavy running costs and expensive extensions which may be avoided by a plan with higher initial costs to give a more economic overall solution. Investment appraisal is therefore concerned with the evaluation of costs and benefits taking account of the time value of money with the general aim of securing best value from investment. Methods of investment appraisal are described in Chapter 19. Traditionally revenue has been ignored, because it was common to optional schemes for meeting a common demand. Competition and the introduction of new services aimed at particular market

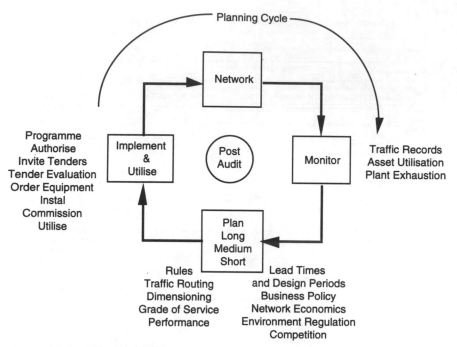

Figure 20.4   The planning cycle

segments of customers make revenue considerations increasingly important.

Thus, it must be recognised that network planning is inextricably tied to business planning and is sometimes constrained by policy considerations concerning availability of capital, procurement of plant and manpower aspects, as shown in Figure 20.5. Since investment in the network and its operating costs, together with revenue from network services, dominate the finances of telecommunication companies, there is a strong linkage between long-term network and business plans. It is therefore necessary to express the network plans in financial terms. A convenient way of doing this is to produce capital- and operating-cost profiles and benefits for the various network sectors such as the local loop, local exchanges and junction network and the trunk network.

Planning can also be heavily influenced by government regulation, which can control the flow of capital; impose growth, economic and social targets; introduce changes to the management structures; dictate equipment-purchasing policy; and determine the monopoly or competitive environment for the network and the services it provides. Where the telecommunications administration is a private limited company, corporation-tax legislation can also impact on planning, for example, by influencing the cost of capital. The strong profit motive of privatisation can also impact on planning by, for example, dictating the priorities to serve high-revenue markets. The conflicting forces on the network are illustrated in Figure 20.6.

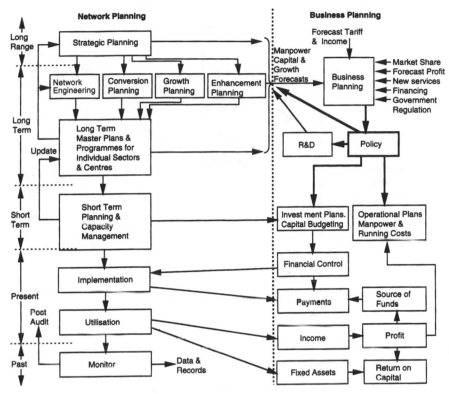

Figure 20.5   Relationship of network planning to business planning

As shown in Figure 20.5, the main phases of planning are strategic planning, network engineering, conversion planning, growth planning, enhancement planning, long-term planning and short-term planning.

20.2.2 Strategic planning

The strategic network plan provides the overall framework within which planning of individual networks and centres must conform. It charts the direction in which a network should move within the next 10–20 years, taking account of the long-term growth in telephone traffic and its distribution, the future services and facilities that may be required, the impact of new technology, the objectives of the business (both financial and service), the need for flexibility, international requirements and standards laid down by international bodies (ITU-T, ETSI etc.) [4].

The strategic network plan may consist of separate plans for the various networks of the system, but they must be coherent. Also, plans must not only cover technology and network architecture; they should also embrace the coherent development of network-management systems, network perform-

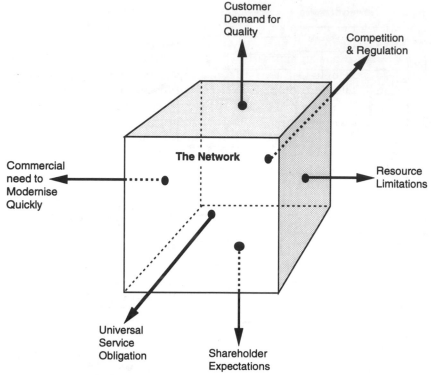

Figure 20.6   Conflicting forces on the network

ance necessary for potential new services, customer-terminal equipment, network standards, business organisation and the skills necessary to operate the network. Once the strategic plan has been produced, it must periodically be updated to ensure that a relevant validated strategy is available for use in formulating shorter-term plans.

20.2.3 *Conversion planning*

There is a requirement to produce conversion plans for evolving from the existing network to the future network as specified in the strategic network plan, especially if the proposals require a radical change in the existing network. Examples include the modernisation of the network by the replacement of existing equipment by new technology, such as digital switching and transmission equipment [5], or the introduction of a radically new technique such as a managed synchronous transmission network using SDH (synchronous-digital-hierarchy) equipment.

20.2.4 *Network (or service) engineering*

This is planning the network to support specific new services or products. Launch of a new service starts with a product or service definition which describes the features of the service as required by customers, together with a forecast of the growth of connections and traffic. Increasingly, service descriptions have to meet international standards so that new services can be offered on an international basis with common customer operating procedures. An example is the European telecommunications standards promulgated by the European Telecommunications Standards Institute (ETSI) for services provided on the ISDN.

Network realisation requires determination of the features necessary in switching nodes and links, call-control requirements of signalling, interface and network standards and support-system requirements, together with estimates of equipment and costs. The latter are required to derive cost-based tariffs and to construct a business case for the product or service. When the business case is approved, technical specifications are required for the purchase of new or enhanced equipment and the relevant network and customer-support systems.

20.2.5 Growth planning

Although it is self-evident that long-term forecasts will not be accurate, it is necessary to create long-term as well as short-term plans to cater for growth. Long-term forecasts will identify site and building exhaustions, but will also influence the topology of the network in terms of the size and disposition of nodes and links or the cost-effective technology that should be developed. For example, it could trigger the development of high-capacity transmission systems.

20.2.6 Enhancement planning

An important aspect of network planning is identifying opportunities to enhance the network to reduce its cost, improve its performance and increase its flexibility to respond to growth and new service opportunities. For example, a periodical inspection of how operating costs are spread over the various sectors of the network may indicate areas which should be studied to identify opportunities for cost reduction. If a high-cost area is identified, an investigation should be carried out to determine the 'cost drivers' that have a major influence on the sector cost. This will enable appropriate planning or organisational action to be undertaken to reduce the cost. For example, the operating cost of the local loop is traditionally a high proportion of the total network cost, due to maintenance and provision-of-service costs. Analysing the cause of, say, high-fault-rate areas could indicate where both network and organisational improvements could reduce costs [6].

20.2.7 Long-term planning and programme management

By necessity, the strategic and implementation plans are 'broad brush' and do not take account of the more-detailed requirements of individual centres. There is therefore a need for a continuously updated long-term 'master plan'

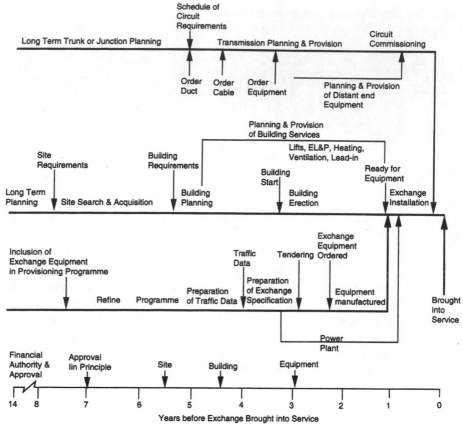

Figure 20.7   Planning a new exchange

for each exchange, which details its development over the next 10–20 years. This may seem a long time, but it must be borne in mind that the lead times necessary for planning new exchanges and the design periods for which plant and buildings must be provided are extremely long. Thus, to acquire a site, erect a building, equip it and bring it into service takes 5–10 years and the building will then take growth for 20 years or more (see Figure 20.7).

The master plan should indicate the need and timing for additional sites for buildings, the replacement of existing plant and the strategy for the provision of new types of exchange equipment. It therefore sets out the constraints within which all short-term planning is exercised.

20.2.8 Short-term planning

This consists of the continuous review and replanning of the system to meet its growth, to implement conversion and enhancement plans and to introduce new services.

20.2.9 Implementation of planning

The last step is implementation and it must be realised that, however good the previous steps were, they can be entirely negated if implementation is not carefully controlled. It can be seen from Figure 20.7 that the successful completion of a major telecommunications project requires a large number of activities to be carefully dovetailed.

Some form of control mechanism is therefore necessary to ensure that the project is completed to time and in the most efficient manner. For large projects, Critical Path Analysis (CPA) or Programme-Evaluation-and-Review-Technique (PERT) methods may be used. Although merely refinements of bar-chart methods, they are amenable to computer assistance and hence rapid response.

20.2.10 Post audit

Formal post-audit investigations to examine schemes after they have had time to mature, to see whether they have turned out as originally envisaged, are an essential follow-up to planning. Planning is not an exact science; it is the application of rules and standards together with estimates and judgements. Thus, post audit is an essential discipline on those responsible for planning and its implementation, to ascertain whether or not expectations have been realised and to see what lessons can be learned to improve the planning of subsequent projects.

20.3 Budgeting

The end product of planning is substantial capital investment in the network, which has to be funded. It is therefore important to convert the plans into individual switching and transmission-equipment orders and then determine the outturn of expenditure to create an annual budget. The budget will be authorised by management and is subsequently used to monitor the outturn of expenditure. Typically, there is an authorisation for the first year plus a forecast for the next four years. If funds cannot be obtained, then plans have to be adjusted to reduce the capital requirements.

The fall-out of expenditure from an order generally occurs in instalments over a few years and is determined by the payment terms for the particular type of equipment, involving the lead times between placing an order, the manufacture of equipment, delivery to site and the 'contractual' completion of a job. Thus, the budget expenditure for a particular year consists of committed expenditure arising from orders placed in previous years plus uncommitted expenditure from orders which will be placed during the year.

20.4 Planning standards

20.4.1 General

To meet the objectives of the telephone system, network planning is constrained to meet various planning standards, namely numbering, transmission, quality of service, switching, signalling, routing and charging. These considerations are not independent; they are inter-related. For example, charging, routing and numbering are closely related. The directory number of a called customer defines both the route for the call and its charging rate. The flexibility with which these can be handled is determined by the capabilities of the switching equipment and signalling systems employed.

It should be recognised that the planning standards and rules have a major impact on the capital cost and they must therefore be determined by appropriate economic studies.

20.4.2 Numbering plan

A national numbering plan must identify each customer's installation uniquely. (In the future intelligent network, personal numbering may well identify the customer, whatever installation is being used.) It must permit an efficient charging plan and must be flexible to meet inevitable changes of circumstances. Numbering is considered in more detail in Chapter 9.

20.4.3 Charging plan

The tariff structure and means of charging customers have a considerable influence on network billing functionality, market demand and planning. Charging methods are described in Chapter 9.

20.4.4 Network-performance plan

In order that information can be passed between customers connected to the network without unacceptable degradation, it is necessary for the network to meet certain performance standards and these must be taken into account when planning the network. For example, the transmission performance plan defines the maximum permissible end-to-end attenuation in decibels and how it should be apportioned between nodes and links. The apportionment is made to minimise network cost while giving satisfactory transmission on all connections. Other performance parameters include error rate, availability (i.e. proportion of time that the network is operational), call-failure rate, call-setup time etc. Such parameters must be closely aligned with customer quality-of-service expectations. Network performance is discussed in Chapter 17.

20.4.5 Routing plan

The traffic-routing plan details the economic routing of individual parcels of traffic between specific exchanges. The plan takes account of the network configuration and has regard to constraints imposed by the transmission plan, the end-to-end and distribution of grade of service, the setting-up time and routing-digit limitations. It is influenced by the types of switching and signalling used and the need for network resilience against unexpected variations and surges in traffic. Routing methods are considered in Chapter 9.

20.4.6 Signalling plan

The prime function of signalling is to pass information through the network to establish, supervise and clear down calls. The signalling plan is thus an essential ingredient to network planning. It specifies the type of signalling to be used on individual traffic routes in order that signalling equipment at each end of a circuit is compatible and that the system chosen meets the constraints imposed by the transmission plan. It must also take account of signalling information to be passed, i.e. called number, call supervision, metering etc., requirements for special services and facilities and the type of traffic route (i.e. unidirectional, both-way, terminal or tandem etc.).

20.4.7 Switching plan

The switching plan defines the method of switching in terms of type of exchange (Strowger, crossbar, electronic, digital, SPC etc.), the mode of switching (two-wire or four-wire) and its function (local, trunk, junction etc.) at each of the nodes of the network. It is therefore directly related to both routing and signalling plans.

20.5 Other planning parameters

20.5.1 Design periods

To meet the growth requirements of a network, it is usual to provide plant in instalments in step with the demand. The period of time for which a plant increment will meet demand is known as the design period. The factors which determine the optimum length of the design period are the fixed and variable costs, the traffic-growth rate, interest rates on capital, the size modules in which plant is provided and practical constraints such as the time needed to install the plant.

The provision of plant in large increments tends to give rise to low unit costs and infrequent interruptions to service from installation work, and provides a buffer against growth requirements higher than forecast. However, high initial capital cost is incurred before it is necessary and this leads to a large burden of spare plant not generating revenue. On the other hand, short design periods give small plant increments, resulting in higher unit costs, greater risk of interruption to service and reduced flexibility to meet unexpected traffic growth. However, there is a lower burden of spare plant.

Examples of design periods employed in UK networks are:

Sites	20–40 years
Buildings	10–20 years
Switching and transmission plant	2–3 years
Cables	5–10 years
Duct	20 years

20.5.2 Planning lead times

Planning lead times are the standard periods of time allocated to the planning, procurement and installation processes which allow the timing of major events in the planning cycle (see Figure 20.7) to be determined so that the activities dovetail to meet the required 'brought-into-service' date. The lead times for a particular plant installation are based on the size, cost, complexity, manufacturing processes and installation procedures.

20.5.3 Forecasting

Good forecasting is fundamental to efficient network planning. The need arises because:

(*a*) The logistics of telecommunication-equipment provision is such that there is a considerable time lag (the planning lead time) between identifying a requirement and satisfying it. Thus, it is necessary to forecast demands sufficiently far ahead for the necessary action to be taken. This, in turn, requires information on when resources are likely to be exhausted and how much should be ordered for replenishment to cover their design period. For this, it is useful to provide, annually, a 'rolling' forecast of traffic and connections for each of up to five years ahead. Exceptionally, for building planning, with its long lead times and design periods, forecasts of up to 15–20 years ahead may be required.

(*b*) The preparation of strategic and long-term plans requires forecasts for at least 20 years ahead, broken down into intervals of, say, five years.

Although, for planning purposes, forecasts need to be as accurate as possible, particularly those on which short-term action is planned, it should be recognised that the only certainty about forecasts is that they will not materialise as predicted. It is therefore important that sufficient flexibility is built into the network plan to cater for the unexpected.

Forecasting is becoming more complex with the introduction of new services where there is no past history from which to extrapolate. In such cases, it is necessary to carry out market research to determine probable demand which, although small in comparison with basic telephony traffic, could have a considerable impact on particular nodes of the network, for instance where value-added-network services (VANS) are accessed. It is also important to realise that the traffic characteristics of new services (e.g. calling rate, holding time, busy periods etc.) could be significantly different from those for the basic telephony service. In a competitive environment there is the additional complication of taking into account the market-share prediction in the forecast [7,8,9].

20.5.4 Network topology

Network topology concerns the arrangement of the nodes, which act as the sources and sinks of traffic and carry tandem traffic, and their inter-connecting traffic links. The determination of the structure of a network must take account of originating traffic and its geographical dispersal, the location of nodes, the pattern of links connecting these nodes, the routing system for finding a path between the nodes and the required number of circuits on each route, together with the resilience of the arrangements to unexpected traffic variation. The main variants are star, mesh, hierarchical and grid-type networks.

20.6 Network optimisation

The end product of network planning should result in good availability in the network, satisfactory service to the customer, low cost, flexibility to meet traffic variations and deviations from forecast, and high resilience against breakdown and traffic surge. Some of these characteristics are incompatible; hence the need for network optimisation. This involves the study of combinations of switching and transmission plant, together with the routing of traffic, to choose the solution that best meets the optimisation criterion (i.e. the objective function). This is usually the minimum-cost solution to meet certain quality-of-service criteria and a given demand forecast. However, other optimisation possibilities include maximising quality for a given demand requirement or cost, or maximising capacity for a given quality cost. Although always implicit in the network-planning process, more precise optimisation procedures stem from the use of mathematical operational-research concepts (e.g. modelling and graph theory) and the use of computers. Network optimisation is the basis of all short- and long-term network planning; however, its application to a total network is generally confined to strategic planning in view of the size of the network to be studied and the complex inter-relationship of plant and network parameters. Short-term planning usually requires the application of rules and procedures to suboptimise various elements, such as the physical trunk transmission network, traffic-routing plan, local-cable network etc., which, when brought together, give a reasonably well dimensioned network. However, the application of computer assistance to planning gives scope for much better optimisation procedures in the future.

An example of network optimisation is the determination of the number, locations and catchment areas of exchanges. For the local network, a small number of exchanges results in large catchment areas; the cost of the local

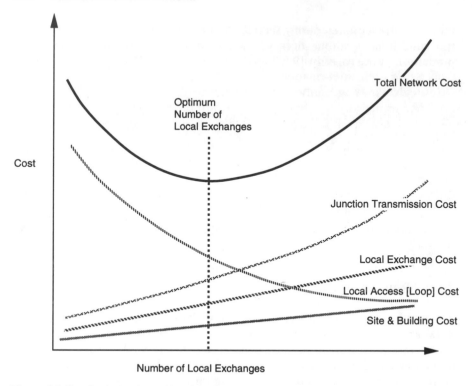

Cost

Total Network Cost

Optimum
Number of
Local Exchanges

Junction Transmission Cost

Local Exchange Cost

Local Access [Loop] Cost

Site & Building Cost

Number of Local Exchanges

Figure 20.8   Optimisation of local and junction networks

access network is therefore high, because the average line is long and circuit costs are high to meet transmission and signalling limits. However, the cost of exchange equipment, sites and buildings and the junction transmission network is comparatively low. If the number of local exchanges is increased, the cost of the access network decreases because there are shorter less-expensive circuits. At the same time, the cost of local exchanges, sites and buildings and the junction transmission network increases. As shown in Figure 20.8, the optimum solution is a compromise between these two extremes. Similar considerations are obtained for the optimisation of trunk exchanges and the trunk network interconnecting them.

20.7 Planning methodology

20.7.1 General

The methodologies adopted for long-term strategic planning and shorter-term planning for growth of local, junction and trunk networks are broadly similar in that they commence with a knowledge of the existing network and a forecast of traffic or connection requirements, determine routing and physical disposition of plant, dimension the network (or subnetwork) and evaluate costs. It is also necessary to consider both the total network or subnetwork and its individual nodes and links, i.e. the macro and micro aspects and their inter-relationship.

Strategic planning is inherently 'broad brush', but it is essential in order to provide a framework for shorter-term growth planning which will ensure that the network, as a whole, develops in a coherent fashion towards a strategic goal. A number of 'scenarios' should be postulated which indicate possible technically-viable ways in which the network may develop within social and technical environments. The consequences are then evaluated in terms of cost, resource requirements and financing to determine the optimum solution. The main planning steps and their interaction are shown in Figure 20.9. Since the only certain thing about the future is its uncertainty, it is important to carry out sensitivity testing to test the credibility of costs and assumptions and highlight critical variables requiring closer examination. This is carried out by evaluating the effects of separately varying one or more of the major parameters to test the strength of the conclusions. Risk analysis is more thorough, because it assumes that variables are interactive and therefore carries out re-evaluations with different probabilities assigned to the variables.

As shown in Figure 20.10*a*, the initial stage of a 'greenfield study' would be to determine the optimum size, location and boundaries of the local exchanges. From the growth forecast of connections for each exchange, its originating traffic can be derived by applying appropriate calling rates to the mix of residential and business sectors served by the exchange as shown in Figure 20.10*b*. The next stage is to determine the distribution of traffic between the exchanges according to the community of interest between customers served by the exchanges, as shown in Figure 20.10*c*. This is often related to the distances between exchanges and their relative sizes, but it will also be influenced by relationships between businesses or industry in the exchange areas.

The total network can now be dimensioned and costed. The essential element of the dimensioning phase is determining the most economic routing of traffic. This is carried out by converting the traffic between all exchanges into circuits routed direct or via intermediate switching points (tandem exchanges), taking account of routing and transmission constraints, as shown in Figure 20.10*d*. A traffic-route matrix of circuits between all switching points can now be assembled, as shown in Figure 20.10*e*. From this, the physical line-plant layout is determined, as shown in Figure 20.10*g* and the total line-plant and switching quantities calculated and costed, as shown in Figure 20.10*f*. It is important to include, for each network scenario, the annual manpower costs for operating the network. These can vary considerably for each scheme, e.g. a network of few but large exchanges requires fewer maintenance staff. This 'matrix' approach can facilitate the use of computer spreadsheets for simple planning studies.

Ideally, the costs should be in terms of present value of annual charges or discounted cash flow, to take account of the increments of plant installation throughout the study period. However, for studies of large networks it would not be feasible to handle such a vast number of calculations; therefore, the usual method is to cost the network, in terms of annual charges, at a number

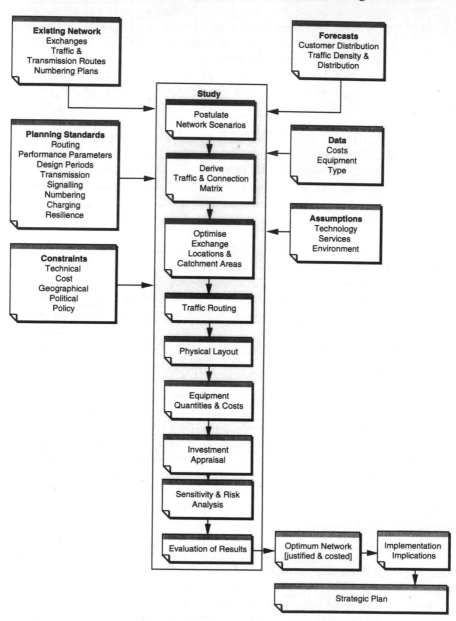

Figure 20.9   Methodology for strategic planning

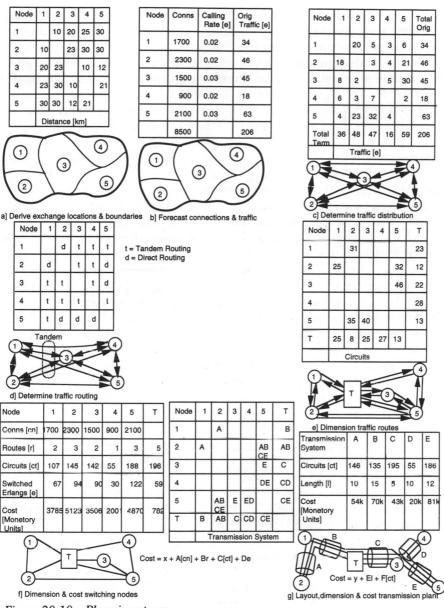

Node	1	2	3	4	5
1		10	20	25	30
2	10		23	30	30
3	20	23		10	12
4	23	30	10		21
5	30	30	12	21	
	Distance [km]				

Node	Conns	Calling Rate [e]	Orig Traffic [e]
1	1700	0.02	34
2	2300	0.02	46
3	1500	0.03	45
4	900	0.02	18
5	2100	0.03	63
	8500		206

Node	1	2	3	4	5	Total Orig
1		20	5	3	6	34
2	18		3	4	21	46
3	8	2		5	30	45
4	6	3	7		2	18
5	4	23	32	4		63
Total Term	36	48	47	16	59	206
	Traffic [e]					

a] Derive exchange locations & boundaries b] Forecast connections & traffic

c] Determine traffic distribution

Node	1	2	3	4	5
1		d	t	t	t
2	d		t	t	d
3	t	t		t	d
4	t	t	t		t
5	t	d	d	d	

t = Tandem Routing
d = Direct Routing

Node	1	2	3	4	5	T
1		31				23
2	25				32	12
3					46	22
4						28
5		35	40			13
T	25	8	25	27	13	
	Circuits					

d] Determine traffic routing

e] Dimension traffic routes

Node	1	2	3	4	5	T
Conns [cn]	1700	2300	1500	900	2100	
Routes [r]	2	3	2	1	3	5
Circuits [ct]	107	145	142	55	188	196
Switched Erlangs [e]	67	94	90	30	122	59
Cost [Monetory Units]	3785	5123	3506	2001	4870	782

Node	1	2	3	4	5	T
1		A				B
2	A				AB CE	AB
3					E	C
4					DE	CD
5		AB CE	E	ED		CE
T	B	AB	C	CD	CE	
	Transmission System					

Transmission System	A	B	C	D	E
Circuits [ct]	146	135	195	55	186
Length [l]	10	15	5	10	12
Cost [Monetory Units]	54k	70k	43k	20k	81k

Cost = x + A[cn] + Br + C[ct] + De

f] Dimension & cost switching nodes

Cost = y + El + F[ct]

g] Layout, dimension & cost transmission plant

Figure 20.10   Planning stages

 a Derive exchange locations and boundaries
 b Forecast connections and traffic
 c Determine traffic distribution
 d Determine traffic routing
 e Dimension traffic routes
 f Dimension and cost switching nodes
 g Lay out, dimension and cost transmission plant

of 'snapshot' base dates, say at 5, 10, 15 and 20 years. This planning procedure is repeated for each of the network scenarios postulated to indicate the optimum network configuration.

A similar method can be used for planning the trunk network. The location and catchment areas of trunk exchanges are determined by optimising to minimise the cost of junction circuits to the local exchanges, plus the trunk-transmission costs between trunk exchanges and the trunk-exchange costs, as shown in Figure 20.8. Likewise, the method can be used to determine the growth requirements of an existing network.

Having determined the network strategy, it is necessary to construct a long-term master plan for each centre or subnetwork (i.e. a trunk exchange and its catchment area) outlining the course of development to meet the needs of the centre and the network. This is particularly essential for site and building planning in view of the protracted lead times and long design periods.

It must be appreciated that a network is, by nature, dynamic, responding to short-term demands which may result in unexpected changes in size and shape. Also, the predictions of the future are revised with changes in circumstances. It is therefore important to re-examine strategic and long-term master plans regularly and revise them to reflect more up-to-date information.

Short-term growth planning is the regular (at least annual) survey of plant, to determine exhaustion against forecast demand, and programming for the provision of plant increments to afford relief. Although the planning cycle involves examination of individual nodes and links, this should be co-ordinated to ensure that growth of switching and transmission is matched to produce a coherent network. The most suitable vehicle for co-ordination is an annually produced forecast of requirements for each traffic route, which should preferably lay down requirements in annual steps for at least five years ahead. Thus, the forecast traffic carried on the routes can be amalgamated to produce switching requirements at each node and the route-circuit forecasts translated into transmission-plant requirements.

The organisational issues related to network plans can contribute significant costs or savings and must not be ignored. For example, a plan which concentrates switching into large nodes could result in a more efficient maintenance organisation due to centralisation, the higher utilisation of expensive support tools and sharing of overheads.

In telecommunications, the scale of the problems to be solved is often such that it is not possible to produce exact solutions. In such cases, 'heuristic' methods are used which produce a good solution but without guaranteeing that it is optimum. This is not unreasonable, since the data are not precise and an exact solution can never be obtained.

Modelling techniques can be used to assist network planning in a number of ways. Examples are the use of econometric models to assist forecasting of telephone growth by relating demand to various political, economic and social indicators, the prediction of traffic distribution and network optimisation [10].

For network optimisation the model uses:

(*a*) input data, such as the forecasts of traffic and its distribution, equipment costs etc.;
(*b*) the laws describing the behaviour of the network, e.g. technical characteristics and growth modules of equipment, traffic routing rules, dimensioning procedures etc.;
(*c*) constraints imposed by quality-of-service criteria etc.; and
(*d*) optimisation objectives, such as a minimum-cost solution etc.

The output can include the dimensions of the network (e.g. the number and size of traffic routes, switching nodes etc.), the utilisation of equipment types, the total cost (broken down in terms of plant type, network segments etc.), overall grade of service etc.

20.7.2 *The local (loop) network*

The long-term planning of the local distribution network (the access network or 'loop') involves the determination of the optimum number of local exchanges together with their locations and catchment areas. It is usually necessary to examine existing exchanges in an established network against forecast growth to determine measures necessary to meet plant exhaustions. The detailed requirements depend on whether a metropolitan, urban or rural situation is being examined. However, the basic approach, shown in Figure 20.11, is as follows:

(*i*) Optional layouts are prepared for exchanges and their catchment areas, as shown in Figure 20.11*b*. This is best carried out by examining the development forecast for the territory being studied. The forecast should be plotted on a map broken down into summary blocks (i.e. small portions of the territory such as primary cross-connection points), as shown in Figure 20.11*a*. A map showing forecast customer density is a convenient way of postulating exchange locations, which should be in high-density areas, and boundaries, which should run through low-density areas coinciding with physical features (roads, rivers etc.). The options are costed to find the optimum arrangement taking account of junction costs.
(*ii*) When the boundaries of an exchange area have been fixed, the economic location of the exchange is determined using a two-stage procedure. First, the *theoretical centre* is found. This is the centre that minimises the total length (and hence cost) of customers' lines ignoring existing plant and topological features. The map is divided into equal squares containing the forecast of customers and the total for each row and column is now calculated, as shown in Figure 20.11*c*.
The area is then divided by two lines (vertical and horizontal) such that the numbers of customers on both sides of each line are equal. The intersection

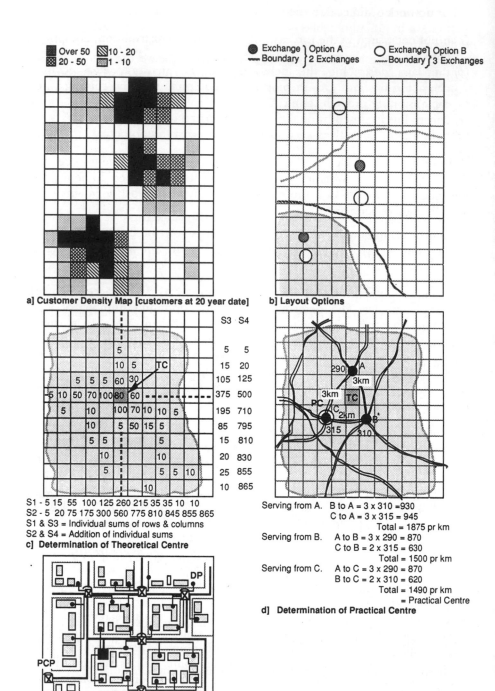

a] Customer Density Map [customers at 20 year date]

b] Layout Options

c] Determination of Theoretical Centre

S1 - 5 15 55 100 125 260 215 35 35 10 10
S2 - 5 20 75 175 300 560 775 810 845 855 865
S1 & S3 = Individual sums of rows & columns
S2 & S4 = Addition of individual sums

d] Determination of Practical Centre

Serving from A.   B to A = 3 x 310 = 930
                   C to A = 3 x 315 = 945
                         Total = 1875 pr km
Serving from B.   A to B = 3 x 290 = 870
                   C to B = 2 x 315 = 630
                         Total = 1500 pr km
Serving from C.   A to C = 3 x 290 = 870
                   B to C = 2 x 310 = 620
                         Total = 1490 pr km
                       = Practical Centre

e] Layout of Main and Distribution Cable Network

Figure 20.11   Planning a local network

of the two lines is the centre of telephone density or the *theoretical centre*. The distribution of junction cables will also affect the exchange location and can be taken into account by adding the appropriate number of connections to the squares at the boundary where junction routes leave the area. A more accurate method is to take moments (connections×distance) from arbitrary vertical and horizontal axes to find the centroid. However, this is generally only used for trunk-exchange locations.

(*iii*) The theoretical centre serves as a guide to the likely area in which the actual economic location, i.e. the *practical centre* of the exchange, may be. This centre is obtained by considering the cable layout and the summary blocks grouped to be served by each main cable. A knowledge of the development in the blocks enables the size and length of cables to be determined; the pair distance can therefore be calculated for each main cable route. A straight-line diagram of cable lengths and weighted development forecasts is helpful in carrying out these calculations. The practical centre for an area is usually most conveniently located at the point of intersection of roads which constitute existing or potential cable routes.

It will often be found that several such points of intersection will exist in the vicinity of the theoretical centre, each constituting a potential practical centre. Each common point is considered in turn and the pair distance calculated to route the cable-pair concentrations from the other points are calculated, as shown in Figure 20.11*d*. A more accurate method is to determine the plant required (cable and duct) to route to each point taking account of existing plant and carry out a discounted-cash-flow (DCF) study over a 20-year costing period. It should be realised that telephone growth is generally not uniform and parts of an area may develop at different rates. Thus, practical-centre studies should determine the minimum cost point at the 10 and 20 year dates.

(*iv*) The determination of the practical centre does not take account of the availability of a site: hence an 'area-of-search' map must be prepared as a guide to the purchase of a suitable site. This shows the practical centre and lines of search along existing and potential main cable routes. These also indicate the additional cost in departing from the practical centre (the 'out-of-centre' cost) and therefore allow the final choice of site to be made on an economic basis, e.g. it may be more economic overall to purchase a site further from the practical centre than one nearer but more expensive.

(*v*) The detailed structure of the local network, i.e. the type (overhead or underground) and layout of distribution, is planned by establishing the location and catchment areas for distribution points (DPs), grouping DPs into areas served by secondary cross-connection points (pillars) and grouping pillars into areas served by primary cross-connection points (cabinets), as shown in Figure 20.11*e*. Then the main- and distribution-cable networks interconnecting these points are designed, ensuring that the cable gauge is sufficient to meet signalling (loop-resistance) and transmission limits. The choice of DPs depends on the forecast penetration. Since augmentation and rearrangements to the diffused distribution network are expensive, a design period of 10–20 years is used. For the larger higher-growth main cables, the design period is generally 5–10 years.

Although the local loop is dominated by copper-pair cables, opportunities have now emerged for the economic introduction of optical-fibre cables and electronics in conjunction with digital exchanges. Initial applications are to give flexible access to major customers which have large numbers of access lines for PABXs, private circuits etc. Such bundles of circuits can be multiplexed to use higher-order digital transmission systems carried over the fibre. At the exchange end, groups of 30 PSTN circuits may access the digital exchange via 2 Mbit/s ports. Further penetration of fibre in the loop could become cost effective over copper cables through the use of passive optical technology. Considerable interest is being shown in the introduction of fibre in the loop [11], particularly in the USA where the broadband capability is particularly interesting for cable television.

A recent breakthrough in the exploitation of copper cables in the loop came with the development of low-cost digital speech processors for basic-rate access (144 kbit/s) to the ISDN. The lowest-capacity pair-gain system, known as the digital-access carrier system (DACS), is a $0+2$, which allows two independent 64 kbit/s channels to be carried over a single copper pair. It can be deployed in a customer's premises to provide an additional line or externally to serve two premises. 15 terminals can be combined at the exchange end directly to interface a digital exchange at 2 Mbit/s, thereby removing the need for line cards.

Multichannel pair-gain systems are now being developed to exploit the loop further; high-bit-rate digital subscriber line (HDSL) was originally developed in the USA for the delivery of 1.5 Mbit/s services over two copper pairs. The European version operates at 2 Mbit/s over three copper pairs. HDSL can provide 2 Mbit/s or $n \times 64$ kbit/s private circuits or pair-gain individual customer PSTN lines. It can also be partially equipped at 384 kbit/s for, say, video conferencing. A variant of HDSL is asymmetric digital subscriber line (ADSL) which, over a single copper pair, could be used to transmit 1.6 Mbit/s from exchange to customer with a 16 kbit/s return channel. This would be useful for those asymmetrical services which require a high-capacity channel in one direction and a control channel in the reverse direction, e.g. interactive services such as video and database access. Development is currently being carried out on very-high-speed digital subscriber line (VHDSL) which has a potential bandwidth of up to 10 Mbit/s.

In established areas, local-network planning is mainly relief planning to augment exhausted cable routes, and it usually involves considerable rearrangements. This often requires establishment of new cross-connection areas and the taking over of cross-connection areas by other cables [12].

20.7.3 The junction network

Planning the junction network requires:

(*i*) determination of tandem switching points, which may be dedicated exchanges or combined with trunk or local exchanges;

(*ii*) deciding the routing of junction traffic between local exchanges, which may be direct or via the tandem exchange; and

(*iii*) planning the physical transmission network.

In rural and urban situations, planning is the straightforward application of planning rules and the location of the tandem switching point is generally self-evident. Planning becomes more complex in metropolitan areas, where numerous options exist for the number, location and catchment areas of tandem exchanges and there is scope for the economic application of complex traffic-routing strategies such as automatic alternative routing.

20.7.4 Trunk network

Planning the trunk network is complicated by the complexity of its structure, the large number of routing options and the variety of transmission media. Long-term planning involves determination of the optimum layout from a number of technically viable solutions. This involves defining the number of levels in the hierarchy, the number, location and catchment areas of exchanges in each level, traffic-routing arrangements, the dimensioning of routes and exchanges and the physical realisation.

The study is carried out by postulating a number of network scenarios. These are dimensioned and costed to identify the optimum solution, which should take account of network-resilience aspects together with resource and financial constraints. The process is illustrated in Figure 20.10.

20.7.5 The transmission network

Trunk exchanges are interconnected by the trunk transmission network which consists of transmission links interconnecting transmission nodes (repeater stations). The transmission media can be cable or radio systems of various circuit capacities which are made up of modules of circuits multiplexed by frequency division or time division as described in Chapter 4.

The transmission nodes provide multiplexing and demultiplexing facilities and they form flexibility points for the interconnection of transmission links or modules of circuits between transmission systems, as shown in Figure 20.12*a*. The planning process consists of routing blocks of capacity (usually 480 circuits at 34 Mbit/s) via the various transmission systems (140 or 565 Mbit/s) to meet the source–destination-demand matrix as shown in Figure 20.10*e*. The recent introduction of synchronous multiplexing, with its drop-and-insert multiplexers and digital cross-connects, has facilitated the creating of managed synchronous networks with a great deal more flexibility

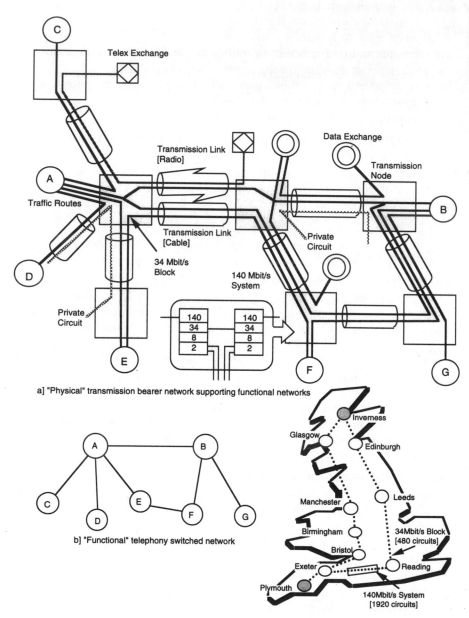

Figure 20.12   Structure of transmission network

　a Physical transmission-bearer network supporting functional networks

　b Functional switched telephone network served by transmission network in (*a*)

　c Diverse routing between Plymouth and Inverness (using 34 Mbit/s blocks) and management capabilities [13].

As shown in Figure 20.12*a* and *b*, the physical transmission network can be considered separately from the switched network (i.e. the functional network of exchanges and interconnecting traffic routes). It requires optimisation to:

(*a*) minimise cost by minimising the length of the circuits interconnecting exchanges, maximising the number of circuits sharing the same transmission links to achieve economies of scale and selection of the most economic transmission media; and

(*b*) maximise network resilience, for example by diversifying physical routing of transmission media and providing back-up equipment.

The planning procedure is as follows:

(i) examination of the matrix of traffic routes and their capacity requirements to produce a matrix of circuit demands between the source and destination transmission nodes (repeater stations) to serve the particular exchanges; adding the additional circuit demands for private circuits, other functional networks (e.g. telex, data, visual etc.) and circuits to the international network and the networks of other licensed operators;

(ii) grouping the circuit demand into appropriate modules or blocks consistent with the multiplexing hierarchy, e.g. 2 Mbit/s (30 channels), 34 Mbit/s (480 channels) etc., and deciding which modules require to be diversely routed for security reasons;

(iii) determining the least-cost transmission routing between source and destination. Clearly, it would be uneconomic to provide separate transmission routes between all sources and their destinations (roughly 400 for the BT network); hence, the network is a mesh of high-capacity transmission systems (e.g. 140 Mbit/s or 565 Mbit/s) interconnecting the nodes, each of which carries the blocks of capacity for a number of source-destination routes;

(iv) examining scale economies which occur by aggregating as many demands as possible to justify high-capacity systems, as shown in Figure 22.12*a*. However, this may result in individual point-to-point blocks (e.g. 34 Mbit/s) traversing a large number of systems, but cost and fault liability occur each time a block has to be demultiplexed and remultiplexed to be routed from one system to another. Hence, the optimum network solution must achieve the shortest source–destination routings, balancing the economies of large systems against the increased cost of intermediate multiplexing. Generally, low-capacity networks have short-distance systems with large numbers of intermediate multiplexers. However, as capacity increases, longer-distance systems are justified.

Growth planning of transmission networks is carried out by applying the incremental point-to-point block forecast (usually every year for each of five years) to the existing transmission network, routing the demands in the most economic manner using spare capacity in existing systems, identifying those systems which would be exhausted by the demand and planning the additional capacity required. In some cases, it will be preferable to create capacity on existing systems by off-loading existing blocks to other systems to give more efficient routings; however, the tradeoff between deferring expenditure on new capacity against the cost of network rearrangement must be considered carefully.

Because the capacity of individual transmission systems is shared between numerous 34 Mbit/s blocks interconnecting a number of different source–destination routings, implementation of plans must be carefully scheduled to satisfy the dates by which the additional blocks are required. The implementation plans are generally known as works planning, works execution and utilisation.

As the network increases in size, resilience against breakdown has to be constantly re-evaluated. Transmission resilience is generally provided by diverse routing of point-to-point capacity, for example as shown in Figure 20.12c, and by provision of a service-protection network of standby systems which can be patched in to restore failed systems [14].

A large transmission network is a mesh of hundreds of nodes and transmission links and thousands of potential routes for the point-to-point demands. Computer assistance is essential to plan such a network.

20.7.6 Other networks

Although other functional networks, such as the telex network, public data network, mobile network and visual-services network, have different technical and customer requirements, the approach to network planning is essentially the same as for the PSTN. Consideration must also be given to planning the support networks, namely the synchronisation network, the signalling network and administration network.

Planning of private networks differs from that for public networks, since the main objective of the customer is to minimise the charges. Hence, the mix of private circuits interconnecting PABXs, computer centres, video-conference centres etc. and public circuits depends on the tariff structure of the telecommunications company providing the facilities. The position is complicated if access is available to more than one competing public operating company. It may be advantageous to access different networks by time of day to gain the most economic tariffs.

20.8 Use of computers

The high capital intensity of the telecommunications system, the extreme complexity of the network, the need to manipulate large quantities of data and the variety of possible planning options make network planning a prime candidate for profitable computer application. The following advantages are offered:

(*a*) Extensive numerical calculation can be carried out which may not be realistic manually.

(*b*) Various network or subnetwork options can be evaluated with extensive sensitivity analysis to produce optimum solutions, giving a better foundation for planning.

(*c*) Consistency of treatment helps to ensure that results are on a comparable basis.

(*d*) Plans can be kept up-to-date with the minimum of effort by simply making the appropriate changes to input data and resubmitting the problem to the computer.

(*e*) Statistics can be produced and analysed more easily and accurately.

Typical computer applications in network planning include:

(i) *Modelling*
Models can range from econometric models for system forecasting (by relating growth of the telecommunications system to external economic factors) to network models to simulate actual networks in their development (thereby assisting network-optimisation studies).

(ii) *Data storage*
Extensive records can be stored and can be analysed and updated easily to provide an accurate common database which can be accessed from a variety of sources for numerous planning purposes. This ensures consistency of treatment; e.g. the national estimated traffic-route requirements used for switching, signalling and transmission planning are the same.

(iii) *Exhaustion surveys*
Details of existing plant quantities, on file, can be compared with forecast requirements to indicate exhaustions and produce a programme of plant augmentations, to dimension nodes and links and to produce detailed specifications of equipment requirements.

(iv) *Financial control*
Different forecasts of demand can be postulated, together with procurement strategies, to indicate funding and resource implications. This assists the formulation of financial policy.

20.9 Conclusions

Network planning gives rise to the largest proportion of capital expenditure incurred by a telecommunications business and to significant running costs. The economic health of the business is therefore critically dependent on good network planning. Although the network is a configuration of nodes and links, it is only effective as a network. Therefore, planning of the individual links and nodes must ensure that the network grows in a coherent manner within the framework of a long-term plan designed to meet a strategic objective.

The network-planning process is becoming ever more complex, with the introduction of new technology, sophisticated services, profit pressures of privatisation, global networking and regulatory requirements to allow network interconnection to competitors' networks and independent service providers. However, the fundamental planning processes will endure and need to be understood.

20.10 References

1. MEYER, F.: 'The planning process in a competitive environment'. Sixth international *Network Planning Symposium*, 1994, pp. 27–32
2. D'SA, D., and LOWE, A.: 'Planning the international network', *J. Inst. Br. Telecom. Eng.*, 1993, **12**, pp. 165–175
3. WARD, K.E.: 'Basic network planning', *J. Inst. Br. Telecom. Eng.*, Structured information Programme, Issue 7, January 1993
4. ROSENBROCK, K.H.: 'Telecommunications standards for Europe', *FITCE J.*, 1993, pp. 14–19
5. SUTCLIFF, K., HAYES, A.E., and NEWBEGIN, K.: 'The modernisation of a rural network', *J. Inst. Br. Telecom. Eng.*, 1991, **9**, pp. 291–298
6. PRIOR, J., and CHAPLIN, K.: 'The St. Albans study'. Sixth international *Network Planning Symposium*, 1994, pp. 283–288
7. LITTLECHILD, S.C.: 'Elements of telecommunications economics' (Peter Peregrinus, 1979), pp. 23–43
8. JENNINGS, D.L., WILLIAMSON, M.A., and GILSTIEN, C.Z.: 'Techniques for estimating demand for new services and assessing forecast uncertainty'. Third international *Network Planning Symposium*, 1986, pp. 152–155
9. WHEELWRIGHT, S.C., and MAKRIDAKIS, S.: 'Forecasting Methods for management' (Wiley, 1980), 3rd ed.
10. NIVERT, K., and WARD, K.E.: 'Methods for planning and optimisation of telecommunications networks—a European project'. *Telecommunication Networks Planning Symposium*, 1980, pp. 14–18
11. LISLE, P., MARSHALL, J., and PARSONS, T.: 'Access network technology; design principles', *J. Inst. Br. Telecom. Eng.*, Structured information Programme, issue 8, April 1993
12. McLACHLAN, R.: 'Planning the access network', *J. Inst. Br. Telecom. Eng.*, Structured information Programme, issue 7, January 1993
13. REID, A.B.: 'Defining network architectures for SDH', *J. Inst. Br. Telecom. Eng.*, 1991, **10**, pp. 116–125
14. DILASCIO, M., GAMBARO, A., and MOCCI, U.: 'Protection strategies for SDH networks'. Sixth international *Network Planning Symposium*, 1994, pp. 387–391

Chapter 21
Case studies

21.1 Flat Earth Island telecommunications project
N.F. Whitehead and A.R. Allwood

21.1.1 Introduction

Flat Earth Island is a square-shaped island with a side length of 100 km. The population density is exactly uniform at 100 persons per square km. The island economy is based on agriculture and cottage industries. There are no cities or significant hills or mountains (hence the island's name). The island is located in the Atlantic Ocean approximately 400 km west of France and 400 km south-west of The Lizard. The existing telecommunications facilities are negligible.

21.1.2 Objective

A firm of consulting engineers has been approached to plan a completely new telecommunication network. The firm is required to propose a network design including exchange topology and interconnection strategy. It should explain the factors which influenced its design proposal, discussing their relative importance and giving any additional assumptions needed. The objective is to achieve the most efficient solution over the operational life of the network, which may not necessarily be the solution offering the lowest first cost.

The consultants' report should also consider how the network is likely to develop over the next 10–15 years and give some indications as to how this might affect their proposals.

This example is artificial; however, it illustrates real factors affecting the planning of a network. As can be seen from Chapter 20, a real planning study of a national network for an inhomogeneous territory would require more pages than the editor can provide.

21.1.3 Costing information

Processor + core-switch costs
Switching: £400 per switched erlang

Processor:

£1 800 000 for the control of up to 200 000 lines or 50 000 erlangs (maximum permissible exchange size), or

£500 000 for the control of up to 50 000 lines or 12 500 erlangs, or

£100 000 for the control of up to 5000 lines or 1250 erlangs.

Units cannot be extended from one size to the next without an interruption of service.

Line-unit costs

£100 per line

For line units remote from the switch, 30 lines may be served from a 2 Mbit/s PCM system, using a 64 kbit/s-to-2 Mbit/s multiplexer at the line-unit end at an additional cost of £1500.

Concentrator costs

Concentration-switching costs are small relative to the core-switching costs, and can be ignored in calculations.

For remote concentrator units (RCU), each RCU serves up to 2048 lines and requires $n \times 2$ Mbit/s transmission systems (depending on the traffic concentration) to the processor and core switch. 64 kbit/s-to-2 Mbit/s multiplexers are not needed. An RCU requires a minimum of two 2 Mbit/s transmission systems, each of which will carry up to 20 erlangs of concentrated both-way traffic. Add £5000 per RCU for remotely provided services, eg. alarms, test access, isolation working etc.

Trunk/tandem switching

The same core switch can be used to provide international, trunk, tandem and local switching (if there is available switching and processor capacity), or a separate unit may be built. Costs for a stand-alone trunk exchange are as given for the processor and core switch.

Multiple exchanges

When two or more exchanges (*not* concentrators) are colocated to serve the same customer base, traffic carried between the exchanges must be switched twice, once in each unit (known as double switching). 2 Mbit/s transmission systems will also be required between the units.

Local access network

Copper-pair local access may be costed at £30 per km per pair. This figure includes assumptions to include all external plant costs. The total local-loop costs for Flat Earth Island can be estimated by making the following assumptions.

The average straight-line distance from any point in a square of side 1 to the centre can be shown to be about 0.4. The average distance following a path which remains parallel to either of the axes of the square is 0.5. This is a more realistic approximation to the physical routing of telecommunications cables, and has been used in Figure 21.1 to calculate the local-loop costs on Flat Earth Island.

Transmission systems

Cost all transmission systems to operate at 2 Mbit/s. A 2 Mbit/s transmission system will carry 30 channels on optical fibre, supporting up to 20 erlangs of concentrated traffic.

Each 2 Mbit/s system costs
£1000 per terminal (2 per system)
£200 per km (includes fibre, repeaters, and higher-order mux).

In general, a minimum of two 2 Mbit/s systems should be provided for any route carrying concentrated traffic to meet security considerations.

Building costs

Large:	£1 000 000	e.g. to serve up to 200 000 lines+2 Mbit/s trunks
Medium:	£250 000	e.g. to serve up to 50 000 lines+2 Mbit/s trunks
Small:	£100 000	e.g. to serve up to 5000 lines+2 Mbit/s trunks.

Installation costs

All the above figures (including the exchange costs) are inclusive of installation.

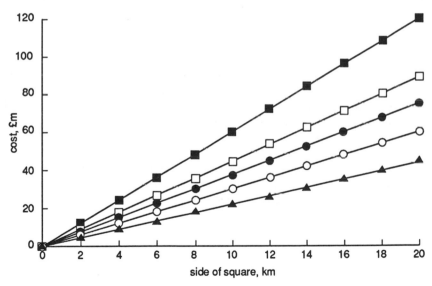

Figure 21.1   Local-loop costs

■ 40 lines/km^2
□ 30 lines/km^2
● 25 lines/km^2
○ 20 lines/km^2
▲ 15 lines/km^2

Depreciation costs

Assume that local-network costs are written off uniformly over 20 years, and that all electronics (including transmission systems) are written off uniformly over 10 years. Buildings are written off over 40 years. Thus:

depreciation of local cable network: 5% per annum
depreciation of electronics: 10% per annum
depreciation of buildings: 2.5% per annum.

Operations and maintenance costs

switches, processors and other electronics: 2% of capital cost p.a.
copper pair network: 5% of capital cost p.a.
buildings: 1% of capital cost p.a.

Note: These figures do not take into account manpower or transport costs.

21.1.4 Marketing report

A marketing report has been prepared giving the expected demand for telecommunications services from the population of Flat Earth Island. The following is an extract from that report.

Forecast demand for lines

The forecast demand for lines during the first 10 years of operation of the Flat Earth Island Telecommunications Project is given in Figure 21.2.

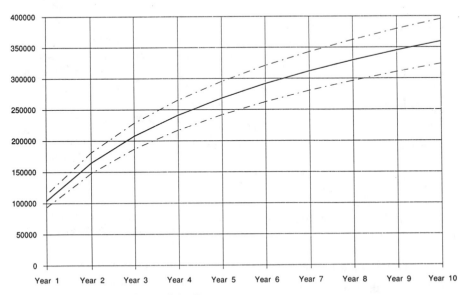

Figure 21.2   Forecast demand for lines

———— median
—·— confidence limits

Forecast traffic volume

The busy-hour traffic generated by the customers of the Flat Earth Island Telecommunications Project is estimated to be a constant 0.1 erlang per line, both-way. Average call-holding time is expected to be 90 s.

21.1.5 *Consultants' report*

(1) *Assumptions*

The majority of the island's telecommunications needs are assumed to be met by the provision of a conventional circuit-switched voice-telephony network. The economic base of the island is agriculture and cottage industry and, in addition, the location of the island makes it an ideal holiday area. There is thus no evidence of a significant need for data or wideband services. It is assumed that any provision of such facilities would not significantly affect the design of the network topology.

The network has been conservatively dimensioned to meet the fourth-year demand forecast of 250 000 lines. In addition, any swift expansion in demand, owing for example to the creation of a tourist industry, can be accommodated by the design without the need to provide additional major capital items for several more years.

(2) *Initial evaluation*

It can readily be seen that a single-level network serving the whole of the island from a central switch by means of a copper-pair local network would be both impractical and prohibitively expensive.

Two-level-network design options are therefore considered, and for this purpose the island's area may be divided into a number of cells. The telecommunications needs for the customers in each cell may then be served from the cell centre by means of a copper-pair local network. The units at the centre of each cell may then be connected to an exchange at the island centre using a number of 2 Mbit/s transmission systems.

The key issues raised by such a topology are:

(*a*) How much switching and processing should be done at each cell?
(*b*) What is the optimal number of cells required to serve the whole island?

(3) *Cell-unit design options*

The island is a square, 100 km by 100 km, with a uniformly distributed population. It is sensible therefore to divide it into equally-sized square cells, and a range of cell sizes between 2.5 km square (1600 cells) and 25 km square (16 cells) was chosen.

Three switching-system design options were evaluated as follows:

(*a*) Remote-multiplexer option: Each cell is served by remote multiplexers which permanently connect all the lines by 2 Mbit/s transmission systems to the central local exchange unit where all processing and switching functions are carried out.

(*b*) Remote-concentrator option: Each cell is served by a remote concentrator unit. The traffic is concentrated for transmission to the central exchange where the remaining local-exchange switching and processing functions are carried out. No trunk exchange is required.

(*c*) Switching option: Each cell is served by a fully-equipped local exchange, with transmission links to a central trunk-exchange unit.

(4) *Network design*

The total capital costs for each of the three design options were then examined over the chosen range of cell sizes. At this stage, the differential costs of depreciation were taken into consideration by relating all costs to those of the electronic equipment. As depreciation is taken to be linear over the plant lifetime, this may simply be done by dividing the building costs by four, and the local-cable-network costs by two. The total costs for buildings and local networks are shown in Figure 21.3.

The cost tradeoffs resulting from the various options may then be evaluated. For the sake of completeness, we have carried out this evaluation by means of a spreadsheet, and the results are shown in Figures 21.4, 21.5, 21.6 and 21.7.

Figure 21.7 shows that the remote-concentrator option will provide the optimal solution, with the island divided into 256 cells. Each cell is a square of side 6.25 km and is served by a single remote concentrator unit, equipped to serve an average of 977 lines. An approximate alternative method of deriving this result is attached to this report as an appendix. This choice of cell size results in a relatively small copper network, and hence low local-network maintenance costs. The choice of the RCU option as opposed to the

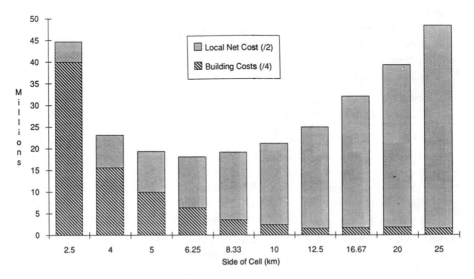

Figure 21.3   Local network and building costs (allowing for depreciation)

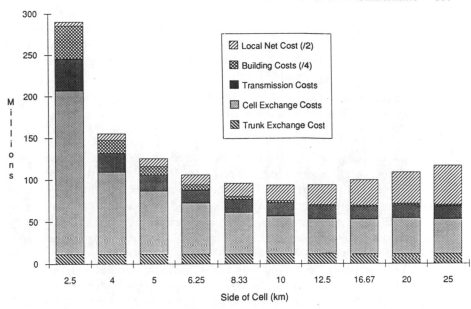

Figure 21.4   Costs for remote-multiplexer option

switching option results in less remotely-housed equipment, and hence lower operations, maintenance and management costs.

For the chosen design, an average of 97.7 erlangs is generated in each cell.

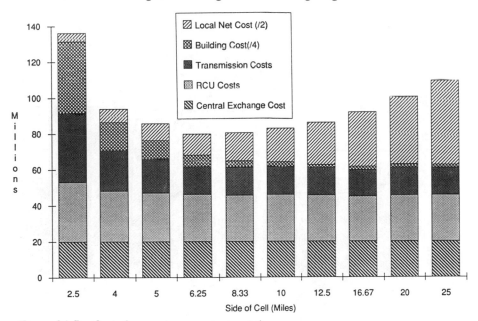

Figure 21.5   Costs for remote-concentrator option

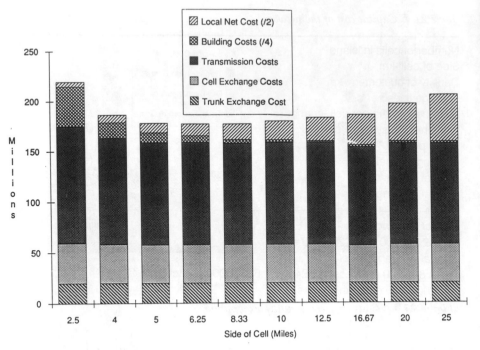

Figure 21.6   Costs for switching option

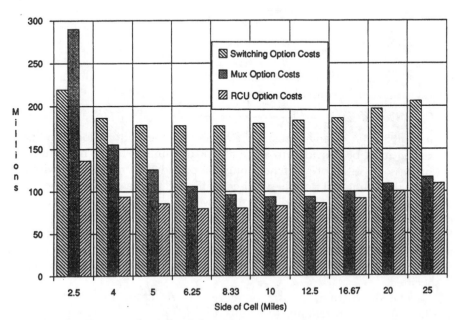

Figure 21.7   Comparative costs of the options

Table 21.1 Capital cost of the network

Number of cells in island	256
Side of cell, km	6.25
Density of customers, km^{-2}	25
Average lines per cell	977
Busy-hour traffic per cell, erlangs	97.7
Transmission systems per cell	Five (2 Mbit/s)
RCUs per cell	One
Building per cell	One (small)
Central exchange	Two large
Central buildings	Two medium

	£ million
Cost of Buildings	
256 × £100 000 + 2 × £250 000 =	26.1
Central-exchange cost	
2 × £1 800 000 (processor) + £400 × 1.5 × 25 000 (switching)	
+ £2200 × 640 (inter unit trunks) =	20.0
Local-loop costs	
£30 × 6.25/2 × 250 000 =	23.4
Transmission costs	
£12 000 × 256 × 5 =	15.4
Remote-concentrator costs	
256 × (£5000 + 250 000 × £100) =	26.3

Grand total	£111.2 million

This traffic may be concentrated onto 5×2 Mbit/s streams, making a total of 1280 systems to be terminated at the island centre.

(5) *Central-unit design*

A total of 25 000 erlangs must be switched at the central location, and processing must be provided to serve all 250 000 lines. Thus, a minimum of two processors (with core switches) must be provided. If more core switches are used, the volume of double-switched traffic will increase. Hence, the total switching cost will rise. In addition, there will be an increasing requirement for interswitch transmission systems.

The central exchange equipment will have adequate switching and processing capabilities to handle international traffic. International-gateway software and appropriate international transmission equipment will need to be specified.

It can be shown that, for two or more core switches, the advantages of distributing the central-unit functions closer to the cell exchanges are outweighed by the increased costs of the inter-unit transmission systems. The optimal solution is therefore to provide two cosited central switches. This solution also carries significant administrative advantages and, if the units are physically located in two separate buildings, an adequate level of security is provided. It is assumed that medium-sized buildings may be used because the concentrator switches are to be housed remotely.

(6) Capital cost

The composition of the proposed network is summarised in Table 21.1 and the capital costs of its components are also listed in the Table. The total capital cost of the network for Flat Earth Island is £112.2 million.

(7) Future development of the network

Network growth for years 4–10 is likely to require the addition of some 100 000 lines. This may be achieved without the need to provide additional remote concentrator units, processors or buildings. The additional equipment required will be: local copper cables (£9.5M), line units (£10M), switching equipment (£4M) and transmission equipment (£1M). These equipments can all be provided incrementally in response to demand. At the time the network is required to serve 350 000–400 000 lines, the economically advantageous choice of a relatively large number of cells and the consequentially small local-network diameter is even more apparent. It can be shown that the choice of a cell size of 6.25 km and the use of remote concentrator units remain optimal.

During the period to year 10, it is possible that a demand will develop for ISDN services and for intelligent-network features (e.g. electronic directories, dial-by-name etc.). These features may readily be accommodated within the proposed network-design concepts. Customers who require basic rate ISDN (2B+D) will need to be provided with an appropriate ISDN line unit in place of their analogue-services line unit. The service may be delivered with no difficulty over the copper-pair network. ISDN software will be required at the two central exchanges. An intelligent-network database may be provided centrally to serve the whole island. Additional software will also be needed at the two central exchanges.

By year 10, alternatives to the chosen network topology might be envisaged as a result of developments in technology and the availability of new types of equipment. In particular, more-advanced transmission systems can be employed, exploiting drop-and-insert multiplexers and point-to-multipoint passive optical networks. Such systems may find a ready application in the junction network (connecting the RCUs and central exchange). If the copper-pair local network on Flat Earth Island is to be enhanced by the use of advanced transmission systems, low-cost multiplexers will need to be provided, coupled with the use of low-cost and environmentally-secure remote electronic-equipment housings. The added operations and maintenance problems of such a solution would need to be carefully assessed.

Beyond year 10, replacements for the major switching and remote-concentrator equipment may be considered. At that time, the book value of the original equipment will have been discounted to zero, and the cost of equivalent replacements is likely to be less than 50% (in real terms) of the year-1 figure. However, the real drive for replacement switching equipment may only emerge when the market demand develops for new multibit-rate services coupled with the availability of technologies, such as asynchronous transfer mode and optical switching, capable of realising them.

21.1.6 Appendix

A quick way to optimise the cell size

Except for networks utilising very large or very small cell sizes, the optimal size is dominated by the cost of buildings (one per cell) and the cost of the local-line plant (more for larger cells). The optimal cell size will occur when the sum of these is at a minimum.

Let the number of cells per side of island = B

$$\text{Length of cell side} = \frac{100}{B}$$

Number of cells = B^2

$$\text{Lines per cell} = \left\{ \frac{100}{B} \right\}^2 \times 25$$

$$\text{Cost per line} = \frac{100}{B} \times \frac{1}{2} \times 30$$

$$\text{Local-loop cost} = \left\{ \frac{100}{B} \times 15 \right\} \times \left\{ \frac{100}{B} \right\}^2 \times 25 \; B^2$$

$$= \frac{375}{B} \times 10^6$$

Building costs = $10^5 \times B^2$

Building costs depreciate at half the rate of the local-loop costs, and so weight the costs accordingly.

We need to solve:

$$\frac{d}{dB} \left\{ \frac{3750}{B} + \frac{1}{2} B^2 \right\} = 0$$

$$\therefore B^3 = 3750$$

$$\text{and } B = 15.536$$

To the nearest integer, $B = 16$, giving a cell side of 6.25 km and a total of 256 cells on the island.

21.2 Introduction of ISDN into an existing public switched telecommunication network

M.J. Duff

21.2.1 Objective

The Utopian Telecommunications Authority has appointed consultants to prepare a plan for the phased introduction of ISDN into the existing public switched telecommunication network of the province of Happy Valley. This province has been chosen by the Authority as it is a good representation of the whole of the country.

Information has been provided by the Utopians concerning the existing network, the projected requirements of the service and the costs involved. However, this information is their view of the situation and the consultants will be expected to supplement this with information from other sources.

21.2.2 Network information

The national trunk network is fully digital and digital links exist between all local exchanges and the trunk network. Some analogue links still exist between adjacent local exchanges, but these are being phased out.

All routes have been provisioned for the demand of normal telephony traffic. The normal provisioning rule is three 2 Mbit/s circuits per 1000 customers.

The network contains two types of analogue exchanges (A and B) and one type of digital exchange (C). The principal characteristics of these systems are as follows:

Exchange type A

Size:	All sizes from small to large local
Switch type:	Analogue
Control:	Automatic path setup under direct subscriber control
General comments:	These exchanges are the oldest in the network and are in the process of being replaced. Initially they were replaced with exchanges of type B, but they are currently being replaced with type-C exchanges.

Exchange type B

Size:	Medium and large local

General comments: This exchange type was developed some 20 years ago, but has been through a number of technology updates and enhancements. They are no longer being used to replace type A exchanges, but the equipment is still being manufactured to provide extensions to existing exchanges. These exchanges have been enhanced to provide call-logging facilities and common-channel signalling (CCITT no. 7).

Exchange type C
Size: Medium and large local
Switch type: Digital
Control: Stored-program control
General comments: This is the newest type of exchange in the network and is being installed as replacements for the aging type A exchanges. It is still being developed and new enhancements are being added as available.

A map showing the locations of the exchanges in the Happy Valley Province is given in Figure 21.8 and details of these exchanges are listed in Table 21.2.

21.2.3 Cost information

The following are the current prices quoted to the Authority for the supply of various items. Also included is any other information relating to the quoted price known at this time. All prices quoted are in Utopia Government Handouts (Ugh).

Costs specifically related to the provision of ISDN

ISDN-line multiplexer (Imux)
  Price 6500 Ugh.   This price is for an Imux.
    Input: 15 circuits at 144 kbit/s (I.420).
    Output: 1 2 Mbit/s channel.

Line plant
  Price 100 Ugh.   This price is per kilometre per circuit and includes all installation charges (a circuit is one 2 Mbit/s link).

Flexible-access-mux
  Price 10 000 Ugh.   This is the projected price for a single flexible-access multiplexer. Each multiplexer will terminate 15 circuits (I.420).

Exchange type B ISDN enhancement package
  Price 60 000 Ugh.   This is the price of the enhancement processor which can handle up to 3840 ISDN customers. If this number is exceeded, another processor is required.

  Price 400 Ugh.   Per-line cost associated with the provision of interface

Table 21.2 Local-exchange information

Exchange no.	Type	Total lines (1000)	Business lines (1000)	Private lines (1000)	No. 7 signalling	Replacement year
1	A	11	5.0	6.0	no	1
2	A	21	3.0	18.0	no	1
3	A	15	2.0	13.0	no	2
4	A	20	7.0	13.0	no	2
5	A	17	5.5	11.5	no	3
6	A	18	15.0	3.0	no	3
7	A	21	7.0	13.0	no	4
8	A	10	1.5	8.5	no	4
9	A	12	1.6	10.4	no	5
10	A	11	4.8	6.2	no	5
11	A	21	2.9	18.1	no	6
12	A	11	8.6	2.4	no	6
13	A	15	2.0	13.0	no	7
14	A	21	10.0	11.0	no	8
15	A	26	9.0	15.0	no	8
16	A	7	0.9	6.1	no	9
17	B	33	26.0	7.0	yes	10
18	B	27	8.7	18.3	yes	10
19	B	14	2.1	11.9	yes	11
20	B	28	14.0	14.0	yes	11
21	B	38	30.0	8.0	yes	12
22	B	23	7.5	15.5	yes	12
23	B	12	3.6	8.4	yes	13
24	B	26	3.7	22.3	yes	13
25	B	22	16.0	6.0	yes	14
26	B	12	3.6	8.4	yes	14
27	B	29	24.0	5.0	yes	15
28	B	22	7.1	14.9	yes	15
29	B	25	14.0	11.0	yes	16
30	B	15	2.2	12.8	yes	16
31	B	16	5.4	10.6	yes	17
32	B	31	17.0	14.0	yes	17
33	B	38	18.0	20.0	yes	18
34	B	11	3.8	7.2	yes	18
35	B	15	4.1	10.9	yes	19
36	B	26	20.0	6.0	yes	19
37	C	16	13.0	3.0	yes	20
38	C	25	13.0	12.0	yes	21
39	C	24	19.0	5.0	yes	22
40	C	30	14.0	6.0	yes	23

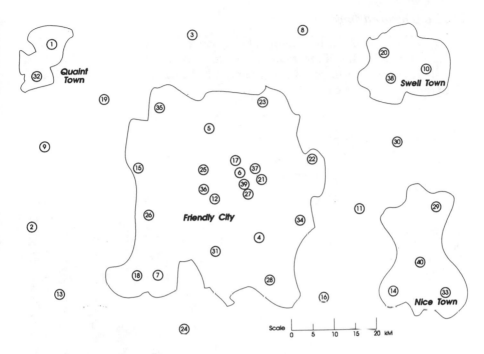

Figure 21.8   Happy Valley province: exchange-location plan

equipment (to be supplied in multiples of 15).

Price 5000 Ugh. Price of replacement junction multiplexer. Each multiplexer provides one circuit at 2 Mbit/s (CCITT no. 7).

Basic-rate-access line cards for type-C exchange
Price 400 Ugh. This is the projected price of the new line card
containing four circuits.

Additional cost information
Type A exchanges
Price 300 Ugh. This is the current equivalent price per line.

Type B exchanges
Price 260 Ugh. This is the current equivalent price per line.

Type C exchange
Price 200 Ugh. This is the current price per line.

Exchange type B enhancement
Price 300 000 Ugh. This is the price paid per exchange for the enhancement to type-B exchanges to provide call logging and
common-channel signalling.

21.2.4 Demand forecast

There is no reliable information about the potential demand for ISDN services in Utopia. It will therefore be assumed that, if ISDN is introduced, its growth rate will be similar to that experienced in Europe. This assumption gives the forecast shown in Figure 21.9.

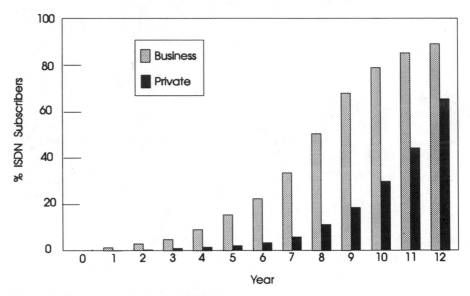

Figure 21.9   Projected growth of ISDN in Utopia

21.2.5 Consultants' report

(1) *Possible methods of implementation*

For those customers already connected to the network via a digital switch the service is automatically available by the replacement of the line-interface circuits. The following relates to the provision of the service to the remaining customers.

On analysing the problems there seems to be the possibility of four methods of achieving the service for these customers. They are:

(*a*) Disassociation
(*b*) Grooming
(*c*) Integration
(*d*) Replacement

These methods are illustrated in Figure 21.10.

The following explanations of the four methods include their advantages

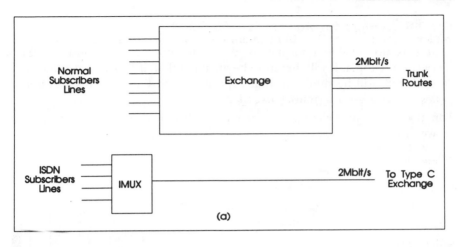

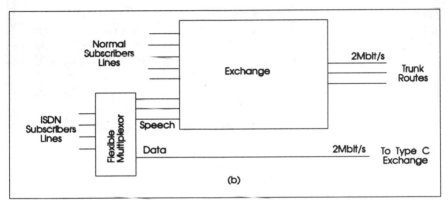

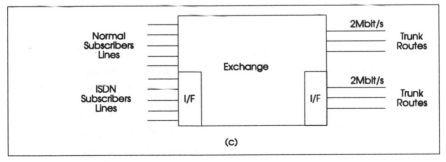

Figure 21.10   Methods of implementing ISDN on existing exchanges

    a Disassociation
    b Grooming
    c Integration

and disadvantages as seen by the consultants. The final plan for the implementation of ISDN could contain more than one of these methods.

Method (a) Disassociation

This method requires all customers requiring ISDN access to be disconnected from their existing exchange and reconnected to the nearest digital switch, as shown in Figure 21.10*a*.

Advantages:
(*a*) simple implementation;
(*b*) low cost;
(*c*) available in areas where type C exchanges exist; and
(*d*) immediate provision of the service.

Disadvantages:
(*a*) requires the customer to have a number change;
(*b*) limited to the number of spare lines available at the nearest exchange of type C. (In most cases the number of spare lines is limited, as the exchanges were originally dimensioned without considering this service);
(*c*) limit to the service penetration possible;
(*d*) no provision in this method should the demand exceed the availability of adjacent spare digital lines; and
(*e*) requirement to provide additional line plant.

Equipment required:
(*a*) IMUX, one for every 15 customers to connect to the nearest type C exchange; and
(*b*) additional line plant.

Availability:
Available now.

Risks:
No technical risks; however, if this solution is used throughout the whole of the network, it would severely restrict the penetration of the service.

Consultants' opinion:
Only suitable if the penetration of the service is small; could be inconvenient to customers to change numbers, therefore reducing the service take-up.

Method (b) Grooming

As shown in Figure 21.10*b*, this method requires the insertion of a flexible-access multiplexer between the customer line and the exchange to separate the data traffic from the telephony traffic. The telephony traffic will then be routed via the existing exchange connection and the data via the nearest digital exchange.

Advantages:
(*a*) low cost; and
(*b*) available in areas where type-C exchanges exist.

Disadvantages:
(*a*) requires the customer to have two numbers, since a new number on the digital switch will be required;
(*b*) limited to the number of spare lines available at the nearest exchange of type C. (In most cases, the number of spare lines is limited, as the exchanges were originally dimensioned without considering this service);
(*c*) limit to the service penetration possible;
(*d*) no provision in this method should the demand exceed the availability of adjacent spare digital lines;
(*e*) requirement to provide additional line plant; and
(*f*) requires the development of new type of flexible-access multiplexer.

Equipment required:
(*a*) a flexible-access multiplexer, one for every 15 customers; and
(*b*) additional line plant.

Availability:
Flexible multiplexers are available now, but they would require the development of additional software to provide this facility.

Risks:
Some risk associated with the development of the new software. This method would also limit the penetration of the service, which would become available with the provision of digital switches.

Consultants' opinion:
Only suitable if the penetration of the service is small. Could be inconvenient to customers to have two numbers, therefore reducing the service take-up. This method, although feasible, is not technically very good. This method is also more costly in terms of exchange ports. As both ISDN channels could be used for either speech or data, each subscriber would therefore require two ports on each of the switching machines.

Method (c) Integration

This method is shown in Figure 21.10*c*. It requires the development of a new enhancement package to control new ISDN interfaces. To transmit digital signals across the switch, the interface circuits on both sides will have to be changed. The subscriber interface will terminate the I.420 line signalling and the junction multiplexers will terminate the CCITT no. 7 signalling. Both will contain the necessary circuits to transmit at 64 kbit/s through the switch.

Advantages:
(*a*) no number changes required;
(*b*) no limit to the possible penetration of the service;

(*c*) no requirement to have connections to an adjacent type-C exchange; and
(*d*) exchanges can be fitted with the control and the line interfaces only added as required by service demand.

Disadvantages:
(*a*) can only be done on type-B exchanges; and
(*b*) requires the replacement of junction multiplexers.

Equipment required:
Exchange ISDN enhancement package.

Availability:
Would be available within two years of the start of development.

Risks:
Normal risks associated with a major system development.

Consultants' opinion:
An attractive method which gives unlimited penetration while retaining type-B exchanges.

Method (d) Replacement

This method requires the replacement of the existing type-A and type-B exchanges with new type-C exchanges.

Advantages:
(*a*) no number changes required; and
(*b*) no limit to the possible penetration of the service.

Disadvantages:
(*a*) requires the replacement of all type-A and B exchanges; and
(*b*) exchanges will require new line interfaces for basic rate access.

Equipment required:
(*a*) replacement type-C exchange; and
(*b*) basic-rate-access line cards.

Availability:
Available now, new basic-rate-access line card due to be available within two years.

Risks:
None.

Consultants' opinion:
The cleanest and most straightforward method.

(2) *Implementation costs*

(a) *Disassociation*

IMUX   6500 Ugh = 433.33 Ugh per line
Line plant   100 Ugh =   6.66 Ugh per line per km
          Cost of method = $n(433.33 + 6.66d)$ Ugh

(b) *Grooming*

Primary mux   10 000 Ugh = 666.66 Ugh per line
Line plant   100 Ugh =   6.66 Ugh per line per km
          Cost of method = $n(666.66 + 6.66d)$ Ugh

(c) *Integration*

Processor   60 000 Ugh
I/F equipment   400.00 Ugh per linc
Replacement mux 5000 Ugh =   15.00 Ugh per line
          Cost of method = $60{,}000 + 415n$ Ugh

(d) *Replacement*

Type-C cxchange 200.00 Ugh per line
Line cards   400 Ugh = 100.00 Ugh per line
          Cost of method = $200N + 100n$ Ugh

where n = Number of ISDN lines
          N = Total number of exchange lines
          d = Distance to nearest type-C exchange

(3) *Timing of introduction of ISDN*

It is recommended that the ISDN service be introduced in five years' time. It will take that long to implement enough of the plan to be able to introduce the service. By that time, as shown in Figure 21.9, some 12% of business customers should be requesting the service. This is a large enough proportion to make the service viable.

In five years' time, 10 of the existing type-A exchanges will have been replaced by type-C exchanges. The network will then contain the mix of exchanges shown in Table 21.3.

(4) *Cost comparisons*

In five years' time, exchanges 1–10 will have been replaced by type-C exchanges and so be able to offer ISDN services. Exchanges 37–40 are already type C. Cost comparisons between the different methods for the remaining exchanges are shown in Table 21.4. The numbers of additional lines required on type-C exchanges are shown in Table 21.5.

Table 21.3 The network as planned in five years' time

Type	Number Exchanges	Number Lines	Value (Approx.)
A	6	95000	20 M Ugh.
B	20	469000	130 M Ugh.
C	14	251000	50 M Ugh.

(5) *Recommendations*

Four methods for providing ISDN were given and we have chosen to recommend two of them: integration and replacement.

The methods rejected, disassociation and grooming, are primarily rejected on two accounts:

(*a*) In both cases the customers requiring the ISDN service would be required

Table 21.4 Cost comparisons between methods

Exch. No.	No. ISDN Business Lines	Distance to Type C (km)	Disass'	Groom'	Integ'	Repl'
11	348	19	195	275	-	4235
12	1032	9	510	750	-	2305
13	240	16	130	185	-	3025
14	1200	9	590	870	-	4320
15	1080	19	605	855	-	5310
16	108	21	60	85	-	1411
17	3120	3	1415	2140	1355	6915
18	1044	5	485	730	490	5505
19	252	18	140	200	165	2825
20	1680	7	805	1195	757	5770
21	3600	3	1630	2470	1554	7960
22	900	13	470	680	435	4690
23	432	14	230	330	240	2445
24	444	14	235	340	244	5245
25	1920	9	950	1395	855	4595
26	432	14	230	330	240	2445
27	2880	3	1310	1980	1255	6090
28	852	11	430	630	415	4485
29	1680	14	885	1275	750	5170
30	264	16	140	200	170	3025
31	648	11	330	480	330	3265
32	2040	9	1010	1485	905	6405
33	2160	9	1065	1570	955	7815
34	456	11	230	340	230	2245
35	492	13	260	375	265	3050
36	2400	10	1200	1760	1055	5440

Table 21.5 Numbers of additional lines required on type-C exchanges

Exchange	Number of lines	Exchange	Number of lines
1	2292	7	1921
2	240	37	4500
3	648	38	1944
4	1416	39	6312
5	2004	40	5388
6	5040		

to change their existing numbers. For most people, especially businesses, this would be hard to justify and would hinder the take up of the service.

(b) Both methods are generally more expensive than integration, which can be applied to type-B exchanges. The number of type-A exchanges does not justify the inclusion of these in the overall network solution.

It can also be seen from Table 21.5 that, in order to be able to provide a full service, if either of the above methods were used and the demand follows the predicted growth, all of the city and town type-C exchanges would have to be extended to provide the required circuits. This would add additional cost and would require the redimensioning of the network to take account of the redistribution of customers.

(6) *The plan*

The proposed plan allows for the major launch of the service in five years' time, which allows for the equipment to be installed. During that time, exchanges 1–10 will be replaced by exchanges of type C at their planned change-out dates. All type-B exchanges will be fitted with the ISDN-enhancement processor package and enough per-line equipment for the initial demand. Additional per-line equipment will be added as the demand increases.

This plan covers the majority of the exchanges in the network. However, there are a number of anomalies.

(a) Three of the remaining type-A exchanges (12, 14 and 15) could have a significant requirement for the service; therefore it is recommended that their replacement be brought forward. This could be done at the expense of other less-important exchanges, for example exchanges 8 and 9.

(b) Some of the type-B exchanges could be excluded from the programme of enhancement. For example, exchanges 19 and 30, being rural exchanges, have smaller proportions of business customers making it harder to justify the expense of the enhancement. Also, exchange 17 is not due for replacement for ten years and requires a large investment to enhance. This exchange should be considered for early replacement.

(c) If any customer on an exchange which does not have the ISDN capability wants the service, then it could be provided using disassociation. Since this would result in the requirement for a number change, the demand could probably be provided from spare capacity on a type-C exchange.

(7) *Service introduction*

The service should be introduced in two phases, as follows.

Phase 1: A limited trial/pilot service to be introduced in Friendly City, to customers already connected to the type-C exchange. This can then be extended to the three towns (exchange 1 when replaced will provide the opportunity to offer the service in all three towns at the same time).

   The pilot scheme will provide valuable feedback to enable demand to be confirmed, and it allows time for the enhancing of the type-B exchanges. As the number of type-C exchanges increases and the type-B-exchange enhancement programme gets underway, the pilot service could be expanded.

Phase 2: Launch of the full ISDN service requires facilities to be available for any customer in the network. This could be achieved by a major launch or by the expansion of the pilot scheme. If early replacement of type-A exchanges is not made, then this will limit the effectiveness of the launch; therefore the only option will be to expand the pilot service.

21.3 Corporate private telecommunication network project

R.K. Bell

21.3.1 The company

The Telnet Group consists of two group-headquarters sites and three operating divisions:

Division A manufactures bottles and distributes soft drinks. It consists of:

Divisional headquarters
Two manufacturing sites
Two distribution depots
One manufacturing-plus-distribution site.

Division B manufactures and distributes confectionery goods. It consists of:

Divisional headquarters/factory/distribution depot
Two distribution depots
Two sales offices.

Division C is a transport company. In addition to distributing Division B's goods to retailers, it also transports products for other organisations. It consists of:

Divisional headquarters

Six distribution depots/warehouses.

Fuller details about all Group sites are given in Tables 21.6–21.10.

21.3.2 Objective

The object of the study is to ascertain if a financially and technically viable integrated private telecommunication network can be designed.

The network will be considered financially viable if tangible savings resulting from provision of the network produce a payback of capital expended in setting up the network in less than three years.

The network will be considered technically viable if it retains all existing communication capability at a lower cost and also allows interconnection of at least 70% of information terminals.

21.3.3 Terms of reference

Existing systems and terminals may be either retained, or changed, if more appropriate for the network proposed. If they are changed, the cost of doing so should be included in the network cost case, unless a separate cost case for their change can be made.

Existing point-to-point analogue private circuits may either be incorporated into the new network or replaced. If they are replaced, any resulting costs must be included in the network cost case and their present annual costs claimed as a saving.

The design parameters given in Section 21.3.4 must be strictly adhered to unless a valid reason for deviation can be presented. The cost information detailed in Section 21.3.7 should be used. The effect of inflation can be ignored. The cost of providing any necessary accommodation can be ignored.

The cost of managing the system can be ignored, although, if the complexity of the proposed network is sufficient, the provision of a network-management system may be recommended.

The cost of charging out the running cost of the network to its users can be ignored.

21.3.4 Design parameters

(i) The overall worst-case grade of service of the switched network is to be better than that provided by the PSTN, i.e. 0.06.
(ii) The network design date is to be two years after commissioning, which will be the end of the current year.
(iii) The overall worst-case transmission loss, or group return delay, should be significantly better than the worst case in the PSTN.
(iv) A unified numbering and addressing scheme is essential, so that the number and address to gain access to any device is the same from any user-access point of the network. This ruling need not apply if both devices are

Table 21.6 Site details

Site	Site function	Site text facilities
London	Group HQ (main board directors, group finance, group planning)	5 PCs, 1 group 3 facsimile, 1 telex line
Bristol	Group HQ (personnel, legal, management services)	3 PCs, 1 group 3 facsimile, 1 teletext adapter
Birmingham	Division B HQ/factory/distribution depot	20 PCs, 1 group 3 facsimile, 2 telex lines
Brighton	Division A HQ	10 PCs, 1 group 3 facsimile, 3 telex lines
Worcester	Division C HQ	2 telex lines
Reading	Division A factory/laboratory	4 PCs, 1 group 3 facsimile, 1 telex line
Portsmouth	Division A distribution depot	1 telex line
Leeds	Division A factory	1 PC, 1 telex line
Liverpool	Division A distribution depot	1 telex line
Edinburgh	Division A factory/distribution depot	1 telex line
Peterborough	Division B distribution depot	1 group 3 facsimile
Manchester	Division B sales office	1 group 3 facsimile
Newcastle	Division B sales office	1 group 3 facsimile
Glasgow	Divison B distribution depot	1 group 3 facsimile
Wigan	Division C warehouse	1 telex line
Milton Keynes	Division C warehouse	1 telex line
Cambridge	Division C garage/warehouse	1 telex line
Cardiff	Division C garage/warehouse	1 telex line
Norwich	Division C warehouse	1 telex line
Paisley	Division C garage/warehouse	1 telex line

situated on the same site.

(v) Automatic user-to-user access should be provided wherever possible; i.e. the intervention of a manual operator should be avoided.

(vi) All special apparatus provided as part of the network should be located at existing established sites.

21.3.5 The present system

The company's sites are listed in Table 21.6, which also gives the text-communication facilities at each site. The mainframe and minicomputer sites are listed in Tables 21.7 and 21.8. Details of the present telephone systems and their exchange lines are given in Table 21.9 and the existing point-to-point private circuits are listed in Table 21.10.

21.3.6 Traffic records

Accurate records of traffic (speech, data and text) were taken in June 1994, December 1994 and June 1995. Results of the June 1995 figures are given in

Table 21.7 Mainframe and minicomputer sites

Mainframe sites	
Site	Applications
Birmingham	1 Online order entry
	2 Interactive personal computing
	3 Database on food products
Brighton	1 Online order entry
	2 Viewdata
	3 Group salary

Minicomputer sites	
Site	Applications
London	1 Chief-executive information system
	2 Database on food and drink
	3 Group accounts
Bristol	1 Share register
	2 Personnel records

Note: Birmingham mainframe application 2 will be superseded by the provision of personal computers within the next 12 months. When this occurs, access to the Birmingham mainframe application 1 will be required to obtain appropriate file information.

Tables 21.11–21.14. Figures shown are call minutes per day. The other records indicated that a growth rate of 6% per annum was prevailing. No major reorganisations of the Group structure are envisaged, but the consumption of confectionery per capita has shown a 15% decline over the past 20 years.

50% of daily calls occur between 09.00 and 13.00 h and the remainder between 13.00 and 18.00 h.

The day-to-busy hour ratio for dial-up traffic is 1:5.

The observed usage of data terminals is given in Table 21.15.

Table 21.8 Mainframe terminals

MAINFRAME TERMINALS

	SITES	APPLICATION 1		APPLICATION 2		APPLICATION 3	
		Qty.	Comms Type	Qty.	Comms Type	Qty.	Comms Type
Divn. A Main frame	BRIGHTON	5	Direct Connection	5	1200/75 bit/s Dial-up	4	Direct Connection
	READING	4	9.6 kbit/s Synch.	6	1200/75 bit/s Dial-up		
	PORTSMOUTH	6	9.6 kbit/s Synch.	1	1200/75 bit/s Dial-up		
	LEEDS	2	9.6 kbit/s Synch.	2	1200/75 bit/s Dial-up		
	LIVERPOOL	6	9.6 kbit/s Synch.	1	1200/75 bit/s Dial-up		
	EDINBURGH	8	9.6 kbit/s Synch.	2	1200/75 bit/s Dial-up		
	LONDON			1	1200/75 bit/s Dial-up	2	1200 bit/s Dial-up
	BIRMINGHAM					4	1200 bit/s Dial-up
Divn. B Main frame	WORCESTER					2	1200 bit/s Dial-up
	BIRMINGHAM	10	Direct Connection	10	300 bit/s Dial-up	2	1200 bit/s Dial-up
	PETERBOROUGH	4	4.8 kbit/ Synch.				
	MANCHESTER	2	4.8 kbit/s Synch.				
	NEWCASTLE	2	4.8 kbit/s Synch.	1	300 bit/s Dial-up		
	GLASGOW	4	4.8 kbit/s Synch.	1	300 bit/s Dial-up		
	LONDON					1	1200 bit/s Dial-up
	BRISTOL			2	300 bit/s Dial-up		

MINICOMPUTER TERMINALS

SITES	APPLICATION 1		APPLICATION 2		APPLICATION 2	
	Qty.	Comms Type	Qty.	Comms Type	Qty.	Comms Type
LONDON	8	Direct Connection	3	9.6 kbit/s Dial-up	4	9.6 kbit/s Dial-up
BRISTOL	1	9.6 bit/s Dial-up	1	2 9.6 bit/s Dial-up		
READING			1	9.6 bit/s Synch.		
BRISTOL	4	Direct Connection	2	1200 bit/s Dial-up		
LONDON			1	1200 bit/s Dial-up		
BRIGHTON			1	1200 bit/s Dial-up		
BIRMINGHAM			1	1200 bit/s Dial-up		

21.3.7 Cost information

(*a*) The capital cost of stand-alone circuit-switching equipment will be:

Two-wire switching: £400 per trunk
Four-wire switching: £500 per trunk

(*b*) The capital cost of enhancing existing private branch exchanges (PBXs) to provide tandem-switching capability will be:

Two-wire switching: £300 per trunk
Four-wire switching: £400 per trunk

These prices will also apply to enhancing PBXs for the connection of private

Table 21.9 Telephone-system details

SITES	TELEPHONE SYSTEM		PUBLIC EXCHANGE LINES	
	TYPE	ANNUAL COST (£)	QUANTITY	ANNUAL COST (£)
LONDON	Digital PABX	3,500	12	1,128
BRISTOL	Digital PABX	10,600	30	2,820
BIRMINGHAM	Digital PABX	20,090	50	4,700
BRIGHTON	Digital PABX	1,620	20	1,880
WORCESTER	PMBX	1,060	12	1,128
READING	Digital PABX	8,500	16	1,504
PORTSMOUTH	Key and Lamp Units	496	8	752
LEEDS	Strowger PADX	2,080	10	940
LIVERPOOL	Strowger PABX	2,550	15	1,410
EDINBURGH	Strowger PABX	2,550	15	1,420
PETERBOROUGH	Key Telephone System	260	8	752
MANCHESTER	Key Telephone System	260	8	752
NEWCASTLE	Key Telephone System	290	9	846
GLASGOW	Strowger PABX	1,820	7	658
WIGAN	4 x Plansets	176	4	376
MILTON KEYNES	PMBX	248	4	376
CAMBRIDGE	Key Telephone System	464	2	188
CARDIFF	Key Telephone System	200	6	564
NORWICH	Key Telephone System	400	4	376
PAISLEY	Key Telephone System	348	2	188

circuits.

(c) The cost of private circuits leased from British Telecom or from Mercury Communications and of call charges will be in accordance with the published tariffs.

(d) The cost of multiplexers will be:

Type	Purchase cost	Annual maintenance
Time division	£2500	£125
Statistical (analogue)	£1400	£70
Statistical (64 kbit/s)	£4000	£200
Statistical (2 Mbit/s)	£12 000	£500
Drop and insert	£9000	£450

(e) If a connection needs to have more than one high-speed link i.e. 9.6 kbit/s analogue or 64 kbit/s digital or 2 Mbit/s digital, these costs will increase as follows:

Two high-speed links: plus 30%
Three high-speed links: plus 40%
Four high-speed links: plus 50%

Table 21.10 Existing point-to-point private circuits

SITE A	SITE B	QTY. OF CIRCUITS	TYPE OF CIRCUIT	CIRCUIT FUNCTION
LEEDS	BRIGHTON	3	Tariff T	Online Order Entry
EDINBURGH	BRIGHTON	2	Tariff T	Online Order Entry
LIVERPOOL	BRIGHTON	1	Tariff T	Online Order Entry
READING	BRIGHTON	1	Tariff T	Online Order Entry
PORTSMOUTH	BRIGHTON	1	Tariff T	Online Order Entry
BRISTOL	BRIGHTON	1	Tariff S1	Inter PBX P.W.
BRISTOL	LONDON	30 x 64 kbit/s	2.0 Mbit/s	Various
BRISTOL	BIRMINGHAM	2	Tariff S2	Inter PBX P.W.
LONDON	BIRMINGHAM	2	Tariff S2	Inter PBX P.W.
LONDON	WORCESTER	1	Tariff S2	Inter PBX P.W.
BRISTOL	WORCESTER	2	Tariff S3	Inter PBX EXTN.
PETERBOROUGH	BIRMINGHAM	1	Tariff S3	Online Order Entry
NEWCASTLE	BIRMINGHAM	1	Analogue B	Online Order Entry
GLASGOW	BIRMINGHAM	1	Analogue B	Online Order Entry
MANCHESTER	BIRMINGHAM	1	Analogue B	Online Order Entry
LONDON	BRIGHTON	2	Analogue B	Inter PBX P.W.
READING	LONDON	1	Tariff S3	Database Access

Table 21.11 Telephone communication (call minutes per day)

From \ To	(a)	(b)	(c)	(d)	(e)	(f)	(g)	(h)	(i)	(j)	(k)	(l)	(m)	(n)	(o)	(p)	(q)	(r)	(s)	(t)
(a) LONDON	-	400	100	150	80	200	10	5	5	10	5	20	20	5						
(b) BRISTOL	380	-	120	140	90	300														
(c) BIRMINGHAM	80	300	-	180	180	80									80	100	120	120		
(d) BRIGHTON	140	180	80	-	10	200	100	180	80	120	80	60	70	110	80	80	80	80	80	70
(e) WORCESTER	60	80	100		-															
(f) READING	200	250				-		100		100			100							
(g) PORTSMOUTH				100		180	-													
(h) LEEDS				120		80	20	-	100	80	80									
(i) LIVERPOOL		10		60			80		-											
(j) EDINBURGH		20		60						-										
(k) PETERBOROUGH			280		60						-									20
(l) MANCHESTER			120							60		-				80	60		40	
(m) NEWCASTLE			100										-							
(n) GLASGOW			160										20	-						180
(o) WIGAN			60		80										-	80	40	80	20	60
(p) MILTON KEYNES					60										70	-	80	70	50	20
(q) CAMBRIDGE	10	20	60		80										40	80	-	10	60	10
(r) CARDIFF	10	15	70		60										20	40	10	-	10	10
(s) NORWICH			60		40										40	60	10	60	-	5
(t) PAISLEY	10	10	60		100										80	40	20	50	10	-

Table 21.12 Facsimile communication (call minutes per day)

From \ To	(a)	(b)	(c)	(d)	(e)	(f)	(g)	(h)	(i)	(j)	(k)	(l)	(m)	(n)	(o)	(p)	(q)	(r)	(s)	(t)
(a) LONDON	-	40	10	15		20														
(b) BRISTOL	30	-	10	15		30														
(c) BIRMINGHAM	6	10	-																	
(d) BRIGHTON	6	6		-		30														
(e) WORCESTER					-															
(f) READING				40		-														
(g) PORTSMOUTH							-													
(h) LEEDS								-												
(i) LIVERPOOL			10						-											
(j) EDINBURGH			10							-										
(k) PETERBOROUGH			10								-									
(l) MANCHESTER			10									-								
(m) NEWCASTLE			10										-							
(n) GLASGOW			10											-						
(o) WIGAN															-					
(p) MILTON KEYNES																-				
(q) CAMBRIDGE																	-			
(r) CARDIFF																		-		
(s) NORWICH																			-	
(t) PAISLEY																				-

Table 21.13 *Telex communication (call minutes per day)*

From \ To	(a)	(b)	(c)	(d)	(e)	(f)	(g)	(h)	(i)	(j)	(k)	(l)	(m)	(n)	(o)	(p)	(q)	(r)	(s)	(t)
(a) LONDON	-				40		10											10		
(b) BRISTOL		-	10																	
(c) BIRMINGHAM		10	-																	
(d) BRIGHTON				-																
(e) WORCESTER					-										80	80	100	100	100	140
(f) READING						-														
(g) PORTSMOUTH							-													
(h) LEEDS								-												
(i) LIVERPOOL									-											
(j) EDINBURGH										-										
(k) PETERBOROUGH											-									
(l) MANCHESTER												-								
(m) NEWCASTLE													-							
(n) GLASGOW														-						
(o) WIGAN					40										-					
(p) MILTON KEYNES					40											-				
(q) CAMBRIDGE					40												-			
(r) CARDIFF					30													-		
(s) NORWICH					20														-	
(t) PAISLEY					60															-

Table 21.14 PC communication (call minutes per day)

From \ To	(a)	(b)	(c)	(d)	(e)	(f)	(g)	(h)	(i)	(j)	(k)	(l)	(m)	(n)	(o)	(p)	(q)	(r)	(s)	(t)
(a) LONDON	-	40	20	20		10														
(b) BRISTOL	40	-	10	15		20		10												
(c) BIRMINGHAM	(20)	(20)	-	(10)																
(d) BRIGHTON	20	30	15	-		30		30												
(e) WORCESTER																				
(f) READING																				
(g) PORTSMOUTH																				
(h) LEEDS																				
(i) LIVERPOOL																				
(j) EDINBURGH																				
(k) PETERBOROUGH																				
(l) MANCHESTER																				
(m) NEWCASTLE																				
(n) GLASGOW																				
(o) WIGAN																				
(p) MILTON KEYNES																				
(q) CAMBRIDGE																				
(r) CARDIFF																				
(s) NORWICH																				
(t) PAISLEY																				

NOTES   1. (20) Indicates potential communication.   2. With the exception of London/Bristol all word processors use PSS to communicate.

Table 21.15 Data-terminal usage

TERMINAL TYPE	SESSION DETAILS		PROCESSOR	
	Sessions per day	Average duration of session	Site	Application
9.6 Kbit/s Synchronous	2	4 hours	BRIGHTON	1
4.8 Kbit/s Synchronous	2	4 hours	BIRMINGHAM	1
9.6 Kbit/s Dial-up	4	30 minutes	LONDON	2
1200 bit/s Dial-up	1	30 minutes	BRIGHTON	3
1200/75 bit/s Dial-up	6	10 minutes	BRIGHTON	2
300 bit/s Dial-up	6	40 minutes	BIRMINGHAM	2
1200 bit/s Dial-up	4	30 minutes	BIRMINGHAM	3
9.6 Kbit/s Dial-up	1	10 minutes	LONDON	3
1200 bit/s Dial-up	1	20 minutes	BRISTOL	2
9.6 Kbit/s Synchronous	3	20 minutes	LONDON	2
9.6 Kbit/s Synchronous	1	10 minutes	LONDON	3

Notes:
1 In the busiest hour of the day 50% of terminals of the same type on any site are in use simultaneously.
2 Average time of session includes setting-up and logging-off times.

(*f*) The cost of modems will be:

Type	Purchase cost	Annual maintenance
300 bit/s asynchronous	£400	£20
1200 bit/s asynchronous	£600	£30
1200/75 bit/s viewdata	£300	£15
4.8 kbit/s synchronous	£2000	£100
9.6 kbit/s synchronous	£2000	£100
X.25 PAD	£3000	£150

(*g*) The cost of a 2.048 Mbit/s link is the same as 16 64 kbit/s links.

21.3.8 The project report

(1) Proposed network plan

Tables 21.11–21.14 show that most traffic is to or from London, Birmingham and Liverpool. These sites are thus the natural choices for the locations of the tandem exchanges to switch traffic between other sites. The switched network shown in Figure 21.11 has therefore been designed. Figure 21.11 also shows the numbers of circuits required between the nodes of the network.

It is proposed that this functional switched network shall use the digital transmission-bearer network shown in Figure 21.12. It is cheaper to aggregate circuits to use a single 2 Mbit/s link than to provide separate 64 kbit/s circuits if more than 16 circuits are required on a route. A 2 Mbit/s link from London to Cambridge therefore also carries London-to-Norwich traffic. A 2 Mbit/s

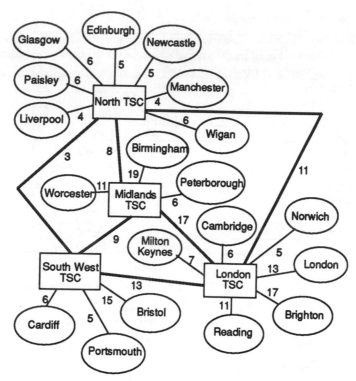

Figure 21.11   Telnet Group private network

☐ tandem switching centre
○ PABX

link between London and Reading also carries traffic between London and Bristol, Portsmouth and Cardiff. A 2 Mbit/s ring connects all sites in Scotland and the north of England with Manchester. To provide resilience, this ring also includes Birmingham. Dual routes are also provided between Birmingham and London.

(2) *Costs*

The estimated capital cost of implementing the proposed network and its estimated annual running costs are shown in Table 21.16. These costs are based on the information in Section 21.3.7. The Table also shows the annual running costs of the present system.

It can be seen from Table 21.16 that introduction of the new network will produce an annual cost saving of over £390 000. This cost saving will pay for the capital expenditure in little more than two years.

(3) *Conclusion*

The network is technically viable and its costs more than meet the financial target of a pay-back time of less than three years. It is therefore recommended that the plan be implemented.

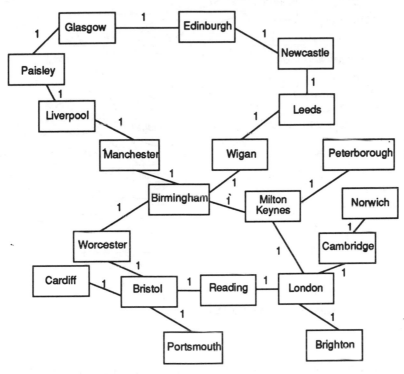

Figure 21.12   Telnet Group digital bandwidth infrastructure

☐ multiplexer
— 2 Mbit/s line

Table 21.16   Costs of Telnet Group private network

1 Existing running costs	Cost, £ per year
PSTN call charges	568 553
Private-wire rental	163 700
PBX rental or maintenance	57 512
Public-exchange-line rental	22 748
Modem maintenance	11 900
Total	824 413

2 New-network capital cost	Once-off cost, £
Line-connection charges	104 740
New PABXs for TSCs	66 000
Upgrade PABXs for TSCs	58 000
Upgrade PABXs for private lines	65 200
Provision of digital multiplexers	480 000
Total	773 940

3 New-network running costs	Cost, £ per year
PSTN call charges	-
Private-line rental	372 850
PBX rental or maintenance	71 942
Public-exchange-line rental	10 152
Total	454 944

4 Savings	£ per year
1 minus 3 =	369 469

5 Payback period	2.1 years

21.4 Rural-network modernisation

D.F. Onions

21.4.1 Introduction

The cost of providing a telecommunication network in a rural area is high because of its scattered population, which results in small exchanges and long customers' lines. Moreover, the quality of service obtained is often poor. The high cost and low quality of service are generally associated with small rural exchanges (e.g. below 500 lines), which often cover very large geographical areas. Nevertheless, rural customers have a high dependence on telecommunications; thus the line penetration and the revenue per line are often higher than average. Rural networks are not necessarily unprofitable, but they provide an average per-line contribution to the public-telecommunications operator (PTO) which is significantly lower than for urban lines.

Telecommunication networks have immense inertia and PTOs have usually concentrated their modernisation programmes on urban areas, where greater cost savings can be made. In the rural areas of Arcadia, there has been little change in the number and size of exchanges over the years and the exchanges are predominantly electromechanical. The fault rates in rural areas are almost double those in urban networks. There is also a significant tail of exchanges, mainly those below 500 lines in size, with very high fault rates in their access networks. The PTO therefore has a requirement for modernising its rural networks and exchanges, particularly for its very small exchanges of a few hundred lines and below.

The PTO is modernising its urban networks by installing digital exchanges and introducing optical-fibre transmission on junction routes. However, it has found that, on a per-line basis, the average cost of modernising a rural-exchange area is twice that of modernising an urban network. This average contains a very wide spread; modernising a 50-line exchange can cost 20 times more per line than modernising a 2000-line rural exchange. The ongoing costs of providing and maintaining rural networks are also about twice those of urban networks.

The economies of scale enjoyed by urban networks are not available to rural networks, using the standard solutions currently adopted. The PTO therefore decided to study the possibility of providing radio access to rural customers. One option is to replace a rural network by a cellular-radio system. This has the potential for improving customers' quality of service and eliminating the maintenance and provision costs of the access network. Also, its reach would provide the opportunity for serving several exchange areas from one system, which would enable economies of scale to be achieved.

Another option is to use a form of point-to-multipoint radio system. The system considered provides up to 120 analogue lines, or 60 ISDN2 links, from a base station to fixed antennas at customers' premises. It can be used for the complete replacement of a small exchange and its access plant, or to supplement a larger rural exchange and replace all its long lines. It would also enable ISDN2 demand to be satisfied where it cannot be provided over existing metallic pairs (i.e. those longer than 4 km). Where the multipoint system is used as a complete replacement for a small exchange, the network would be trunked (over optical fibre or a point-to-point radio link) to a digital exchange or remote concentrator unit (RCU) which can host the multipoint system.

The PTO has decided to carry out a study to compare the costs of conventional modernisation and each of the above radio options for a representative rural area in its territory. The area chosen for this study is the Bucolia region of Arcadia.

21.4.2 The Bucolia region

The Bucolia region of Arcadia is a coastal area which also contains several islands off its shore. It represents a substantial rural area, with a mix of exchange sizes, and cost information on the present network is readily obtainable because it is managed as a discrete area. Some details of the present network are as follows:

(i)   Number of working lines: 9500

(ii)   Number of exchange sites: 23
(iii) The largest exchange has 4000 lines and the smallest has only 65 lines.
(iv)   The average exchange size is 415 lines.
(v)   20 people are employed to provide and maintain the network.

21.4.3 The study

Three different plans are to be produced for modernising the Bucolic telecommunication network and a comparison made of their costs and the return obtained on the capital investment. The options to be studied are:

(*a*) conventional modernisation;
(*b*) replacement of the network by a cellular-radio system; and
(*c*) introduction of point-to-multipoint radio-access systems.

In each case, it will be assumed initially that no network exists. Means for migration from the current network to the selected option will then be considered.

Typical capital costs for switches, junction plant, transmission equipment and access networks for a range of exchange sizes are shown in Table 21.17. All costs are given in Arcadian Units of Currency (ACU).

For depreciation purposes, the standard plant lives and depreciation rates (based on straight-line depreciation) shown in Table 21.18 are to be used for

Table 21.17 Capital costs of modernising switches, junctions and access networks

Exchange Size (Lines)	Switch Cost Per Line (ACU)	Transmission Cost Per Line (ACU)	Junction Cost Per Line (ACU)	Access Cost Per Line (ACU)	Total Cost Per Line (ACU)
100	2350	505	2955	2860	8670
200	1095	215	973	2605	4887
300	730	130	760	2075	3675
400	570	95	545	2055	3285
500	480	70	420	2055	3025
600	425	60	345	1615	2445
700	386	50	290	1615	2341
800	355	40	282	1615	2292
900	330	35	220	1615	2200
1000	325	35	210	1615	2185
1500	265	20	135	990	1410
2000	240	15	95	990	1340
5000	240	10	25	990	1265
10000	240	5	15	555	815

Table 21.18 Plant lives and depreciation rates

	Plant Life	Depreciation Rate
Switch	13	7.7%
Junction duct	25	4%
Junction cable	20	5%
Junction transmission	10	10%
Radio equipment	10	10%
Access plant	20	5%
Power plant	12	8.3%

the study.

The revenue obtained is to be assumed to be the same for all three options and based on the annual revenue from the present network. This is 625 ACU per line per annum.

21.4.4 Option (a): conventional modernisation

Method

The exchanges are listed in Table 21.19. The new network is to be based on a digital exchange at site K, with remote concentrator units (RCUs) at the other 22 exchange sites. The switches are to be connected by optical-fibre cables or radio systems as appropriate. Customers will have traditional metallic-pair access.

Assumptions

  (i) Switch, junction and access costs are to be assumed as given in Table 21.19.
 (ii) Accommodation and facility charges are to be based on current charges, totalling 714 000 ACU per annum (of which 234 000 ACU is for rates and energy).
(iii) Staff costs for maintenance and provision are to be based on the current charges of 915 000 ACU per annum.

Result

The capital cost of providing the network and the interest and depreciation charges, derived using the above assumptions, are given in Table 21.19.

21.4.5 Option (b): cellular-radio network

Method

Customers will use mobile cellphones at their premises. Three base stations will provide full coverage of the area. The cellular-radio system will have a digital ISDN interface to a centrally located exchange.

Assumptions

(i) Cost of the cellular equipment will be 2000 ACU per customer. This includes both mobile phones and base stations.

(ii) Accommodation and facility charges will be 20% of those for option (*a*). This is based on the requirement for only three buildings instead of 23.

(iii) Staff costs can be reduced to 25% of those for option (*a*). This is because the majority of staff required for option (*a*) are involved in maintenance and provision activities for the access network.

Table 21.19 Option (a): capital costs

Exchange	Number of Lines	RCU Cost (ACU)	Standby Power Cost (ACU)	Junction Cost (ACU)	Transmission Cost (ACU)	Access Cost (ACU)	Total Cost (ACU)
A	140	145460	18000	190000	32400	359630	745490
B	215	158210	18000	190000	32400	437930	836540
C	150	147160	18000	190000	32400	385690	773250
D	145	146650	18000	190000	32400	377870	764920
E	85	136110	18000	190000	32400	237545	614055
F	220	158890	18000	190000	32400	446150	845440
G	475	203090	18000	190000	32400	980710	1424200
H	325	177420	18000	190000	32400	670255	1088075
I	230	160930	18000	190000	32400	470825	872155
J	115	141550	18000	190000	32400	299690	681640
K	4000	800130	18000	190000	32400	3949110	4989640
L	230	161100	18000	190000	32400	472880	874380
M	65	133220	18000	190000	32400	188890	562510
N	185	153110	18000	190000	32400	476900	870410
O	65	132880	18000	190000	32400	183170	556450
P	160	148690	18000	190000	32400	409140	798230
Q	230	161270	18000	190000	32400	474935	876605
R	240	162460	18000	190000	32400	489330	892190
S	385	187450	18000	190000	32400	791560	1219410
T	1180	322430	18000	190000	32400	1167210	1730040
U	170	151410	18000	190000	32400	450840	842650
V	170	150730	18000	190000	32400	440415	831545
W	320	176740	18000	190000	32400	662030	1079170
Total Capital	9500	4417090	414000	4370000	745200	14822705	24768995

Capital/line		465	44	460	78	1560	2607
% Depreciation		7.7	8.3	4.5	10	5	
Depreciation (ACU)		340116	34362	196650	74520	741135	1386783
Depreciation/Line		36	4	21	8	78	146
Interest/ annum (ACU)							1609985

Table 21.20 Option (b): capital costs

Exchange Sites	Number of Lines	RCU Cost (ACU)	Standby Power Cost (ACU)	Junction Cost (ACU)	Transmission Cost (ACU)	Cellular Cost (ACU)	Total Cost (ACU)
Cellular costs	9500	0	0	0	0	19000000	19000000
Central Site	9500	1281500	0	0	0	0	1281500
3 Base Stations	0	0	54000	570000	97200	0	721200
Total Capital		1281500	54000	570000	97200	19000000	21002700

Capital/line (ACU)		135	6	60	10	2000	2211
% Depreciation		7.7	8.3	4.5	10	10	
Depreciation (ACU)		98676	4482	25650	9720	1900000	2038528
Depreciation/Line		10.3	0.5	2.7	1.0	200	216
Interest/annum (ACU)							1365176

Result

The capital cost of providing the network and the interest and depreciation charges, derived using the above assumptions, are shown in Table 21.20.

21.4.6 Option (c): multipoint-radio access

Method

The multipoint-radio solution will be adopted for all exchanges below 240 lines and a traditional metallic access network used for all exchanges above 240 lines. There are 17 exchanges of less than 240 lines (total 3000 lines) and six exchanges of more than 240 lines (total 6400 lines). There will be no reduction in the number of exchanges. The multipoint-radio system is to provide basic telephone service rather than ISDN, thus doubling the system capacity.

Assumptions

(i) The cost of a fully-provided multipoint system for 60 ISDN lines is 120 000 ACU. This cost is made up of a fixed cost of 30 000 ACU for the central station and a variable cost of 1500 ACU for every line. Thus, the per-line cost for a fully-equipped system is 2000 ACU, but 7500 ACU if there are only five lines.

(ii) For basic telephone service, one ISDN link can serve two customers, thus

Table 21.21 Option (c): capital costs

1.Exchanges with conventional RCUs

Exchange	Number of lines	RCU Cost (ACU)	Standby Power Cost (ACU)	Junction Cost (ACU)	Transmission Cost (ACU)	Multipoint Cost (ACU)	Total Cost (ACU)
G	475	203090	18000	190000	32400	980710	1424200
H	325	177420	18000	190000	32400	670255	1088075
K	4000	800130	18000	190000	32400	3949110	4989640
S	385	187450	18000	190000	32400	791560	1219410
T	1180	322430	18000	190000	32400	1167210	1730040
W	320	176740	18000	190000	32400	662030	1079170
Total Capital (ACU)		1867260	108000	1140000	194400	8220875	11530535

Capital/line (ACU)		280	16	170	29	1230	1725
% Depreciation		7.7	8.3	4.5	10	5	
Depreciation (ACU)		143780	8965	51300	19440	411044	634529
Depreciation/Line		22	1.5	8	3	61	95
Interest/annum (ACU)							749485

2.Exchanges with Multipoint radio system

Exchange	Number of lines	RCU Cost (ACU)	Standby Power Cost (ACU)	Junction Cost (ACU)	Transmission Cost (ACU)	Multipoint Cost (ACU)	Total Cost (ACU)
A	140	0	18000	190000	32400	220800	461200
B	215	0	18000	190000	32400	340800	581200
C	150	0	18000	190000	32400	236800	477200
D	145	0	18000	190000	32400	232000	472400
E	85	0	18000	190000	32400	132800	373200
F	220	0	18000	190000	32400	347200	587600
I	230	0	18000	190000	32400	366400	606800
J	115	0	18000	190000	32400	184000	424400
L	230	0	18000	190000	32400	368000	608400
M	65	0	18000	190000	32400	105600	346000
N	185	0	18000	190000	32400	292800	533200
O	65	0	18000	190000	32400	102400	342800
P	160	0	18000	190000	32400	251200	491600
Q	230	0	18000	190000	32400	369600	610000
R	240	0	18000	190000	32400	380800	621200
U	170	0	18000	190000	32400	276800	517200
V	170	0	18000	190000	32400	270400	510800
RCUs		501960					501960
Total Capital (ACU)		501960	306000	3230000	550800	4478400	9067160

Capital/line (ACU)		178	109	1147	196	1591	3221
% Depreciation		7.7	8.3	4.5	10	10	
Depreciation (ACU)		38650	25400	145350	55080	447840	712320
Depreciation/Line		14	9	52	20	159	253
Interest/annum (ACU)							589365

Grand Total

Total Capital		2369220	414000	4370000	745200	12699275	20597695
Depreciation (ACU)		182430	34365	196650	74520	858884	1346849
Interest/annum							1338850

reducing the per-line cost of a fully-equipped system to 1000 ACU. This lower cost is unlikely to be fully achieved in a rural situation, because suitable pairs of closely located premises may not exist to share the ISDN link. It will therefore be assumed for this study that the cost is 1600 ACU per line.

(iii) Junction-network costs are assumed to be the same as for option (a).

(iv) Accommodation and facility costs are assumed to be the same as for option (*a*).

(v) Staff costs are assumed to be 67% of those for option (*a*) because of the reduction of the metallic access network by 33%.

Result

The capital cost of providing the network and the interest and depreciation charges, derived using the above assumptions, are shown in Table 21.21.

21.4.7 Conclusions

The financial outcomes for each of the three options are shown in Table 21.22.

Option (*b*), the cellular system, offers the greatest potential. Its contribution is 36% of the revenue, compared with 22% for option (*a*). This contribution would increase further if the depreciation life of the equipment is increased to that of switching equipment. This option also has the lowest annual running costs.

Option (*c*), multipoint radio for small exchanges, shows a contribution of 32%. The return on capital is 9%, compared with 10% for option (*b*), but the capital required is slightly less.

The costings assume that no network currently exists. However, since a network does exist, the full advantages of options (*b*) and (*c*) cannot be obtained in practice. The existing access network, switches and buildings are not yet fully depreciated and represent an additional cost to carry. However, the existing network could evolve towards a cellular-radio-based network by

Table 21.22 Summary of costs and contributions

	Option A	Option B	Option C
Annual Revenue (ACU 000's)	5938	5938	5938
Annual Costs (ACU 000's)			
Interest	1610	1365	1339
Depreciation	1387	2038	1347
Staff	918	230	596
Accommodation	714	142	714
Total Costs	4629	3775	3996
Annual Contribution (ACU 000's)	1309	2163	1942
Capital Employed (ACU 000's)	24769	21003	20598
Contribution	22%	36%	32%
Return on Capital Employed	5%	10%	9%

the addition of a radio overlay for:

(a) meeting new customer demand; and
(b) the early replacement of the access networks (in whole or in part) of specific exchanges where there is a high fault liability or where line limits are exceeded.

The cellular system can then be extended to cover further customers as other exchanges become due for replacement. This strategy provides the opportunity for the PTO to migrate over a period of time towards a higher-quality lower-cost network. It is therefore recommended that the cellular-radio strategy be adopted.

Index